Studies in Computational Intelligence

Volume 627

About this Series

The series "Studies in Computational Intelligence" (SCI) publishes new developments and advances in the various areas of computational intelligence—quickly and with a high quality. The intent is to cover the theory, applications, and design methods of computational intelligence, as embedded in the fields of engineering, computer science, physics and life sciences, as well as the methodologies behind them. The series contains monographs, lecture notes and edited volumes in computational intelligence spanning the areas of neural networks, connectionist systems, genetic algorithms, evolutionary computation, artificial intelligence, cellular automata, self-organizing systems, soft computing, fuzzy systems, and hybrid intelligent systems. Of particular value to both the contributors and the readership are the short publication timeframe and the worldwide distribution, which enable both wide and rapid dissemination of research output.

More information about this series at http://www.springer.com/series/7092

George A. Tsihrintzis · Maria Virvou
Lakhmi C. Jain
Editors

Intelligent Computing Systems

Emerging Application Areas

Editors
George A. Tsihrintzis
Department of Informatics
University of Piraeus
Piraeus
Greece

Maria Virvou
Department of Informatics
University of Piraeus
Piraeus
Greece

Lakhmi C. Jain
Faculty of Science and Technology
Data Science Institute, Bournemouth University
Poole
UK

ISSN 1860-949X ISSN 1860-9503 (electronic)
Studies in Computational Intelligence
ISBN 978-3-662-49177-5 ISBN 978-3-662-49179-9 (eBook)
DOI 10.1007/978-3-662-49179-9

Library of Congress Control Number: 2015959587

Printed on acid-free paper

This Springer imprint is published by SpringerNature
The registered company is Springer-Verlag GmbH Berlin Heidelberg

Foreword

When dealing with real-world problems, we are often faced with mathematical intractability in pursuing their solution. Nonlinearities, chaotic phenomena, the presence of a plethora of degrees of freedom, inherent uncertainties, noise and time variability of the context often prevent us from reaching precise mathematical solutions. In such problems, Intelligent Computing Systems may be of assistance. The core of the Intelligent Computing System is based on computational methodologies that have been inspired from processes that arise in nature and address high-complexity problems very efficiently.

Some of the most common intelligent computational methodologies are artificial neural networks, evolutionary computation and genetic algorithms. Other computational paradigms that have been applied successfully include artificial immune systems, fuzzy logic, swarm intelligence, artificial life and virtual worlds. Finally, it is common for intelligent systems to use hybrid computational methodologies based on combinations of the previous.

In terms of application areas, Intelligent Computing Systems have appeared in virtually all modern scientific disciplines, including engineering, natural, computer and information sciences, economics, business, commerce, environment, health care and life sciences. The successful application of Intelligent Computing Systems in some areas has been driving their applications in new and emerging areas.

The book at hand represents a significant contribution to the field of Intelligent Computing Systems. The editors present a number of emerging areas in which Intelligent Computing Systems may have an impact in the years to come. In particular, the editors have selected chapters on intelligent systems to assist (1) government, business and communities; (2) smart city technologies, intelligent energy systems and environmental monitoring; (3) equipment; and (4) search through and analysis of big/massive data.

The book audience includes graduate students, practitioners and researchers in Intelligent Computing Systems and other overlapping areas. As such, it is self-contained and its chapters are appropriately grouped to provide a better coverage of emerging areas of application of Intelligent Computing Systems.

I believe that the editors have done an outstanding job in addressing the pertinent topics and associated problems. I consider the book to be a great addition to the area of Intelligent Computing Systems. I am confident that it will help graduate students, researchers and practitioners to understand and explore further Intelligent Computing System methods and apply them in real-world problems and applications.

Prof. Dr. Andreas Spanias
Arizona State University, USA

Preface

Intelligent Computing Systems make use of computational methodologies that mimic nature-inspired processes to address real-world problems of high complexity for which exact mathematical solutions, based on physical and statistical modelling, are intractable. Common intelligent computational methodologies include artificial neural networks, evolutionary computation, genetic algorithms, artificial immune systems, fuzzy logic, swarm intelligence, artificial life, virtual worlds and hybrid methodologies based on combinations of the previous.

Intelligent Computing Systems have appeared in most modern scientific disciplines, including engineering, natural, computer and information sciences, economics, business, commerce, environment, health care and life sciences.

The book at hand is an attempt to explore emerging scientific and technological areas in which Intelligent Computing Systems seem to provide efficient solutions and, thus, may play a role in the years to come.

The book includes an editorial (Chap. "Intelligent Computing Systems") and an additional fifteen (15) chapters, covering such application areas of Intelligent Computing Systems as government, business and communities, smart cities, intelligent energy systems and environmental monitoring, Intelligent Computing Systems incorporated in equipment and advanced theoretical tools for processing data. All chapters in the book were invited from authors who work in the corresponding area of Intelligent Computing Systems and are recognized for their research contributions.

This research book is directed to professors, researchers, application engineers and students of all disciplines. We hope that they all will find it useful in their works and researches.

We are grateful to the authors and the reviewers for their excellent contributions and visionary ideas. We are also thankful to Springer for agreeing to publish this book. Last but not least, we are grateful to the Springer staff for their excellent work in producing this book.

George A. Tsihrintzis
Maria Virvou
Lakhmi C. Jain

Contents

Intelligent Computing Systems

Emerging Application Areas

George A. Tsihrintzis, Maria Virvou and Lakhmi C. Jain

There are many real-world problems of such high complexity that traditional scientific approaches, based on physical and statistical modeling of the data generation mechanism, do not prove effective. Typically, these problems are characterized by multidimensionality, nonlinearities, chaotic phenomena and the presence of a plethora of degrees of freedom and unknown parameters in the underlying data-generating mechanism. Additional difficulties in addressing these problems arise from inherent uncertainties, noise and time variability of the context. As a result, loss of information that is crucial to solve a problem is inherent in the data generation process itself, making a traditional mathematical solution intractable.

On the other hand, biological systems have evolved over the ages to address similar problems in biological organisms in very efficient ways. For example, biological neural networks, i.e., networks of interconnected biological neurons in the nervous system of most multi-cellular animals, are capable of learning, memorizing and recognizing patterns in signals such as images, sounds, smells or odors. Similarly, ants exhibit a collective, decentralized, and self-organized intelligence that allows them to discover the shortest route to food in very efficient ways. A third example of a biological system that exhibits high level intelligence is the vertebrate immune system, i.e., a decentralized system of biological structures and processes within an organism that protects against pathogens that threaten the organism and may cause disease.

Intelligent Computing Systems make use of computational methodologies that mimic nature-inspired processes to address real world problems of high complexity for which exact mathematical solutions, based on physical and statistical modelling,

G.A. Tsihrintzis (✉) · M. Virvou
University of Piraeus, Piraeus, Greece
e-mail: geoatsi@unipi.gr

M. Virvou
e-mail: mvirvou@unipi.gr

L.C. Jain
Faculty of Science and Technology, Data Science Institute, Bournemouth University, Poole House, Talbot Campus, Fern Barrow, Poole, UK
e-mail: Lakhmi.Jain@unisa.edu.au

G.A. Tsihrintzis et al. (eds.), *Intelligent Computing Systems*,
Studies in Computational Intelligence 627, DOI 10.1007/978-3-662-49179-9_1

are intractable. Common intelligent computational methodologies include artificial neural networks, evolutionary computation, genetic algorithms, artificial immune systems, fuzzy logic, swarm intelligence, artificial life, virtual worlds and hybrid methodologies based on combinations of the previous.

Intelligent Computing Systems have appeared in most modern scientific disciplines, including engineering, natural, computer and information sciences, economics, business, commerce, environment, healthcare, and life sciences.

The book at hand is an attempt to explore emerging scientific and technological areas in which Intelligent Computing Systems seem to provide efficient solutions and, thus, may play a role in the years to come.

More specifically, the book at hand consists of an editorial chapter (this chapter) and an additional fifteen (15) chapters. All chapters in the book were invited from authors who work in the corresponding area of Intelligent Computing Systems and are recognized for their research contributions. More specifically, the chapters in the book are organized as follows:

The first part of the book consists of six chapters devoted to intelligent computing systems incorporated in various services provided to government, businesses and communities.

Specifically, chapter "Semantic Tools; Their Use for Knowledge Management in the Public Sector", by Theocharis and Tsihrintzis, is on "*Semantic tools; their use for knowledge management in the public sector.*" The authors present ontology-based intelligent systems towards the development of smart applications, Internet search and knowledge management in government.

Chapter "From Game Theory to Complexity, Emergence and Agent-Based Modeling in World Politics", by Paravantis, is entitled "*From game theory to complexity, emergence and agent-based modeling in world politics.*" The author examines the complexity of world politics with an emphasis on global environmental issues and with use of game theory.

Chapter "A Semantic Approach for Representing and Querying Business Processes", by Kalogeraki, Apostolou, Panayiotopoulos, Tsihrintzis and Theocharis, is on "*A semantic approach for representing and querying business processes.*" The authors present the challenges and benefits associated with the coupling of semantic technologies with business process management and describe a methodology for representing the semantic content of the BPMN specification in the form of ontology.

Chapter "Using Conversational Knowledge Management as a Lens for Virtual Collaboration in the Course of Small Group Activities", by Akoumianakis and Mavraki, is on "*Using conversational knowledge management as a lens for virtual collaboration in the course of small group activities.*" The authors focus on a relatively recent approach to knowledge management and collaborative learning, namely *conversational knowledge management*.

Chapter "Spatial Environments for m-Learning: Review and Potentials", by Styliaras, is on "*Spatial environments for m-learning: Review and potentials.*" The author reviews existing spatial hypermedia interfaces as well as related environments and their potential use in educational platforms for mobile devices.

Chapter "Science Teachers' Metaphors of Digital Technologies and Social Media in Pedagogy in Finland and in Greece", by Vivitsou, Tirri and Kynäslahti, is on "*Science teachers' metaphors of digital technologies and social media in pedagogy in Finland and in Greece.*" The authors discuss and analyze pedagogical decisions and choices when the learning space is enriched with social networking environments, and digital and mobile technologies.

The second part of the book consists of five chapters devoted to intelligent computing systems incorporated in various services provided to smart cities, intelligent energy systems and environmental monitoring.

Specifically, chapter "Data Driven Monitoring of Energy Systems: Gaussian Process Kernel Machines for Fault Identification with Application to Boiling Water Reactors", by Alamaniotis, Chatzidakis and Tsoukalas, is on "*Data-driven monitoring of energy systems: Gaussian-process kernel machines for fault identification with application to boiling water reactors.*" The authors present an approach that adopts a set of Gaussian process-based learning machines in monitoring highly complex energy systems.

Chapter "A Framework to Assess the Behavior and Performance of a City Towards Energy Optimization", by Androulaki, Doukas, Spiliotis, Papastamatiou and Psarras, is on "*A framework to assess the behavior and performance of a city towards energy optimization.*" The authors introduce the Smart City Energy Assessment Framework (SCEAF) to evaluate the performance and behavior of a city towards energy optimization, taking into consideration multiple characteristics.

Chapter "An Energy Management Platform for Smart Microgrids", by Delfino, Rossi, Rampararo and Barillari, is on "*An energy management platform for smart microgrids.*" The authors present a platform to address the important issue of planning and management of a, so-called, smart microgrid, that is a group of interconnected loads and distributed energy resources with clearly defined electrical boundaries that acts as a single controllable entity with respect to the public grid.

Chapter "Transit Journaling and Traffic Sensitive Routing for a Mixed Mode Public Transportation System", by Balagapo, Sabidong and Caro, is on "*Transit journaling and traffic sensitive routing for a mixed mode public transportation system.*" The authors propose transit journaling, a crowdsourcing solution for public transit data collection, and we describe CommYouTer, an Android app for this purpose.

Chapter "Adaptation of Automatic Information Extraction Method for Environmental Heatmaps to U-Matrices of Self Organising Maps", by Markowska-Kaczmar, Szymanska and Culer, is on "*Adaptation of automatic information extraction method for environmental heatmaps to U-matrices of self-organising maps.*" The authors we introduce some dedicated processing steps while trying to minimize the number of changes in previously proposed methods.

The third part of the book consists of two chapters devoted to intelligent computing systems incorporated in equipment.

Specifically, chapter "Evolutionary Computing and Genetic Algorithms: Paradigm Applications in 3D Printing Process Optimization", authored by Canellidis, Giannatsis and Dedoussis, is on "*Evolutionary computing and genetic algorithms: Paradigm applications in 3D printing process optimization.*" The authors present the effective utilization of genetic algorithms, which are a particular class of Evolutionary Computing, as a means of optimizing the 3D printing process planning.

Chapter "Car-Like Mobile Robot Navigation: A Survey", authored by Spanogianopoulos and Sirlantzis, is on "*Car-like mobile robot navigation—A survey.*" The authors review the basic principles and discuss the corresponding categories in which current methods and associated algorithms for car-like vehicle autonomous navigation belong.

In the fourth (final) part of the book, we have included two chapters which provide advanced theoretical tools for processing data in various applications of intelligent computing systems.

Specifically, chapter "Computing a Similarity Coefficient for Mining Massive Data Sets", authored by Cosulschi, Gabroveanu and Sbırcea, is on "*Computing a similarity coefficient for miningmassive data sets.*" The authors analyse the connections and influences that certain nodes have over other nodes and illustrate how the Apache Hadoop framework and the MapReduce programming model can be used for a large amount of computations.

Finally, chapter "A Probe Guided Crossover Operator for More Efficient Exploration of the Search Space", authored by Liagkouras and Metaxiotis, is on "*A probe-guided crossover operator for more efficient exploration of the search space.*" The authors propose a new probe-guided crossover operator for the more efficient exploration of the search space through the recombination of the fittest solutions.

In this volume, we have presented some emerging application areas of intelligent computing systems. Societal demand continues to pose challenging problems, which require ever more efficient tools, methodologies, and integrated systems to de devised to address them. Thus, it may be expected that additional volumes on other aspects of intelligent computing systems and their application areas will appear in the future.

Semantic Tools; Their Use for Knowledge Management in the Public Sector

Stamatios Theocharis and George Tsihrintzis

Abstract In recent years it is globally recorded that there is an increased demand of citizens for greater transparency of operations and accountability of governments to the public. The means to achieve this requirement are the IT technologies through the development of appropriate applications. The reproduction of information has taken massive proportions resulting to the insufficiency of the traditional methods and knowledge management tools. Towards the development of smart applications Internet search and knowledge management, we present in this work related technologies focusing on ontologies.

Keywords Open government data · e-Gov ontology · Reasoning

1 Outlines

In the second part we present critical issues about the opportunities of e-Government through the Semantic Web and the "opening" of public data as an introduction and, in the third part, typical relevant works on the areas of open data and semantic web technologies. In the fourth part we develop the theoretical background on the Semantic knowledge representation using RDF triples and OWL ontologies and fifth place matters on the logical inference and its support languages. In the sixth part we present the concepts and properties that we defined in the ontology developed in the framework of this study while in seventh part we present issues on query submission on ontologies technologies, with relevant examples from our ontology. In the eighth part we present the methodology development and the evaluation of our ontology and in the ninth and final part the conclusions of the relevant study.

S. Theocharis (✉) · G. Tsihrintzis
Department of Informatics, University of Piraeus, Piraeus, Greece
e-mail: stheohar@unipi.gr

G. Tsihrintzis
e-mail: geoatsi@unipi.gr

G.A. Tsihrintzis et al. (eds.), *Intelligent Computing Systems*,
Studies in Computational Intelligence 627, DOI 10.1007/978-3-662-49179-9_2

2 Introduction—Presentation of the Field of Interest

2.1 E-Government—The Opportunities Through the Semantic Web

At its early stage, e-government was considered to be synonymous with the ITC introduction in Public Administration, and at a later stage with the development of informative websites of Public Administration and the maintenance of public data in database locally and separately for each public body. Afterwards, upon completing the first four development levels, a fifth one concerning the personalized provision of e-services was added. In this way over the last years e-government has evolved aiming to the integration of smart e-services for the general public as well as the administrative body itself. These services are related to advanced features such as recommendations to traders, advance completion of applications and intelligent information search. The latter level is particularly important if we take into account thc number of handled information both electronically and physically and the number and complexity of transactions with the public administration.

The current designation of government websites and online applications are mainly anthropocentric, meaning that most of the content is designed to be accessible by people and not automated by machines. It is based on documents written mainly in the environment of HTML, which is used to describe a structured text with emphasis on visualization and presents limited capabilities in classifying the text sections of a page. So far, computers can efficiently analyze websites in appearance and routines, but generally do not have a reliable way to process the semantics of their content.

In recent years there has been much debate about the evolution of the Internet and the evolution into the so-called semantic web. Initially, the semantic web is expected to contribute to more intelligent access and manage information handled on the internet through new technologies and the development of corresponding new applications. The overall vision of the transition from the existing Internet of static pages into a dynamic network of Web services providers that automatically discover thc information sought, negotiate for goods or services that the user intends to purchase or gather information from different sources and unite in homogeneous forms with the ultimate goal of sharing and interaction with other systems based on "common language".

Our interest is focused on the possibilities that can be offered by Semantic Web technologies in the further development of e-government and support open government. The field of e-government, as we believe, may be one of the main consumers of smart services based on the semantic web. This is because the modern trend of open government, is supported by the publication of "open data" and its further connection to with semantic web technologies. By using semantic web technologies, are expected improvements related to both parts of the entire system, i.e. the so-called "front-office" and "back-office":

- As far as the citizens service is concerned: It will greatly facilitate the search of the services provided through the use of government portals (development of semantic portals) and through the physical presence in the service point (in the context of operation of one-stop shop). The search for the appropriate service by the corresponding institution can become more effective and efficient.
- The sharing and searching data among stakeholders in the back-office will be significantly improved by using a common vocabulary and common semantics.

The structural reforms, movements and removal of staff in the public sector has brought great changes and malfunctions in the so-called back-Office of Administration. The reduction of operating costs of public sector is vital and is connected to the efficient functioning of public institutions. This is particularly important if take into account the European economic crisis of recent years. What is required is a corresponding evolution in the overall vision of e-government. Their implementation is associated with the adoption of technologies and methodologies of the semantic web for the benefit of saving resources and infrastructure. Since the technologies of semantic web are not a profitable tool for businesses, essentially one of the key factors for further development of the SW remains the public sector that owns and distributes huge data sizes.

2.2 *Public Open Data for the Transition to 'Open Government'*

The transformation of governance from the classical model to the model of e-government is now directed towards the adoption of principles and policies of the so-called "open" government. Relative is the initiative "Open Government Partnership-OGP" (and the related guide published by) of a large part of European states as well as the USA regarding the "opening"—sharing the acts of government to citizens with the basic objective of achieving transparency and accountability of the government to its citizens and their participation. In particular, with the concept of transparency, is meant that people know and understand the functions of administration. The controllability of citizens in the operations and the administration effectiveness is associated with the concept of accountability of the administration. With the concept of participation is meant the ability of citizens to form together with the administration the various policies and any procedures that affect them.

By the processing of data provided "open" by the administration to the public, it is possible to produce important information and thus knowledge, in the various areas of interest. The sharing of useful open public data and their processing by interested parties—the scientific community and businesses, must be accompanied with corresponding user rights to its citizens and businesses and can have serious economic and developmental benefits to the states. So there is the public open data, which may concern various cases of public activity, such as financial data, geospatial administrative data, public safety issues, etc.

Regarding Greece the first steps have already been made towards open government with its participation to the OGP initiative (http://www.opengovguide.com/) but also with the adoption of basic principles with a series of acts of low with the "Diavgeia" program (Law 3861/2010) constituting the basis. Under the existing legislation (as amended) the information that can be posted online and constitute open data is the following:

- Laws, acts of legislative content, Presidential decrees
- General regulating acts except for the regulating acts concerning the organization, structure, composition, arrangement, supplies and equipment of the Armed Forces of the country, as well as any other acts, the publication of which causes damage to the country's national defense and security,
- Circulars,
- Economic data on expenditure and state revenue, as budgets, accounts, balance sheets of State stakeholders
- Contracts
- Acts of appointment, acceptance of resignation, transfer, termination of appointment concerning regular employees and employees of special categories, acts of appointment concerning single member bodies and formation of collective management bodies of public institutions
- Acts of appointment concerning Committees and project teams
- Contest announcements and results
- Spatial data relating to the Public land,
- Spatial data related to the local economic and business activity and the residential development
- Administrative division of the country
- Data related to public safety and the fight against crime,
- The distribution of the population
- Social data related to job opportunities, education and health care

This venture in the last five years, has undoubtedly led to the unprecedented expansion of mobile data at both local and the online environment. Moreover, the variety of standards and methods of sharing documents, has created a chaotic situation regarding the recovery of data and their processing. The data are so many, scattered and varied that the real knowledge is extremely difficult to be recovered through the methods and tools of traditional knowledge management used until now. The first steps of this initiative were made without much coherence by those involved resulting in a "Babel" of applications and management methods of the new open data.

In recent years there has been a more systematic effort to more homogeneous and compact data exposure methodology to encourage its use and processing. This methodology uses tools and methodologies of the Semantic Web. This is the basic idea of interest, especially in the case of "Linked Open Data—LOD" in conjunction with the evolution of the Internet to the Semantic Web.

3 Related Work

The evolution of e-government in open government in recent years has been the subject of Administrative Science and especially of Computer Science and directly connected with the open public data as a subset of open data. The requirements and conditions for their disposal, the methodologies for their connection and the development of applications for its processing and further consumption for the production of knowledge, is of particular interest, which is reflected to the number of relevant scientific references in various fields. The relevant references concern the concepts, methodologies and tools of the semantic web, knowledge management and information retrieval, as directly relevant themes with the implementation of related open data.

Theocharis and Tsihrintzis (2014) is the previous work of the authors on ontologies for the support of open public data, which are a special tool for modeling knowledge in the field of e-Government. Ontologies are used as a dictionary for the description and classification of concepts and their correlations in specific fields. In this paper, we had presented the technologies used for the implementation of open data, such as the triplets RDF, the RDF schemas, the OWL language and SPARQL. We had presented the Protégé as a tool for building the ontology and the creation and management of knowledge base around public data. This was considered as a critical point, since the scope of public information is quite complex and the information from different operators using different datasets whereas there is no a commonly accepted dictionary for concepts and relationships between them. Among the critical points we had recorded from the experience of recent years by opening public data in Greece, especially through the "Diavgeia" refers to:

- The existence of a large number of different systems that do not communicate with each other since they do not support the existing interoperability standards.
- Many bodies store the same data in different places, ignoring the existence of duplicates, resulting in waste of resources and creating conflicts in their use
- Depending on their structure the bodies use similar—but not exactly the same terminology for naming organizational units having the same object.
- There is confusion as to the correct completion of the thematic fields and the type of decision.
- The search for a published decision in the current search system, is based on the word or key expression either by body or to all the stakeholders but the results are not related.

We have been able to conclude that there is a question of efficiency and overall reliability of the information search system which resulted in delays and depreciation of the system of open data especially from the public information producers themselves.

In this paper we did not fully develop the overall model of public administration, but only the part concerning the description of so-called administrative acts and the concepts associated with them. For the demonstration of the ontology constructed in this work, we considered a real scenario of administrative act concerning the municipal elections. Through this, we introduced the relevant concepts and their relationships. By using the reasoners offered by the Protégé we presented the evaluation process of the ontology on the consistency and search of any logical errors. Also we had presented the two main tools available in the Protégé 4.2 to query the knowledge base that we have created and the results of the submission of relevant questions.

By the development of this ontology we recorded significant findings for the construction of the overall description model of public administration in order to be led to the correct connection of the already open public data. One of the main problems we faced in this work involved the compulsory transfer in English of Greek Public Administration terms as there is no semantic content management software in Greek language oriented to public administration.

In Heath and Bizer (2011) we provided general principles for the creation of linked open data on the Web, and we give an overview of the technologies of the Web. These principles include the use of URI's and the connection among them as well as the use of RDF and SparqL in search for information. The application of the overall architecture of WWW, which includes the HTTP protocol as a generalized mechanism of accessibility and HTML as a general formatting tool for sharing structured data in a specified format worldwide, is presented as a general principle. Reference is also made to two key issues: the publication and consumption of linked open data by presenting basic methods for connecting information that could serve as linked open data. Also presented are standards for connection of open data and methodologies for their implementation. Finally are presented, the applications and architectures on which they are build, so that it will be able to constitute a training guide for researchers involved in developing applications in open data.

In (Educational Curriculum for the usage of Linked Data Project, in http://www.euclid-project.eu/), an educational tool through a European program for the consolidation of related data and methods for their implementation to interested professionals or researchers, is presented. In particular, one of the main objectives of the project is to support different teaching scenarios. Therefore, EUCLID provides an extensive training curriculum, covering the main technologies, tools, use cases and skills that need to be acquired in order to complete both basic, as well as more complex, tasks related to dealing with Linked Data. Among the topic that are covered are: the basics Linked Data principles, the SparqL as a Query tool, the visualization and the search in the linked data and the creation of Linked Data applications. In addition, one of the main features of EUCLID's curriculum is that it not only covers the six core topics related to learning how to use and employ Linked Data through the six developed modules. It also provides a set of diverse materials supporting a variety of learning scenarios.

The page in (http://www.w3.org/standards/techs/gld#w3c_all), summarizes the relationships among specifications, whether they are finished standards or drafts in the domain of Government Linked Data. Each title in this page links to the most recent version of a specific document, such as the RDF Data Cube Vocabulary, Data Catalog Vocabulary etc. It's about the W3C recommendations in the domain of Semantic Web and the Open Data. Through the various links the ability to examine one concrete examples is provided in addition to prototyping in specific themes. For instance, the document in (http://www.w3.org/TR/2014/REC-vocab-org-20140116/) describes a core ontology for organizational structures, aimed at supporting linked data publishing of organizational information across a number of domains. It is designed to allow domain-specific extensions to add classification of organizations and roles, as well as extensions to support neighbouring information such as organizational activities.

There are also several reports for associated open data among which we refer Parundekar et al. (2010) which presents an approach to the production of equivalent relations through research in the environment stating the equivalent classes in various entities. In addition in Jain et al. (2010) we refer to the ontologies behind the Linked Open Data and present a system about finding schema-level links between LOD datasets. A view on the ontology of public administration is given in Goudos et al. (2007) which will form, according to the authors, the cognitive basis for developing semantic web applications in the field of e-Government.

4 Semantic Representation of Knowledge

4.1 The RDF Data Model

The Resource Description Framework—RDF is a W3C standard for simple description of metadata on the internet although there are many who consider it as an ontological language. This model enables you to implement graphs of data and to share between users and machines. For this reason various programming languages such as JAVA and Python, support through appropriate libraries the RDF data. Also many software applications have been developed for the use of such data and the production of these knowledge.

The basic concept of the representation of knowledge is the resource. In this sense, we mean any entity the World Wide Web such as a website, a part or a group of web pages, electronic files or even objects that are not immediately available online, for example a book. The description of the resources of the Internet with the RDF focuses primarily on efficiency of metadata such as title, author's name, creation date, etc. The general idea is that each resource has one or more properties which have specific rates or they may be themselves resources from the Internet. To summarize, the RDF information is given as suggestions in the form of triples subject—properties—object. In this triple, the subject is a resource on the Internet which is identified by a specific unique URI. The property attributed to the subject is specifically described in another resource of the internet (and therefore is itself a resource). The object can either be a literal or is itself another resource. A set of RDF triples can be perceived as a graph. In this graph the objects and subjects play the role of nodes and properties play the role of the connecting edges.

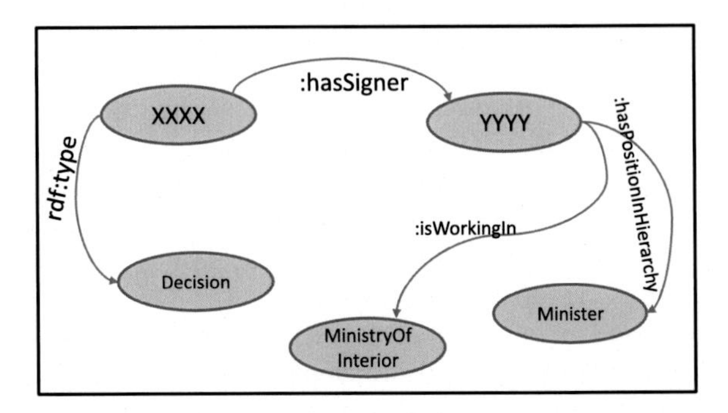

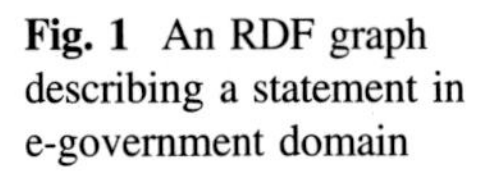
Fig. 1 An RDF graph describing a statement in e-government domain

For example in the model of public administration that we have developed in this paper, we consider the case of the "*XXXX decision is signatory to YYYY who is the Minister of the Interior*". The RDF graph that describes the above statement is shown in Fig. 1.

The syntax used in the RDF representation of the data, is based on the XML standard referred to as RDF/XML syntax. For reasons of generality the RDF uses URI references to identify the entities that are in the position of the subject the property and the object.

4.2 The URI's Use

In order to implement the RDF representation, where basic concept is the resource we use as unique identifiers the Universal Resource Identifiers—URIs. The known URLs (Universal Resource Locators) or the strings that we use to identify each site on the internet is a subset of URIs. The URIs generalize logic of URLs in the sense that anything that can be described and recovered can be identified in a manner analogous to websites. The URI that is used in an RDF statement is characterized as URI reference—URIRef. A URIref comprises a URI and an optional fragment identifier. For example the URIRef: http://www.example.org/index.html#doc1 consists of the URI http://www.example.org/index.html and the fragment identifier *doc1* which distinguish from the URI of the use the # symbol. It is used for the first part of the URI to be entrusted to a relative prefix which is then accessible. In this way each person can identify the resources they wish to publish electronically uniquely worldwide. So for example, a ministry can use for one of its resources the URI: http://www.ministryofinterior.org#path1/docl, which is a component of a unique URL assigned to the institution (http://www.ministryofinterior.org#) followed by the corresponding internal path (*/Path1/doc1*) with which each resource is identified in the internal system featured. With RDF statement: "*PREFIX mint:* <http://www.ministryofinterior.org#>", the URI could be accessible in the form of mint: *Path1/doc1*. Because the URI are used in the manner mentioned as unique identifiers of resources worldwide they are considered as strong identifiers.

4.3 RDF Schema Specification Language

To RDF data model does not offer the possibility of restrictions on reporting relationships that exist between the properties and resources. For example in RDF there is no statement to define that "*signatory of a decision can only be a minister*" or that "*the MinistryOfInterior is the individual of the class Ministry*". So apart from the RDF descriptions of resources, the RDF data model is also supported by a schema definition language (sets of classes and properties), the RDF Schema Specification Language (RDFS). With the help of RDFS, it is possible to identify mechanisms for determining classes of resources and the limitation of possible relationships between classes using appropriate correlations. An RDF schema consists of statements classes, attributes and relationships between classes. Similar resources are grouped under the same heading. Based on the above, we can distinguish three different levels of abstraction in the data model RDFS: In the lower level there are the same resources (documents, websites, persons, or whatever).

The next level of abstraction is the data level, which is the description of information resources using specific vocabularies, which are described in the last level, the shape level. The shape level is the abstraction level where RDF shapes are developed to facilitate semantic resource description. At this level, the classes represent abstract entities and collectively refer to sets of similar objects. For the graphical display of RDF/S descriptions and figures a directed graph model is used, with labels both to the edges and nodes which can easily combine many different vocabularies and be expanded simply by adding more edges. The nodes represent objects (resources or classes) and edges represent relationships between nodes (properties). As nodes we represent also Literal types, i.e. strings and other basic data types as defined in the standard XML Schema. The nodes are represented as gaps and individual values as parallelograms. The edges can be of three types: Performance attributes, creating instances and employees. The performance attribute edges represent attributes of nodes and relations between them while the subsumption edges used to indicate a shape layer that a node (class) or property, are a subclass of a broader semantic node or property respectively. Finally, the snapshots creating edges form the link between RDF standards and RDFS, allowing the creation of instances of a class and assigning types to information resources described.

For better sharing of resources RDF descriptions and schemas on the web the model RDF "borrows" same syntax as XML. This leads to the RDF/XML syntax of RDF schemas and RDF description of resources shown in the graph representation.

4.4 Web Ontology Language—OWL

OWL was created to satisfy the need for a web ontology language and is one of the W3C's recommendations on implementing Semantic Web technologies (http://www.w3.org/OWL/). OWL is designed to be able to process the content of information. The purpose of OWL (Uschold and Gruninger 1996) is to provide a standard format that is compatible with the Internet architecture and the Semantic Web. The standardization of ontologies in OWL language will make the data on the Web more machine processable and reusable applications. Thus, scalability, convertibility and interoperability possess high priority in language design. Compared with existing technologies, OWL exceed the known and widely used languages XML, XML Schema, RDF and RDF Schema as it supports greater clarification of the Web content than machines, providing additional vocabulary along with a formal semantics.

The OWL adds larger vocabulary for the description of properties and classes such as relations between classes (e.g. disjointness), the number of elements in a set (cardinality), logical combinations of classes (union, intersection), equality, richer typing of properties, characteristics of properties (e.g. symmetry, uniqueness, transitivity), and enumerated classes (enumeration).

An advantage of OWL ontologies is the availability of tools which may draw conclusions about these (reasoners). These tools produce general support related to stated rules or events that follow specific morphology and vocabulary. OWL is developed as an extension of RDF vocabulary. It derives from the web ontology language DAML+OIL and it is written in XML format that can easily be independent of the operating system and language implementation of a computer. An OWL ontology includes descriptions of classes, properties and their instances. Taking into account such an ontology, the formal semantics OWL specifies how to produce the logical conclusions, i.e. facts not literally present in the ontology, but implied by the semantics. These implications can rely on a single document or multiple distributed documents that have been combined using defined OWL mechanisms.

As the Semantic Web is distributed, the OWL should allow to gather information from distributed sources. This is done partly by allowing ontologies to be associated, including clearly the import of information from other ontologies. The practical applications of OWL, include Web portals, which can be used to create the classification rules, in order to improve the search, the Multimedia Collections, which can be used to enable content-based searches of media and Web Services, where it can be used for the discovery and composition of web Services as well as for rights management and access control.

Table 1 shows the correspondence between the abstract syntax of OWL class descriptions and descriptions of concepts of descriptive logic. As we can see, in this match, derives the semantics of the structural elements of the OWL language. A similar correlation can be set in the case of class axioms of OWL language. These axioms may be assigned to axioms of affiliation and equivalency between concepts of a Description Logic. The correlations are shown in Table 2 and in most cases are quite obvious. The only case that is worth comment on is the one of foreign classes. From the set theory we know that two sets are disjoint if their intersection is the empty set. If transferred from the field of sets to the abstract space of concepts then we can define two concepts as foreign if the intersection of their falls into the empty concept. It is pertinent to note that there is an alternative statement that two

Table 1 Correspondence between the abstract syntax of OWL class descriptions and descriptions of concepts of descriptive logic

OWL syntax	DL	Example	Semantic annotation
IntersectionOf	$C \sqcap D$	CivilServant⊓CitizenAttributes	All the Civil Servants that have specific attributes
UnionOf	$C \sqcup D$	CivilServant⊔Professor	Anyone that is civil servant or a professor
Atomic megation	$\neg C$	¬CivilServant	Anyone than is not a Civil Servant
hasValue(o)	$\exists R.\{o\}$	∃hasSigner.CivilServant	The signer of a document must have value CivilServant
allValuesFrom(C)	$\forall R.C$	∀CivilServant.{isWorkingIn some State}	Civil servant is anyone that is working in some unit of State

Table 2 Class axioms in OWL

OWL syntax	DL
Class(A partial C1,…Cn)	$A \subseteq C1 \sqcap \ldots \sqcap Cn$
Class(A complete C1,…Cn)	$A \equiv C1 \sqcap \ldots \sqcap Cn$
EnumeratedClass(A O1,…On)	$A \equiv \{O1\} \sqcup \ldots \sqcup \{On\}$
SubClassOf(C1, C2)	$C1 \subseteq C2$
EquivalentClasses(C1,…Cn)	$C1 \equiv \ldots \equiv Cn$
DisjointClasses(C1,…Cn)	$Ci \sqcap Cj \subseteq \perp$

concepts are disjoint. More specifically the term C is foreign to the meaning D if-f $C \subseteq \neg D$.

Table 3 shows the correlation between the properties of the OWL axioms and axioms of roles descriptive sense. Once again we observe that OWL does not offer anything more than the expressive ability that gives us a Description Logic System even where this is not immediately obvious, as in the case of functional roles explained above. From this table however we observe something interesting. The axioms of definition of range and domain of a property, which we met in language RDF-S, may themselves be encoded by using Descriptive Logic axioms. More specifically a form axiom definition of the range of ObjectProperty (R domain (C)) corresponds to the office of entry for $\exists R.\top \sqsubseteq C$, while an axiom definition of the form of range ObjectProperty (R range (C)) corresponding to the entry office, $\top \sqsubseteq \forall R.C$. Intuitively the first axiom tells us: "*If an object is connected through the relation R to*

Table 3 Property axioms in OWL

OWL syntax	DL
SubPropertyOf(R1,R2)	$R1 \subseteq R2$
EquivalentProperties(R1,R2)	$R1 \equiv R2$
ObjectProperty (R domain (C))	$\exists R.\top \sqsubseteq C$
ObjectProperty (R range (C))	$\top \sqsubseteq \forall R.C$

something, then this object is of type C". So this axiom tells us that all individuals who are in the first position of a pair (a, b) which belongs to the relation R is type C. The second axiom tells us that "*for all objects of the world, whenever they are connected to another object* via *the relationship R then the object is of type C*". So all individuals that are in the second position of a pair (a, b) which belongs to the link.

5 Reasoning Tools

The acquisition of knowledge from a database or from a base of knowledge is often associated with the submission of relevant questions through appropriate interfaces. The answer to a question in semantic web, is not just about data control (such as databases), but it requires the implementation of complex considerations that cover a huge crowd classes and instances. Classes and instances are linked together under declared rules and constraints that describe specific events and roles. For this reason, the combination of declared facts and concepts in an ontology are particularly important so that the correctness or not of another event occurs as a conclusion or so that a new unexpected fact is produced. This process is described as reasoning and is one of the main differences between databases and knowledge bases of the semantic web. Of particular interest is the sub-process of reasoning concerning the production-addition of new facts in ontology referred also as inferencing.

The reasoning depends on the way of representing the events in the ontology. This representation is formulated by using formal languages based on various, such as predicate logic and descriptive logic. The last, are used mainly for supporting the semantic web and based on Open World Assumption. In descriptive logic, we use the concepts, objects and people in order to define the nomenclature relating to the field of interest. A typical architecture of a system is given below (Fig. 2) based on descriptive logic.

The knowledge base is the combination of statements of two ingredients: (a) Terminology Box—TBox comprising the vocabulary of the field of interest

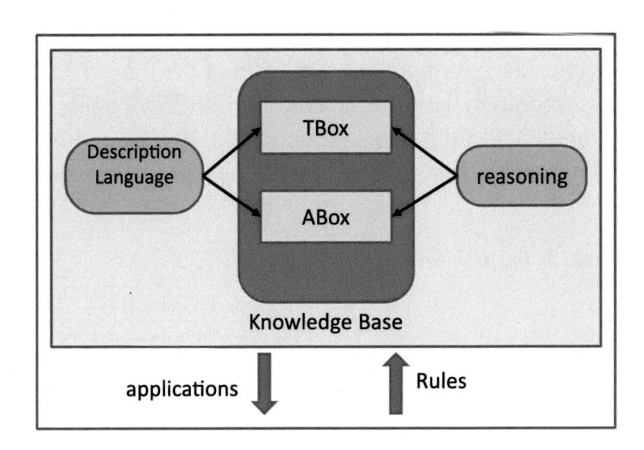

Fig. 2 A typical architecture of a reasoning system

through the concepts and definitions of properties and (b) Assertion Box—Abox comprising all axioms/allegations on individuals included in TBox.

As a vocabulary we mean all the concepts (which usually group together individuals) and the roles referring to binary relations between individuals. The concept and roles description languages for the knowledge base, are based on a basic language, Attributive Language, that provides key manufacturers with common syntax rules.

Special software has been developed under these logic (for example OWL2 Query Tab of Protégé presented below) which support the operation of ontologies management applications by using reasoners such as Pellet, Fact ++ and HermiT used in Protégé. The reasoning (reasoning) provided by the software are particularly useful service and mainly include the following categories of controls: (a) knowledge base consistency checking, (b) concepts classification and (c) instance checking, with the sense of the calculation of classes, which include snapshots of the database.

5.1 SWRL Rules

The SWRL (Semantic Web Rule Language) is a language of the Semantic Web standards and is based on a combination of OWL DL and OWL Lite sublanguage of OWL with Unary/Binary Datalog RuleML sublanguages of the Rule Markup Language. This proposal extends the set of OWL axioms by introducing Horn-like rules (O'Connor 2009). In this way it allows a combination of Horn-like rules with an OWL database. At the same time, it provides a high-level abstract syntax for Horn-like rules, which are expressed in terms of the OWL language (classes, properties, individuals). The rules are in the form of contributory fault between an antecedent and a consequent part. The antecedent part is the main body of a rule and it is known as body, while the consequent part is the consequence of the rule and is known as head. The point at which a rule is intended to be interpreted as follows: whenever applicable conditions laid down in the antecedent of a rule, you should apply the conditions laid down in the consequent part of the rule. The rules are stored as part of the ontology and also it is possible that they interact with reasoners. While the provision of reasoning tools that support SWRL constantly growing, we can distinguish Bossam, R2ML, Hoolet, Pellet, KAON2, RacerPro and SWRLTab. It is known that carries OWL inference capabilities through the features of OWL properties, such as inversion, symmetry and composition of indirect relationships through a combination of direct relations (transitive property). The SWRL carries inference capabilities through the rules. To avoid repetition necessity between OWL extrapolations and SWRL extrapolations would be desirable that rules engines could follow the OWL characterizations. This means that the OWL features should be translated to SWRL equivalents. For this reason, in SWRL is perfectly possible to define rules for attributes that represent symmetry, inversion and transitive possibility.

5.2 The Query Language SQWRL

The library of Protégé-OWL contains a number of built-ins that extend the SWRL in SQWRL. The SQWRL (Semantic Query-Enhanced Web Rule Language) is a language based on SWRL standard for performing queries in OWL ontologies. It has similar methods to SQL to retrieve knowledge from OWL. The determination of SQWRL is done using a library of SWRL built-ins that structure an efficient query language over the SWRL. The built-ins contained in this library are determined by the SQWRL Ontology. The common prefix is sqwrl. The Jess Rule Engine is necessary for the performance of SQWRL questions. The SQWRL questions being built on SWRL can be used for retrieving knowledge which has deduced from SWRL rules. Moreover, SQWRL questions are free to cooperate with other built-in libraries. This free choice and use of built-ins to the questions, provides a continuous expansion means of the expressive power of the query language.

6 Presentation of Our Ontology Through Protégé

6.1 The Ontology Development in Protégé 4.3

Further to our previous work in Theocharis and Tsihrintzis (2014), in this paper, we reviewed some of the structures of the initial model of public administration and added new concepts, properties and instances. The aim of this study was to adapt the ontology so that it describes the actual image recorded today in the public service and not a theoretical model of knowledge. So the concepts and properties that we defined correspond to reality while many of the instances are images of the actual structure of public administration.

In this study we used version 4.3 of Protégé, as one of the most updated, oriented OWL ontologies. As a big advantage of these versions 4.x., we highlighted the friendliness towards the creation of the properties and their limitations on the classes and instances. On the other hand, the concealment of SWRL and the replacement of the corresponding tab that the versions 3.x offered with rules tab, restricts the possibilities for developing complex rules on classes or instances.

6.2 The E-Government Ontology

6.2.1 Defining Classes

Like the previous version of our ontology, the present edition also describes the interaction environment of public administration as a system under the tripolar model comprising three main pillars: the state, citizens and businesses. In our ontology, this is indicated by the definition of the class *Trader* which has as subclasses the: *Citizen*, *Enterprize*, *State*, and an additional two (excluding basic pillars): *NGO* (*Non Governmental Organizations*) and *EuropeanUnion*. The case of subclasses of State interests us particularly, since each country has its own state structure but also internally in each country there are different views as to what is considered as state. Here, the class *State* depicts the Greek reality as it emerges from relevant legal texts e.g. the *Constitution* of our country. By this logic structure created shown in Fig. 3.

We also considered citizens as a member of the system to the same class. Fundamental also is the character of the profession which gives us a way to distinguish citizens. So the *civil servants*, *judicial officials*, *professionals*, *pensioners*, *politicians*, *private employees*, *teachers*, *students*, *the unemployed*, are represented through their respective subclasses of Class *Citizen*.

To describe the properties of the citizens we defined the class *CitizenAttributes* with subclasses the *Gender*, *IllegalActivity*, *Married*. Class *Gender* has two disjoint individuals (*man*, *woman*). The class *IllegalActivity* has additional subclasses according to the type specified by the statements in the General Class Axioms, shown in Fig. 4.

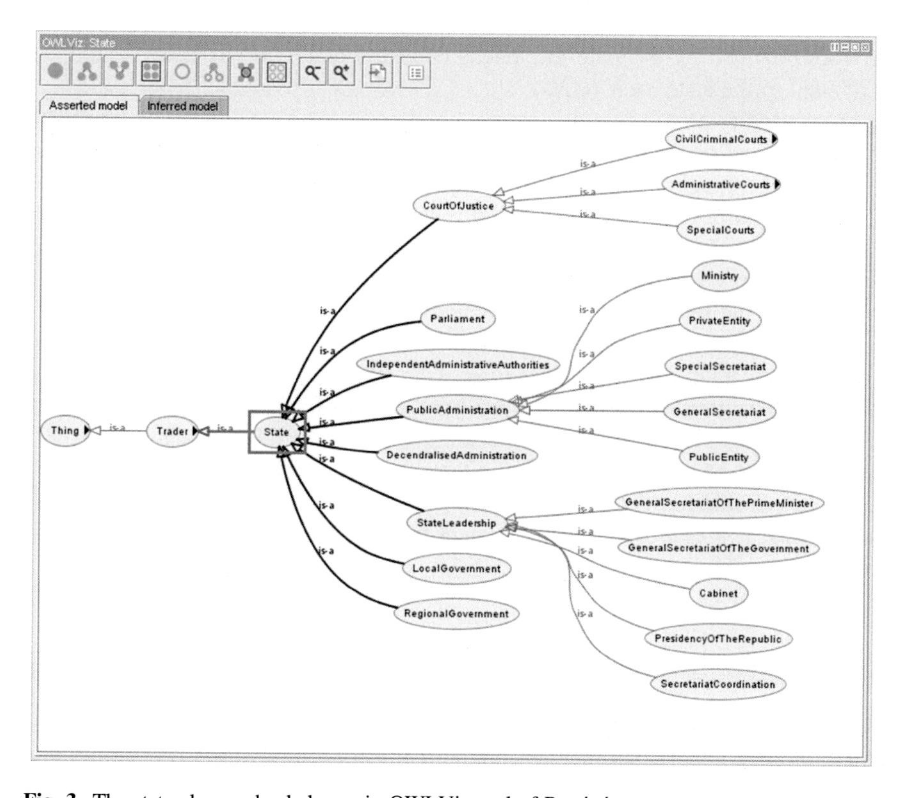

Fig. 3 The state class and subclasses in OWLViz tool of Protégé

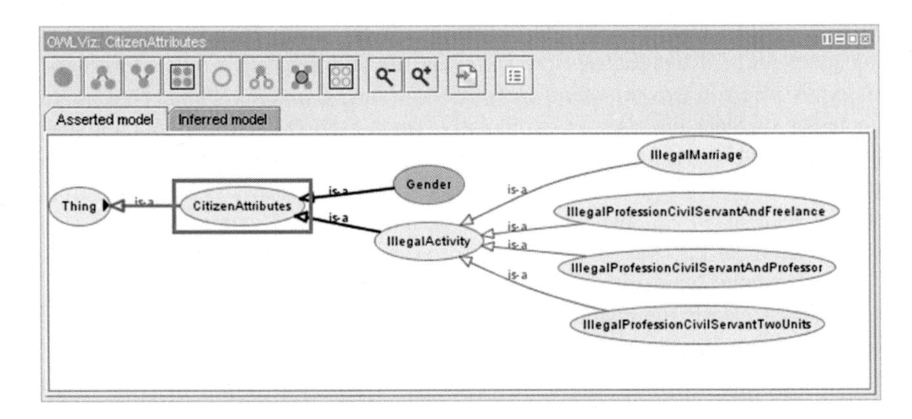

Fig. 4 The CitizenAttributes class and subclasses in OWLViz tool of Protégé

This model supports the so-called "*separation of powers*" of the Constitution into *executive, judicial and legislative*. These concepts are provided by *StateAuthorities* class and subclasses *Executive*, *Judicial* and *StateLegislation* respectively as shown in Fig. 5. At the same basic level with the class *Trader* and *StateAuthorities*, we defined the class *Services* to describe the categorization of the services provided by the state to citizens (subclass *G2C*), to businesses (subclass *G2E*) and to the state itself (*G2G*).

From another perspective, the services and procedures also cover the so-called front-Office and back-Office of Administration. The visual is covered by the class *Process* and respective subclasses that appear in Fig. 6.

In this ontology, we chose to describe the properties of organisms as a class *OrganizationAttibutes* to include two main characteristics of agencies as subclasses: the *bureaucracy* (in the sense of hierarchical organization) and *organizational structure* (in the sense of the units which constitute organizations).

For a description of the operations of the state, we defined the class *ActsAttributes* with subclasses *Gradation*, *Kind*, *Priority*, *Topic* for describing their essential characteristics. In parallel, we have provided a description of the transaction points to the Public Administration with the class *TransactionSpot*. Transaction points can be understood either by their physical nature (*AgentOffice*) or their electronic form (*DiavgeiaSpot*, *WebSite*). The class *Location* and the subclasses *City*, *Prefecture*, *Region* implement the requirement that transaction points are located within a specific spot.

6.2.2 Defining Properties

The protégé enables definition of object properties and datatype properties. The object properties have the role of correlations among individuals and practically implement the triplets RDF. The object properties that we stated in our ontology,

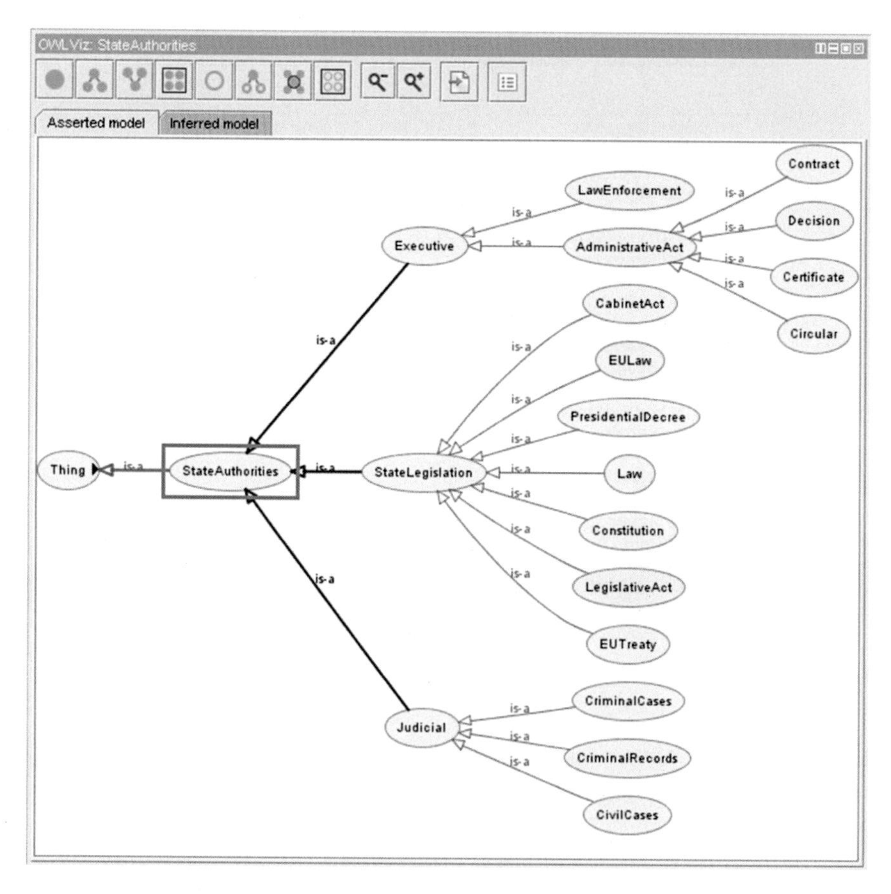

Fig. 5 The StateAuthorities class and subclasses in OWLViz tool of Protégé

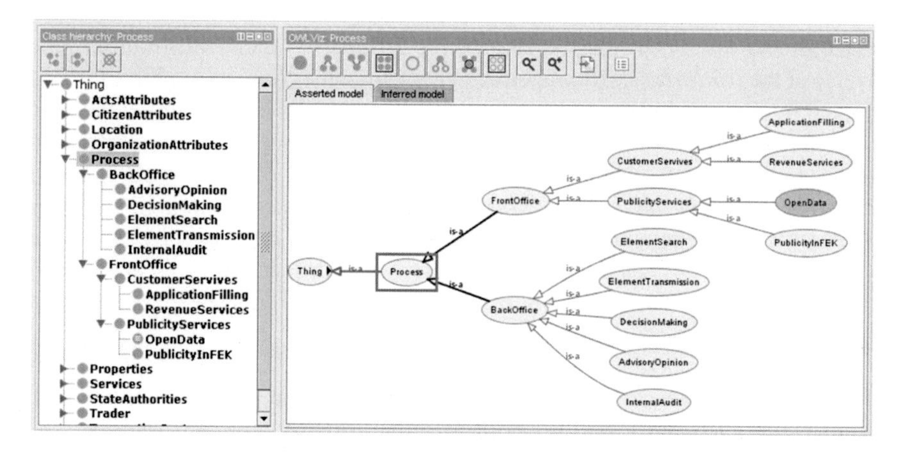

Fig. 6 The process class and subclasses in class hierarchy view and OWLViz tool of Protégé

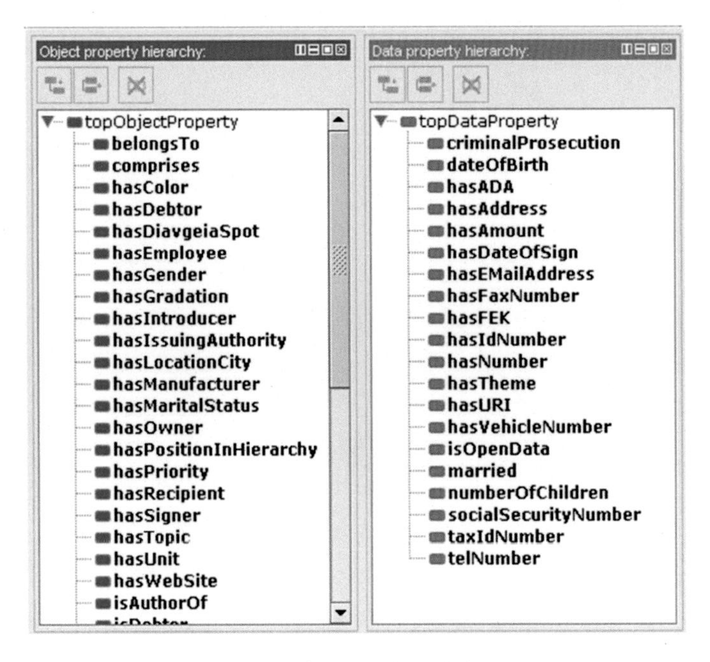

Fig. 7 The ontology properties in hierarchy view of Protégé

are shown in Fig. 7. We note as particularly important the *belongsTo* which is transitional and the inverse of the comprises which we use to attribute semantically that "*a City belongs to a Prefecture*" or "*a Prefecture comprises a Region*". Also for the corresponding requirement that "*a Prefecture belongs to a Region*" or "*a Region comprises a Prefecture*". Also important is the *hasPositionInHierarchy* which has the union of the domain classes *CivilServant* and *PrivateSectorEmployee* and range class *Bureaucracy*. This property is used to describe the requirement that "*an Employee has a specific position in the bureaucracy of the organization*". We also note the *hasSigner* (the inverse of the *isSignerOf*) and *hasIntroduser* (the inverse of the *isIntroduserOf*) properties. These properties have as a subject the acts adopted by the executive authority (*domain* = *Executive*) and as an object the civil servants (*range* = *CivilServant*). With these qualities we implement the description of the requirement that "*an Executive act has signer a civil servant*" or "*an Executive act has introducer a civil servant*". By the property *hasIssuingAuthority* we describe the requirement that every transaction (of the *Executive* or of the *Legislative authority*) has a competent authority which issued it. This is achieved with the statement as a domain of the association of classes *Executive* and *StateLegislation*.

The datatype properties we have set are shown in Fig. 7. The properties of this kind are used to connect the logic of RDF of certain individual with specific literal. For example, every legislative act or some of the administrative acts shall be published in the daily newspaper of the Government and take a unique number (FEK).

This requirement is indicated by the use of the property *hasFEK* which has as a domain the union of *AdministrativeAct, CabinetAct, Decision, Law, LegislativeAct, PresidentialDecree* and as a range type: string. Also, according to the Clarity program, specific categories of administrative acts and instruments are published on the internet, making the so-called open public data. Any document notified on the specific website (www.diavgeia.gov.gr) is characterized by a unique number ADA. This requirement is described in the property *hasADA* which has the same domain and range with the capacity *hasFEK*. Also by using the property *isOpenData* which is type Boolean, we describe the fact that "*a particular kind of the administrative acts or Laws are Open Data*". This is been achieved by the definition of the general class axiom: "*hasADA some string SubclassOf isOpenData value true*".

6.3 The Use of RDF, RDFS, OWL and SWRL Through a Case Study

The RDFS statements involving domain-range are particularly important. This is because the declaration of a class as a domain of a property implements the part of the RDF triplet concerning the subject. In other words, the individual having *rdf:type* the domain of a property is the subject and the particular property is the predicate of a RDF triplet. Correspondingly, the individual having *rdf:type* the range of the property can be the object of the triplet.

For example the property *hasSigner* we have stated to have the domain and range Class *Executive* Class *CivilServant* as shown in the following code section RDF/XML.

```
<owl:ObjectProperty rdf:about="&pa;hasSigner">
  <rdfs:range rdf:resource="&pa;CivilServant"/>
  <rdfs:domain rdf:resource="&pa;Executive"/>
</owl:ObjectProperty>
```

As shown, the *hasSigner* resource is *owl:ObjectProperty* while *CivilServant* and *Executive* resources are *rdfs:range* and *rdf:domain* respectively. In this way we implement the triplet that performs semantic requirement that "*all acts of the executive authority has as signatory a civil servant.*"

We should note at this point that the domain and range do not function as limitations, such as the domain of a mathematical function, but function as a classifier for specific individual. For instance with the statement:

```
<owl:NamedIndividual rdf:about="&pa;Ind_Test1">
  <hasSigner rdf:resource="&pa;Citizen5"/>
</owl:NamedIndividual>
```

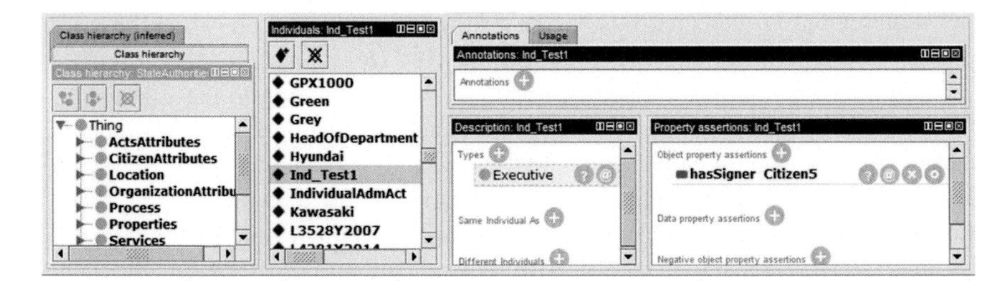

Fig. 8 The reasoner inference about the classification of Ind_Test1 individual

is implied that *Int_Test1* the subject is associated with the object *Citizen5* with the property *hasSigner*. We have stated that that "*Ind_Test1 has Signer Citizen1*". As shown in Fig. 8 using the reasoner's protégé, it is concluded that the *Int_Test1* is type *Executive*, which has not been declared and we do not know if it is a correct conclusion for the classification concerned.

Also important is the function of the features of the properties. Such features are "functional", "transitive", "symmetric" etc. Their use requires attention and understanding of their operation because although they are very useful, they can easily lead to logic errors.

For example, when a property is classified as functional, this means it can get at most one value. That means that the subject in this property can be associated with at most one object. This feature does not work limited as to the subject, but conclusively. For example, if you declare that:

```
<owl:NamedIndividual rdf:about="&pa;Citizen1">
    <hasPositionInHierarchy rdf:resource="&pa;Introduser"/>
    <hasPositionInHierarchy rdf:resource="&pa;Rapporteur"/>
  </owl:NamedIndividual>
```

i.e. if we declare that "*Citizen1 has Position In Hierarchy*" both *Introduser* and *Rapporteur*, (with the use of functional property *hasPositionInHierarchy*), then the reasoner of Protégé concludes that individual *Introduser* and *Rapporteur* is identical, as shown in Fig. 9.

For a highest level of expressivity we can use OWL. Relation between classes can be formally modelled based on description logics (mathematical theory). Because OWL relies heavily on the reasoner, it is possible to express complex constructs such as chained properties for instance or restriction between classes. OWL serves to build ontologies or schema on the top of RDF datasets. As OWL can be serialised as RDF/XML, it is theoretically possible to query it via SPARQL, yet it is much more intuitive to query an OWL ontology with a DL query (which is usually a standard OWL class expression). An example of OWL constructs serialised in RDF/XML is given throughout the object Property *hasPositionInHierarchy*.

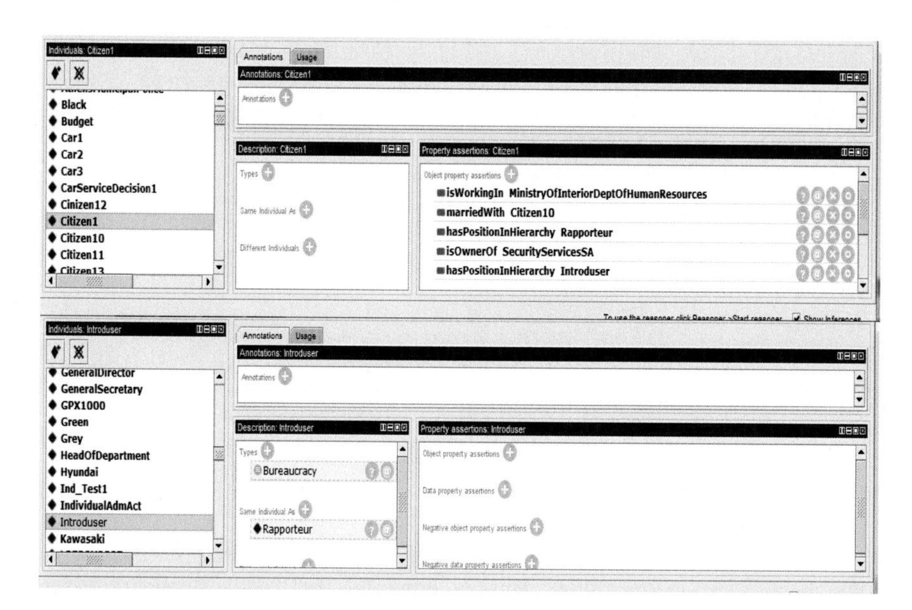

Fig. 9 Classification of Citizen1

The *hasPositionInHierarchy* property connects the *Citizen* classes and *PrivateSectorEmployee* with the *Bureaucracy* which has as domain the union of the first two and the third range from those classes. As shown in the following part of the RDF/XML code, the statements of the domain and range follow the standard RDFS and the concept of classes and their union are attributed with the use of OWL.

```
<owl:ObjectProperty rdf:about="&pa;hasPositionInHierarchy">
   <rdf:type rdf:resource="&owl;FunctionalProperty"/>
   <rdfs:range rdf:resource="&pa;Bureaucracy"/>
   <rdfs:domain>
      <owl:Class>
         <owl:unionOf rdf:parseType="Collection">
            <rdf:Description rdf:about="&pa;CivilServant"/>
            <rdf:Description rdf:about="&pa;PrivateSectorEmployee"/>
         </owl:unionOf>
      </owl:Class>
   </rdfs:domain>
</owl:ObjectProperty>
```

A more complex requirement is presented in the following example. One of the requirements of the Civil Servants Code specifies among others that: “*1. Civil servant is any citizen who works in a public body. 2. Each civil servant holding a position in the bureaucracy of the organization, including: Rapporteur, Head of Department, Director, General Director. 3. A civil servant cannot possess two positions in two public bodies.*"

The requirement 1. is described in the declarations:

```
<owl:Class rdf:about="&pa;CivilServant">
    <owl:equivalentClass>
      <owl:Restriction>
        <owl:onProperty rdf:resource="&pa;isWorkingIn"/>
        <owl:someValuesFrom rdf:resource="&pa;State"/>
      </owl:Restriction>
    </owl:equivalentClass>
    <rdfs:subClassOf rdf:resource="&pa;Citizen"/>
</owl:Class>
```

The statement describing the bureaucracy of an organization that is necessary for requirement 2. is enumeration. This is done by the OWL statement *<owl: equivalentClass>* together with the statement of the members made with the statement *owl:oneof rdf: parseType='Collection' and RDF* statements of the form *rdf:Description rdf:about=NamedIntividual* to give the respective individual.

```
<owl:Class rdf:about="&pa;Bureaucracy">
    <owl:equivalentClass>
      <owl:Class>
        <owl:oneOf rdf:parseType="Collection">
          <rdf:Description rdf:about="&pa;Director"/>
          <rdf:Description rdf:about="&pa;GeneralDirector"/>
          <rdf:Description rdf:about="&pa;Rapporteur"/>
          <rdf:Description rdf:about="&pa;GeneralSecretary"/>
          <rdf:Description rdf:about="&pa;HeadOfDepartment"/>
          <rdf:Description rdf:about="&pa;Minister"/>
        </owl:oneOf>
      </owl:Class>
    </owl:equivalentClass>
    <rdfs:subClassOf rdf:resource="&pa;OrganizationAttributes"/>
</owl:Class>
```

The requirement 3. is described with the class definition *IllegalProfession CivilServantTwoUnits*, with the condition that an individual in order to belong to the class must meet the status *isWorkingIn* at least twice, as seen in the following code fragment:

```
<owl:Class>
      <rdfs:subClassOf
rdf:resource="&pa;IllegalProfessionCivilServantTwoUnits"/>
      <owl:intersectionOf rdf:parseType="Collection">
        <rdf:Description rdf:about="&pa;Citizen"/>
        <owl:Restriction>
          <owl:onProperty rdf:resource="&pa;isWorkingIn"/>
          <owl:onClass rdf:resource="&pa;State"/>
          <owl:minQualifiedCardinality
rdf:datatype="&xsd;nonNegativeInteger">2</owl:minQualifiedCardinality>
        </owl:Restriction>
      </owl:intersectionOf>
    </owl:Class>
```

7 Data Mining Technology from Ontologies

As mentioned above, the knowledge held by experts in each field, can be described based on a common vocabulary with the help of ontologies. Many times, however, the knowledge derived from Data mining tools and the knowledge of experts contained in a ontology is often written in different knowledge representation models or in different shapes. So the first step is the representation of knowledge in a common format using translation mechanisms from different formats to a common one. In the next step we confront issues of resolving any disputes and then processing them.

7.1 SPARQL

Knowledge in Semantic Web is in the form of triplets RDF, to which specific languages are focused. Their main object is the recovery of results upon submission of tailored questions in the form of triplets. Such languages are RQL and SPARQL. With these languages ontological knowledge that has been represented in RDF can easily be questioned, cut, transformed or concentrated. This knowledge often stems from a Web Service and is updated either from a specialist, or it is the result of a Data mining process. In our work we are interested in SPARQL language, which we can use in different ontology management tools such as Protégé.

The SPARQL (SPARQL Protocol and RDF Query Language) is the query language for RDF documents which has become a prototype for the W3C in January 2008 (www.w3c.org). The SPARQL is based on previous query languages for RDF, like RDQL and SeRQL with which they form a family of query languages. Their use depends on the applications developed and run every time. One of their common features is that they comprehend the RDF data as simple triples without some shape or another ontological information unless it is explicitly set to the RDF document. Also, these languages are all SQL-like, i.e. mimic the SQL syntax styles. The basic building block of SPARQL is the triple pattern and is an individual (elementary) question. A standard triple is actually an RDF triple (subject—predicate—object) which may contain a variable in one or more of the three positions. A prerequisite for the question to give a result, in other words for the variable to get some price is the existence of pairing with some RDF triple of the document upon which the question is put. A typical example is the question of the form:

SELECT ?subject ?predicate ?object
WHERE {?subject ?predicate ?object}

which returns all RDF triples contained in the document.

Remarks by the implementation of SPARQL query

The Protégé environment allows the submission of SPARQL queries. In this environment we submitted specific questions shown in Fig. 10.

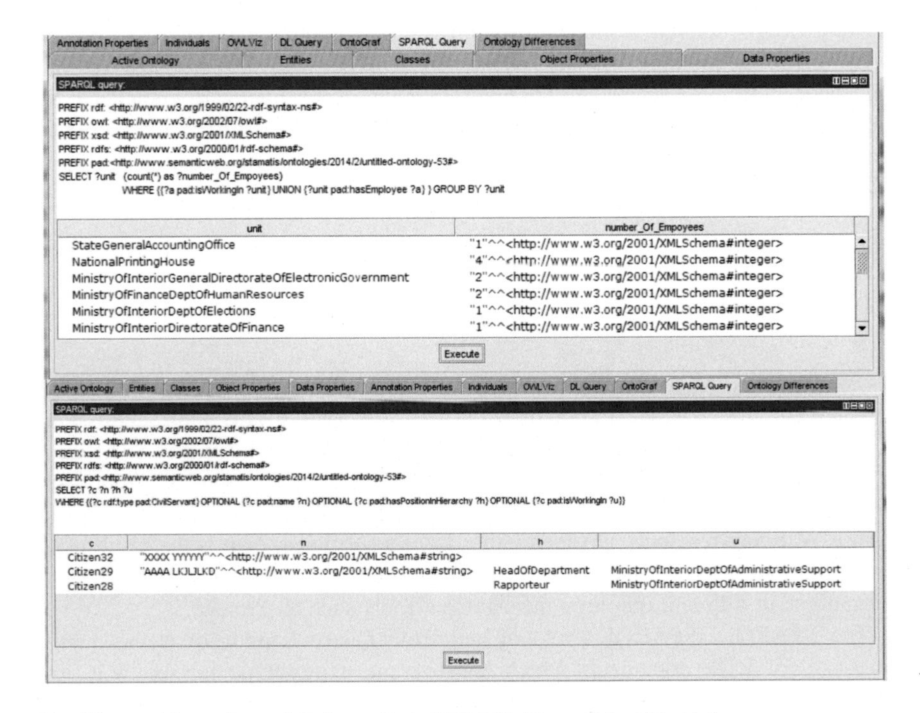

Fig. 10 a and b queries and their results in SPARQL Query Tab of Protégé

a. Number of employees per organization unit:

```
PREFIX   pad:<http://www.semanticweb.org/stamatis/ontologies/2014/2/untitled-
ontology-53#>
SELECT ?unit   (count(*) as ?number_Of_Empoyees)
         WHERE {{?a pad:isWorkingIn ?unit} UNION {?unit pad:hasEmployee
?a} } GROUP BY ?unit
```

b. The data about employees in the units of organizations:

```
PREFIX   pad:<http://www.semanticweb.org/stamatis/ontologies/2014/2/untitled-
ontology-53#>
SELECT ?c ?n ?h ?u
WHERE {{?c rdf:type pad:CivilServant} OPTIONAL {?c pad:name ?n}
OPTIONAL {?c pad:hasPositionInHierarchy ?h} OPTIONAL {?c
pad:isWorkingIn ?u}}
```

From these questions, we note the following important points.

- There is no comprehensive analysis on the performance of the relevant questions. Empirically, we can say the performance depends on the interrogated database in the form of storing information, the complexity of the query itself, the optimizations introduced by the query engine and other environmental factors.
- SPARQL does not support the semantics of RDF Schema. For example, the query

```
SELECT ?s
WHERE {?s rdf: type pad: Citizen}
```

- where prefix pad regards the namespace of our ontology and Citizen argument the corresponding class, will return only instances of Class Citizen and not instances of the subclass, as shown in Fig. 11.
- In its current version it is not possible to modify an RDF document, i.e. there are no structures of the respective INSERT, UPDATE, or DELETE the SQL. The SPARQL is currently a pure query authoring language for knowledge acquisition.
- The SPARQL specification defines the query results based on the RDF implication system. Nevertheless, in the specification there is a general, parameterized definition matching model graph, which can be extended to any implication system. Alternatively, the specification acknowledges that inquiries may be made to a virtual graph which has no explicit definition. Using this feature with suitable software, it is possible to produce a graph based on the rules of RDFS implication which are simple SPARQL questions.

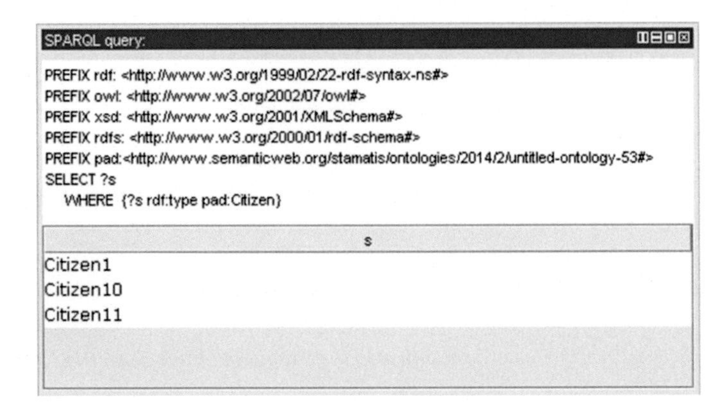

Fig. 11 Members of Citizen class as a response from SPARQL

- As regards the OWL-DL ontology, the OWL-DL axioms do not always lead to a unique graph. Therefore, to answer SPARQL queries to an OWL-DL ontology based on the implication rules of OWL-DL, the implementation of the general customized default template matching securities mentioned is required, suitable for implication based on the rules the OWL-DL. Such an extension of SPARQL is the language SPARQL-DL (which was not developed in this work).

7.2 *SPARQL-DL in OWL2 Query Tab of Protégé*

SPARQL-DL was designed as an extension of SPARQL, since, as already mentioned, was not supported by the discovery of knowledge from OWL-DL documents (Sirin and Parsia 2007). It appears as a query language capable to retrieve information but also as a separate query submission tool on OWL-DL documents (http://www.w3.org/2001/sw/wiki/SPARQL-DL). SPARQL-DL combines TBox/RBox/ABox queries and templates used by SPARQL to ensure an interoperability form the semantic web.

In our research we installed OWL2 Query Tab as a plugin in Protégé 4.3 and we used it (as well as the DL Query tool and SPARQL tool) for querying the Knowledge base that we have developed. It is a conjunctive query and metaquery engine (for SPARQL-DL and negation as failure), and visualization plug-in. It facilitates creation of queries using SPARQL or intuitive graph-based syntax, and evaluates them using any OWL API-compliant reasoner.

It also provides the possibility of submitting additional atom queries related to the hierarchy of classes as: *directSubClassOf*, *strictSubClassOf*, *directSubPropertyOf*, *strictSubPropertyOf*, *directType*. These questions do not represent ontology data but are implemented through the existing questions of SPARQL.

7.3 DL Query Tool of Protégé

The DL Query tab provides a powerful and easy-to-use feature for searching a classified ontology. It comes with the standard distribution of Protégé Desktop (versions 4.x and above), both as a tab and also as a view widget that can be positioned into any other tab. The query language (class expression) supported by the plugin is based on the Manchester OWL syntax, a user-friendly syntax for OWL DL that is fundamentally based on collecting all information about a particular class, property, or individual into a single construct, called a frame (http://protegewiki.stanford.edu/wiki/DLQueryTab).

With this tool, we are able to submit questions which are based on inference of rdfs shapes. As seen above, the questions of SPARQL retrieve information from the RDF statements in the relevant documents without using inferences. This weakness is covered with the DL Query of the Protégé. The search results correlate triples on the basis of logical conclusions generated by the chain (transitional, reverse, etc.) properties and relations between classes and subclasses. For example, it supports the logic that an instance of a class P is at the same time snapshot of superclass Q. With mathematical symbols would you describe this logic as follows:

$$(w \in P) \cap (P \subset Q) \rightarrow w \in Q$$

In Fig. 12 the reasoners show snapshots of Citizen Class following the relevant question. We note here that the corresponding SPARQL query had no results found.

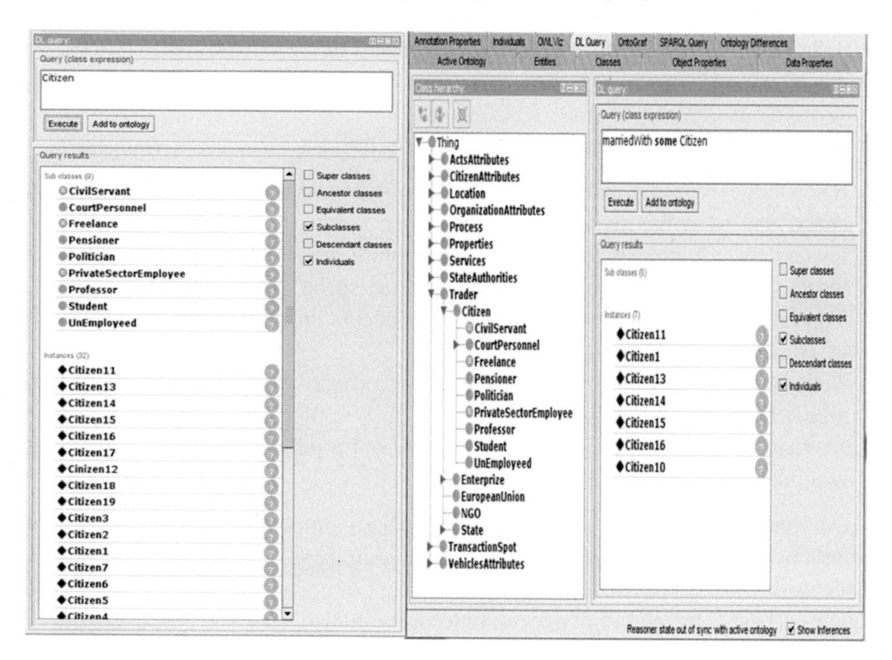

Fig. 12 The DL query results in Protégé

This tool allows the exporting of the results as general axioms in the form of classes. For example the question "*marriedWith some Citizen*" we can define the results in the form of general axiom as shown Fig. 12.

8 Evaluation of Ontology

8.1 Categorization of the Ontology

Generally, an ontology includes a vocabulary of terms and some form of specifications for their importance. Concerning the degree of formality of the representation of an ontology can be:

- Informal, expressed in a natural language.
- Semi-informal, formulated in a limited and structured subset of a natural language.
- Semi-formal, formulated in an artificial and strictly certain language.
- Rigorously formal: definitions of terms with semantics, theorems and proofs of properties such as soundness and completeness.

The ontology developed in this work, is an OWL ontology, i.e. formulated in a certain artificial language. In Protégé environment we have defined a set of rules, axioms and restrictions on classes and properties according to which any stored snapshots are linked. The concepts and properties that we defined follow the semantics based on the actual situation so as listed in the official legal sources. So we could classify our ontology in semi-formal ontologies with several elements from strictly formal ontologies.

A different categorization of ontologies is as follows:

- Knowledge representation ontologies: entities provide representation without specifying what this represent.
- General or common ontologies: aim to capture general knowledge around the world, providing basic concepts such as time, space, events etc.
- Top-level ontologies: provide general concepts under which all terms are associated to existing ontologies.
- Metadata ontologies: provide a vocabulary to describe the information content, which is available electronically.
- Domain ontologies: represent knowledge about a particular field, e.g. medical, administration etc.

Our ontology is at the intersection of general ontologies and of domain of definition ontologies. This is because, in our ontology we have defined various concepts of our everyday life, such as the concept of citizenship, the concept of the state, etc. That is the nature of general ontologies. It also has basic characteristics of domain of definition ontologies, since our main goal was the description of the

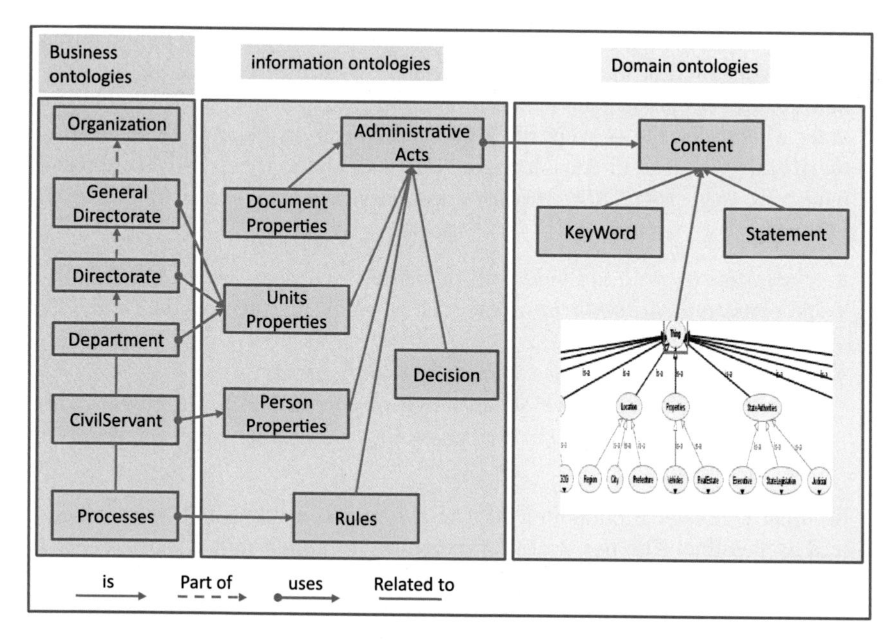

Fig. 13 Categorization of our ontology

scope of public administration and public open data. Finally, our ontology, combines features of business ontologies, information ontologies, as shown below in Fig. 13.

8.2 *Basic Principles of Design*

When designing the ontology of authentic instruments and public administration, we have considered basic principles (Uschold and Gruninger 1996) of ontologies design, among which we mention the following:

- Clarity and objectivity. The ontology is intended to describe a particular subject field. This field should be attributed clearly and if possible objectively. Our ontology was developed based on existing legislation for the separation of powers, the Greek legislation which defines the criteria of open data and concepts set out in the provisions of administrative law.
- Consequency: the concepts and axioms declared do not contradict or conflict with the axioms or conclusions deduced. This principle was tested extensively using reasoners available in Protégé and analyzed in detail in the preceding paragraph.
- Scalability—reuse of concepts: the design of the ontology should provide scalability and expertise at all levels, i.e. the classes, properties and instances.

This point is important, given that the needs required to cover the ontology may change compared to the original design and that it may be necessary to add a new concept or relationship between concepts. For example, in our ontology in order to describe the concept of "*responsible position Head of Department in the Ministry*" instead of considering a class with all the Heads of Department in ministries (e.g. *HeadOfDtpMinistry*) as individual, we chose to define the following:

- class *Ministry* with individual all ministries
- property *hasPositionInHierarchy*, as a predicate for specific individual having the type *Citizen*
- The individual *HeadOfDepartment*, *type:Bureaucracy*.
- With these definitions we achieve two objectives: (a) achievement of the wanted description and (b) the possibility of reusing of the concepts to describe others.

- Minimal Ontological commitment: The axioms declared should be as generalized as possible. The possibility of extending or application to other fields of interest, shall be given without a clear reference to the specific area covered by the ontology. For example, the statement of the rule that "*whoever works in a department -> also works at the corresponding direction*" was implemented by the rule:
  isWorking (? x,? y), isUnit (? y,? z) -> isWorking (? x,? z),
  so that it can also be applied in the case where we want to describe that "*anyone who works in an direction -> work and the corresponding general direction*"
- Minimize coding: because the general idea is to reuse the ontology and systems with different approaches, the coding should be as simple as possible and not to affect the information we want to model.
- Semantic analysis of conditions. For example, the concept of citizen is often associated to the property of professional status. Ie we consider a civil servant or a retired or a student etc. is a citizen of a country. So we considered that kinds of professions are subclasses of the Citizen.
- Analysis is aiming to the reuse of concepts.

8.3 Methodology of the Ontology Development

In the bibliography several models of ontologies' development are recorded, depending on the scope and needs to be covered. Among these models we mention the model of Uschold-King (Uschold and King 1995) and the model of Gruninger-Fox (Gruninger and Fox 1995). The development of the ontology in the context of this work follows the basic steps of the model of Uschold-King enriched with elements of the methodology of Gruninger-Fox (Gruninger and Fox 1995). In more detail, the steps we followed in the development of the ontology were:

1. Identification of the purpose for which the ontology was built in conjunction with the capturing of motivating scenarios. The purpose usually arises from the need to provide solutions to specific problems that either have not been resolved or the solutions found do not correspond to reality. In our example, the sector of public administration with the associated public open data is a field that involves many different aspects and concerns such a huge number of concepts and relationships. As mentioned above, the semantic web is expected to provide solutions in this area.
2. Ontology Building
    a. Standardization of informal competency questions (formulation of competency questions) is considered to be important. These questions are based on motivating scenarios and form a sort of requirements to be met by the ontology. These questions must be answered with the terminology developed in the ontology.
    b. Conception of ontology. At this stage we make the identification of the most important concepts and relationships between them and then we make their registration.
    c. Definition of axioms. The axioms define any restrictions of the concepts combined with the atypical sufficiency questions.
    d. Coding. It is done in the Protégé environment with the help of Description logic.
    e. Integration of existing ontologies. We examined already published ontologies in the field of public administration with a view to incorporating any concepts or relationships. No models that align with our model were found.
3. Assessment of the ontology. Besides texts of the documentation of the ontology, the evaluation record results from the use of reasoner featured by Protégé. Their use enables us to ensure any issues of inconsistency restrictions and general axioms.
4. Maintenance of the ontology. In the final stage, which may be the first of any revision of the ontology, we considered any omissions or the need to adapt to new requirements.

9 Conclusions

In this paper we have presented issues concerning the application of Semantic Web technologies in the area of e-government and open data. This area is characterized by the huge number of primary data which are open public data, by the failure to observe common vocabulary for the concepts and procedures used throughout the public sector, and by the insufficient set of information retrieval tools of the public sector. The technologies that have been introduced aim to support the semantic approach in the seeking of information in the vast area of open public data. They are the infrastructure for the development of new intelligent applications that will use both the public and the same public administration for the management of open public data. From the study which was preceded we can point the following conclusions.

- The Protégé's assistance in the development of the ontology was important. This program which constitutes free open source software, provides the ability to manage ontologies both local on the computer of each user and online. In its last versions it is a very user-friendly tool developing only OWL ontologies, with query reporting possibilities with two main tools: the DL Query oriented exploitation of semantic expressiveness of OWL and SPARQL oriented simple data in RDF format.
- The Protégé comes with different versions with the main 3.4.8 and 4.x. Key differences are recorded to support OWL 2 versions 4.x, non-supported version RDF documents and OWL-Full in newer versions (http://protegewiki.stanford.edu/wiki/Protege4Migration#Side_by_Side_Comparison). However, the existence of different technologies and versions of OWL (OWL-Full, DL OWL 2) creates compatibility problems and confusion between different versions of Protégé and other related tools.
- Significant is the assistance of reasoners for debugging and correcting logical errors in addition to their basic function is to support the logical inference basis of OWL expressive possibilities.
- The methodology followed for the development of our ontology, is a combination of two basic methodologies that are used in numerous ontologies.
- Since the Protégé does not support the Greek language, the concepts we have set have been translated into English. The issue of supporting the concepts in Greek language is open and its resolution in the future will contribute to the further development of semantic web applications.
- There is a clear need to develop online queries applications using SPARQL-DL to support the submission of intelligent queries that exploit the increased semantics offered by OWL ontologies.

References

Goudos, S.K., Peristeras, V., Loutas, N., Tarabanis, K.: A public administration domain ontology for semantic discovery of eGovernment services. ICDIM (2007)

Gruninger, M., Fox, M.S.: Methodology for the design and evaluation of ontologies. In: Workshop on Basic Ontological Issues in Knowledge Sharing, Montreal (1995)

Heath, T., Bizer, C.: Linked data: evolving the web into a global data space. Synth. Lect. Semant. Web: Theory Technol. **1**(1), 1–136 (2011). Morgan & Claypool. Electronic version in http://linkeddatabook.com/editions/1.0/

Jain, P., Hitzler, P., Sheth, A.P., Verma, K., Yeh, P.Z.: Ontology alignment for linked open data. In: The Semantic Web—ISWC 2010, Lecture Notes in Computer Science, vol. 6496, pp. 402–417 (2010)

O'Connor, M.: The semantic web rule language, Stanford University, in http://protege.stanford.edu/conference/2009/slides/SWRL2009ProtegeConference.pdf (2009)

Parundekar, R., Knoblock, C.A., Ambite, J.L.: Linking and building ontologies of linked data. In: The Semantic Web—ISWC 2010, Lecture Notes in Computer Science, vol. 6496, pp. 598–614 (2010)

Sirin, E., Parsia, B.: SPARQL-DL: SPARQL Query for OWL-DL, 3rd OWL Experiences and Directions Workshop (OWLED-2007) (2007)

Theocharis, S., Tsihrintzis, G.: Ontology development to support the Open Public Data—The Greek case. IISA (2014)

Uschold, M., Gruninger, M.: Ontologies: principles, methods and applications. Knowl. Eng. Rev. **11**(2), 93–136 (1996)

Uschold, M., King, M.: Towards a methodology for building ontologies. In: IJCAI95 Workshop on Basic Ontological Issues in Knowledge Sharing, Montreal (1995)

www.w3c.org/owl

From Game Theory to Complexity, Emergence and Agent-Based Modeling in World Politics

John A. Paravantis

The last lesson of modern science is that the highest simplicity of structure is produced, not by few elements, but by the highest complexity.

—Ralph Waldo Emerson, 1847

... in place of the old local and national seclusion and self-sufficiency, we have intercourse in every direction, universal interdependence of nations.

— Marx and Engels, Communist Manifesto, 1848

Look, you've got it all wrong! You don't NEED to follow ME, You don't NEED to follow ANYBODY! You've got to think for your selves! You're ALL individuals!

— From the "Life of Brian" movie (1979), http://www.imdb.com/title/tt0079470/quotes

Abstract This chapter examines the complexity of world politics with an emphasis on global environmental issues. Concepts of game theory are reviewed and connected to international relations (IR). Game theoretic models found in IR, such as the prisoner's dilemma, and global environmental negotiations, such as the North-South divide, are presented and discussed. The complexity of world politics, taking place on a highly interconnected global network of actors organized as agents and meta-agents, is presented and discussed as a multiplayer extension of game theory that should not be regarded as a theory alternative to realism but as a novel approach to understanding and anticipating, rather than predicting, global events. Technology, interconnections, feedback and individual empowerment are discussed in the context of the complex world of global politics. Furthermore, evolution and adaptation are related to the concept of fitness and how it may be approached for the case of actors in world politics. Finally, it is suggested that many events of world politics constitute emergent phenomena of the complex international community of state and non-state actors. The presentation is complemented with a review of research problems from the fields of social science, political science, defense, world politics and the global environment that have been

J.A. Paravantis (✉)
University of Piraeus, Piraeus, Greece
e-mail: jparav@unipi.gr

G.A. Tsihrintzis et al. (eds.), *Intelligent Computing Systems*,
Studies in Computational Intelligence 627, DOI 10.1007/978-3-662-49179-9_3

successfully addressed with agent-based simulation, arguably the most prevalent method of simulating complex systems. This chapter concludes with a recapping of the main points presented, some suggestions and caveats for future directions as well as a list of software resources useful to those who wish to address global problems with agent-based models.

1 Introduction

Lacking a world government, the world is an anarchic community of some 200 sovereign nation-states that strive to survive (Waltz 2008). Wars and conflict erupt as a result of this international anarchy (Waltz 1954). Neorealism (Waltz 2010) asserts that whether the nature of man is flawed (Morgenthau 2005) or not, is of no importance: human nature is not essential to an explanation of conflict. It is the very organization of international relations (IR) rather than the nature of man that determines war (Weber 2009). In response to the international anarchy, states realize that their dominant strategy (to use a game theoretic term) externally is to try to secure their survival by increasing their power; internally, they can afford to focus on quality of life for their citizens (Weber 2009). The Cold War was an example of such a mad spiral in which two global superpowers were trying to outpower one another that as Weber notes, resulted rather surprisingly in a nuclear peace that lasted for almost five decades. Such balance of power arrangements may in fact be more likely to play out in this way when there are only two poles, i.e. the world is a bipolar system. When there are more than two poles, things may become trickier and balances harder to strike.

The unexpected nuclear peace that prevailed in the Cold War and the even more surprising events that have transpired in world politics since then, motivated the author of this chapter to approach the system of world politics from a joint game theoretic and complexity viewpoint. This discussion along with explanation of key elements of game theory is presented in Sect. 2 that contains examples from the field of global environmental diplomacy (Sect. 2.1). In Sect. 3, the author presents complexity science as a multiplayer extension of game theory and argues that many phenomena of world politics may constitute emergent behaviors of the complex global system (Sect. 3.1). Section 4 focuses on simulation and presents agent-based modeling as the premier method of simulating complex adaptive systems (CAS) with a review of key literature publications that have applied agent-based models (ABMs) in problems of social science and world politics (Sect. 4.1). This chapter is rounded up with conclusions (Sect. 5) and a list of resources.

2 Game Theory in World Politics

Cases where a decision making agent is faced with an individual decision and the outcome is not influenced by other agents, are the subject of decision-making. Game Theory examines decision problems where two or more agents (called players) choose between alternative strategies in order to maximize their payoff—this is called rational behavior. Such problems are called strategic games, they can be of simultaneous or sequential movement and they highlight the interactive interdependence of players that together determine the final outcome (Dixit and Skeath 2004). Games are solved when players reach a point of balance, an equilibrium. A small number of equilibrium types constitute frequent solution concepts, including: the dominant strategy equilibrium, consisting of choices that are clearly the best players may achieve; the Nash equilibrium, which is determined by choices that are optimal responses to the selections of the other players; and the focal point equilibrium, which describes an outcome that seems plausible or fair to all players (Schelling 1980). In IR (same source) but also in the business world (Brandenburger and Nalebuff 1996) it has become clear that far fewer games than previously thought are game of pure conflict (called constant or zero sum games), i.e. most games may be directed to outcomes that benefit all players from the distribution of additional value created by smart strategic choices that enlarge the sum of the expected outcomes.

In a newer version of their classic book, Dixit and Nalebuff (2008) present examples from three areas in which game theory is traditionally applied: (a) daily life, (b) business and economics and (c) political science and IR. Attention now shifts to these areas, with emphasis on the latter.

Games in Daily Life

Games are a useful model for the interpersonal relationships of daily life. Stevens (2008) discusses four classic games that he refers to as atomic games, which are particularly suitable for modeling everyday human interaction: the coordination game, the battle of the sexes, the chicken game (in which he who flinches first receives an inferior outcome) and the well-known prisoner's dilemma, a game that illustrates the difficulty of cooperation when individual interests do not match those of the society. Some interesting example games from everyday life that are presented by Miller (2003) include: the interaction between a parent and a rebellious daughter (in which case, it may be advantageous for the parent to pretend that he or she does not love the daughter); the disciplinary style of a young teacher on unruly underage students (where severe disciplinary measures are in conflict with the desire to have them transfer good impressions to their parents); and various negotiation games (in which it may be advantageous, e.g. to discontinue communications temporarily).

Games in Business and Economics

As illustrated in the classic book of Von Neumann and Morgenstern (1944), the broader discipline of business and economic activity is a suitable area for applying game theoretic tools. In their pioneering book, Brandenburger and Nalebuff (1996) analyzed the application of game theoretic models in business and the economy. The basic argument is that when companies try to increase the pie, as when opening up new markets, then it is to their interest to cooperate; when attempting to increase their share of the pie, as when they try to split up existing markets, then it is to their advantage to compete. Hence, the mixed type of business interaction, called co-opetition (i.e. cooperative competition). Other sources analyze numerous examples from the wider field of the economy and businesses at a layperson (Dixit and Nalebuff 1991), professional (McMillan 1992) or college level (Dixit and Skeath 2004). In his video course, Stevens (2008) devotes much lecture time carrying out a game theoretic analysis of pure competition, incentives for increased productivity, oligopolies, bargaining and auctions (a traditional subject of game theory). In an innovative work, Bennett and Miles (2010) show that the approach of one's professional career as a multiplayer game, with the capacity of game theory to model dynamic situations in which the best responses of one player depend on the movements of the other players, allows the extraction of useful conclusions for professionals that are either at the beginning or the end of their career. Finally, in their classic treatise, Fisher and Ury (1999) present a strategic analysis of negotiations as a nonzero-sum game, in which all participants may secure satisfactory payoffs.

Games in International Relations

While Von Neumann and Morgenstern (1944) laid the foundations for the application of game theory in economics, it was Schelling (1980), a Nobel Prize winner that first showed how it may be used as a tool for the analysis of conflict in IR. The Prisoner's dilemma is the best known game theoretic model of transnational cooperation. The original scenario goes like this (Rapoport and Chammah 1965; Axerlod 1985): two suspects, who are not allowed to come in contact, may either keep their mouth shut or confess their crime and give their partner away. If both manage to keep their mouth shut, they are put in prison for a very short time. If only one confesses (the snitch) and the other keeps his mouth shut, the snitch is released while the other goes to jail for a very long time. Finally, if they both snitch, they both go to jail for a moderately long time. Such is the structure of this remarkable game that a powerful outcome emerges in which they both choose to confess, each one in the naïve hope of being released, but since they both snitch on one another, they both go to prison for a moderately long time. An analyst may be pretty confident that they will both select to confess (i.e. snitch) because this is a dominant strategy, and the unfortunate outcome of both snitching on one another is a dominant strategy equilibrium, the strongest solution concept in game theory. This game belongs to the class of social dilemmas (McCain 2004), which are characterized by the existence of a cooperative solution (e.g. both keep their mouth shut) that is distinct from the game equilibrium (e.g. both snitch on one another).

The Cuban missile crisis that transpired in October 1962, is perhaps the international crisis that has been analyzed with game theoretic concepts more than any other crisis in world politics. The governments of John F. Kennedy (from the side of the Americans) and Nikita Khrushchev (on the part of the Soviets) came into conflict for 13 days, which brought the world close to nuclear disaster. In their classical analysis of the Cuban missile crisis, Allison and Zelikow (1999) considered three conceptual policy analysis models for understanding the events: (1) a rational actor model often encountered in the analysis of public policy, (2) an organizational behavior model, which emphasizes the interplay of complex organizational clusters and (3) a governmental policy model that places emphasis on procedures. The authors suggested that all three models are jointly necessary to explain the behavior of the two rival superpowers in the Cuban Missile Crisis. Their innovation lies in the fact that decision-making models are applied to a foreign policy crisis. An elaborate game theoretic analysis of the Cuban missile done by Dixit and Skeath (2004), included a simple-threat model, a model with hard-liner Soviets and a final model with unknown Soviet payoffs. The latter game theoretic model, which was essentially a real-time chicken game (Brams 2005), used Kennedy's estimate that the chances of the blockade leading to war varied from one out of three (0.33) to even (0.5) to calculate the conditions for successful brinkmanship, i.e. gradual escalation of the risk of mutual harm, on behalf of the Americans, rendering a deeper understanding of the crisis and the strategic manipulation of risk it entailed.

Brams (2005) analyzed several cases of national and global politics as bargaining games: the Geneva Conference on Indochina (1954) highlighted the privileged position of the status quo (Zagare 1979); the Cuban missile crisis was regarded a game of chicken (Dixit and Skeath 2004); the Watergate scandal (1973–74) which led to the resignation of President Nixon; the strategy of President Carter at the Camp David negotiations between Israelis and Arabs (September 1978); the role of Kissinger as an arbitrator between Israelis and Arabs during the Yom Kippur war (October 1978); and the role of threats in the conflict between the Solidarity trade union of Lech Walesa and the Government of the Polish Communist Party (1980–1981). Worthy of note is that Noll (2011) favors the use of professional mediation practices in many different types of international conflict. Game theory has also been used to investigate terrorism (Sandler and Arce 2003). Finally, Brams and Kilgour (1988) use game theoretic tools to analyze national security problems, e.g. the arms race (again as a chicken game) or the Star Wars initiative of President Reagan.

2.1 A Game Theoretic Approach of Global Environmental Diplomacy

Having completed the presentation of key game theory concepts, attention now shifts to a game theoretic outline of international negotiations through which the world tries to solve global environmental problems. Such negotiations are usually placed under the auspices of supranational organizations such as the European Union (EU) or the United Nations.

Fiorino (1995) distinguishes three categories of transboundary environmental problems: bilateral, regional and international. As reported by Brandenburger and Nalebuff for the case of businesses (1996), the relationship between national and other agencies of global politics, can best be described as cooperative competition (co-opetition). Transboundary environmental problems such as those related to the anthropogenic contribution to global climate change, conservation of biodiversity, the protection of oceans and the promotion of sustainable development often put countries that are long-term partners and allies, to rival positions on specific environmental issues.

Parties in Global Environmental Negotiations

Susskind (1994) writes that negotiations on global environmental problems take months, years or even decades and eventually culminate in international environmental meetings such as those of Rio de Janeiro (1992, United Nations Conference on Environment and Development, also known as the Earth Summit) or Kyoto (1997, United Nations Framework Convention on Climate Change). Stakeholders in such international environmental meetings include businesses, industries, environmental groups, activists and scientific organizations. Groups not participating in such conferences necessarily rely on representatives to articulate and support their opinion, so meeting participants are under pressure from many interests that want to influence their position. The negotiating committees that represent national agencies receive explicit instructions from the highest levels of their government (such as Brussels or the White House with the assistance of competent environmental government authorities) and often include specialists from these organizations. However, as Susskind mentions, these organizations oftentimes have different agendas and priorities, e.g. Brussels would not want the European negotiating committee to take a position that damages the EU's relationship with allies and partners in other bilateral negotiations such as those on security or economic cooperation. Similarly, an environment ministry wants to ensure that the positions taken by the negotiating committee of its country, comply with applicable environmental laws and regulations and promote the environmental agenda. At the local level, parliamentary or partisan representatives want their opinion to be heard, and may even persist for inclusion in the negotiating committee. When they are present, they may even reject a treaty that could harm their constituency, even if the treaty would benefit the rest of the country or the world, a behavior contrary to the motto of "think globally, act locally".

As pointed out by the same author (Susskind 1994), a host of nongovernmental organizations (NGOs) supplement the actors trying to influence national delegations. Groups coming directly from the grass roots are a strong political force, but rarely exert a direct influence. Extremist environmentalists usually oppose any form of development in areas they consider sensitive. The followers of neoclassical economics and free markets believe that accurate pricing strategies and appropriate financial incentives (rather restrictive regulations) may ensure greater protection of the environment. Consumer protection institutes and left movements seek to ensure that measures to protect the environment do not burden the poor and the weak. Real estate brokers worry that environmental regulations may restrict local investment options in real estate. Bankers are skeptical regarding the impact that actions intended to protect the environmental may have on economic growth. Finally, representatives of scientific groups try to ensure that any political decisions take into account the best and latest research methods and techniques (with a bias towards their own).

To get a sense of the difficulty, it should be kept in mind that global environmental negotiations try to reach an agreement among 170 national delegations, each with its own political agenda, which is in delicate balance with many internal pressures (Susskind 1994). The countries that are represented in the negotiations originate from different geographical regions and political systems: democracies, dictatorships, nations that struggle with poverty and hunger, deprived countries with low per capita income, nations experiencing rapid population growth, newly industrializing countries with little or no enforcement of environmental protection measures and developed countries with elaborate environmental management systems (Rubino 2009). Obviously, the more countries participate, the more difficult it is to reach an agreement. Thus, ambitious plans are often reduced to a small number of real achievements. The conference in Rio, for example, when planned in 1989 by the United Nations General Assembly intended to contain nine individual treaties, which would deal with climate change, transboundary air pollution, deforestation, loss of territories, desertification, biodiversity, protection of the oceans, protection of water resources and, finally, strategies to finance all these measures. Eventually, agreement was reached only on the issues of climate change and biodiversity, with the treaty framework on climate change containing no specific goals or timetable. These are the weaknesses that the 1997 Kyoto summit attempted to correct, although the Kyoto Protocol that was agreed upon, has registered in history as yet another international action in which the world agreed half-heartedly on an extremely costly and inefficient protocol (since the expected delay in global warming by 2100 would be around five years, provided that the Kyoto protocol measures were in full effect until then, as reported by Lomborg 2007), with the United States of America not willing to participate, Canada announcing its retirement on December 12th, 2011 (CBC News 2011) and China and India aware that it was not to their interest to participate. Naturally, the Kyoto accord had strong communication value in local political audiences. Later on, attention shifted to Copenhagen (2009), Cancun (2010) and, lately, Durban (2015), which have undertaken the onerous task of keeping the Kyoto Protocol alive, despite its significant failings (Boehmer-Christiansen and Kellow 2002).

Games in Global Environmental Negotiations

Having presented and described the participants of international environmental negotiations and highlighted the complexity of the relations among them, the focus now shifts to a few appropriate game theoretic models for these negotiations. International negotiations, trying to solve environmental problems of global concern, target the waste of shared natural resources, such as the oceans, the atmosphere or biodiversity. Like the services provided by a lighthouse or national defense, such common natural resources are pure public goods, i.e. (a) they are not traded in markets, (b) the services they provide cannot be allocated to individual portions nor do they decrease with increasing consumption (because of zero marginal cost), and (c) no one can be excluded from using them (Pirages and Manley DeGeest 2004). The overuse of natural resources, a phenomenon named the Tragedy of the Commons by Hardin in his pioneering work (1968), is understood to be a classical prisoner's dilemma game in which players fail to comply with an agreed cooperative solution (which is the most beneficial to society) because breaking their agreement is a dominant strategy for each of them individually (as long as the others comply with it); the problem is that everyone ends up breaking the agreement and the society is worse off for it. In other words, the nature of this game creates free riders, who enjoy the provision of public goods without fulfilling their agreed obligations. Free riding countries pollute the environment, hoping that other countries will comply with international agreements and bear the costs of remediation. Environmental problems such as air pollution, loss of biodiversity and overfishing (e.g. in South America and Asia) arise in this manner.

The task of analyzing global environmental problems with game theoretic concepts is an interesting task (Finus 2001). Susskind (1994) analyzed the diplomacy of international negotiations on global environmental problems (such as those concerning the ozone layer and global climate change) and found that these negotiations included five specific types of confrontations involving: (1) the rich North against the poor South countries; (2) countries having lax environmental regulations and acting as pollution havens; (3) idealists against realists (as regards their expectations on what constitutes reasonable progress in the solution of global environmental problems); (4) optimists against pragmatists (as regards the improvements in environmental quality); and (5) reformers against conservatives (mainly in reference to the structure of the United Nations). These confrontations constitute games, i.e. interactive decision making by actors that try to maximize their payoffs. Of these, the first four are of importance in the context of this chapter and are considered in the following paragraphs.

The Rich North Versus the Poor South

The first and perhaps the most powerful game governing international environmental negotiations is that in which the rich North is pitted against the poor South hemisphere of the earth. Conflict is caused by the difficulty of maintaining a working relationship among the developed countries of the North and the developing countries of the South because the North seems to deny the dream of economic growth and prosperity to the South for the sake of conserving resources and maintaining the quality of the global environment (essentially in order to ameliorate the global climate change). Developed countries have typically resolved many environmental problems through economic growth, as provided by the theory of Environmental Kuznets Curves (EKCs; Grossman and Krueger 1991). In many of the least developed countries, public interest in environmental quality is not seen as a priority, compared to nutrition and survival problems, a reality no doubt made worse by the global economic recession. This divide between the north and the south makes progress on global environmental issues particularly difficult. The challenge lies in recasting this multi-country (i.e. multiplayer) game of conflict in win-win terms so that a cooperative solution may be arrived at for the benefit of the global man-made and natural environment.

Pollution Havens

The second game that may be distinguished in international environmental negotiations is the phenomenon of pollution havens (Neumayer 2001). Some developing countries deliberately loosen up environmental standards in order to attract foreign investments. Thus, these countries gain a competitive advantage over countries that implement stricter environmental regulations. In this complex multiplayer game, businesses and economically developed countries export their wastes to pollution haven countries that are in a position to provide much more inexpensive environmental compliance oftentimes at the cost of the quality of life of their citizens.

Idealists Versus Pragmatists and Optimists Versus Pessimists

The third and fourth game theoretic models that may be recognized in international environmental negotiations, are these in which idealists confront pragmatists and optimists are opposed to pessimists (Susskind, 1994). Actors characterized as idealists are not in favor of environmental agreements or treaties that fail to promote sound environmental management and achieve substantial improvements in environmental quality. They think that a non-deal is better than a weak and abstract deal that essentially allows politicians to continue communicating empty promises, avoid actual obligations and allow problems to remain unresolved, leading to environmental degradation. From another point of view, actors identified as pragmatists believe that every effort, even a moderate one, is an important step in the right direction and consider even the meeting and commingling of official delegations to discuss environmental issues, to constitute progress even if no agreement is reached at; furthermore, they consider even purely symbolic statements from countries to be valuable because they put pressure on reluctant leaders. A great example of the pragmatist viewpoint was given by an Al Jazeera article (Kennedy 2015) in which Professor Carlos Alzugaray of the University of Havana was quoted as saying "*I am an optimist. If you take out all this noise surrounding these talks, the fact remains that there are officials on both sides engaged in discussions on many different issues. Just to have these people in the same room working towards common goals is a good thing and is hopeful.*" Pragmatists believe that a weak agreement can always be enhanced later—the important thing is to have one, anything instead of a failure to reach a deal. As Susskind mentions, sadly, many pragmatic international agreements have failed to produce any substantial improvement because the cooperation of the international community takes so much time that the environmental protection measures originally proposed, no longer make sense, as when efforts to protect a particular habitat become meaningless when the protected species disappear. The other game, of optimists versus pessimists regarding the range and the prospects of global agreements that aim to improve in environmental quality, is of a related nature and concerns the process more than the substance of such global bargaining.

Deviations from Perfect Rationality

Having introduced a game theoretic perspective to international environmental negotiations, some findings are now reported that shed light into deviations from the assumption of (perfect) rationality, a keystone principle of Game Theory.

Fiorino (1995) mentions four models from the field of administrative theory that are particularly useful in understanding how environmental policy is actually produced: the institutional, the systems, the group process and the net benefits model. The institutional model is the oldest one and it is limited to the examination of the legal aspects of public policy, focusing on institutions, laws and procedures. The systems model is inspired by the analogy between social and biological systems, and analyzes the behavior of an organization by examining its inputs, its outputs and the internal processes that transform inputs into outputs. The group process model, which dominated political science for much of the 20th century, considers special interest groups to be the fundamental unit of analysis, politicians to mediate the competition among these groups, and politics to be the product of this competition. Finally, unlike the group process model that is derived from the field of political science, the net benefits model hails from the field of economics and considers politicians to be analysts who take those decisions that maximize utility to society. Further to these models that put the process of global environmental negotiations in a realistic frame of reference, Fiorino further suggests that the concepts of bounded rationality (Simon 1957, 1991), incrementalism (Lindblom 1959, 1970) and the garbage can model (Cohen et al. 1972) are especially useful in understanding how decisions are taken in practice. All of these conceptual approaches suggest that real decisions are made under conditions of limited rationality and incomplete information, a fact confirmed by empirical findings in areas of social psychology and behavioral economics, as outlined below.

On an individual level, it is impressive to consider than decision making is often done unconsciously (Damasio 1994), which evidently means that neither unlimited rationality nor complete information constitute appropriate assumptions in interpersonal communication and negotiations. The picture is further complicated by the fact that communication is often blocked by the intensity of a conversation (Patterson et al. 2002), or that much of the information is not included in the classic channels of oral or written communication, but transmitted through body language (Navarro and Karlins 2008). Such asymmetries are often created in the communication between people of different gender and constitute additional communication channels that transmit information as meta-messages (Tannen 1990). It is even more impressive to consider that decisions are oftentimes shaped not just subconsciously, but within seconds, and may be more informed than a person would tend to think (Gladwell 2005). An interesting example is narrated by Lehrer (2009): a Lieutenant Commander monitoring a radar in the 1991 Gulf War became subconsciously aware that a radar blip was a hostile Silkworm missile (rather than an American fighter jet) and issued the correct command for the target to be fired upon, long before he realized that this was due to an imperceptible discrepancy in the way the blip appeared in radar sweeps that was indirectly related to its altitude. While the above points concern individual decisions, it is reasonable to assume that they also influence decision-making in groups (Baron 1998), such as those of the actors participating in international environmental negotiations.

A final point of interest regarding deviations from perfect rationality, relates to the influence of individual decisions by some rather unexpected external factors predicted by influence science (Cialdini 2009), which shows that people (and not only animals) make heavy use of automatic responses as a way of dealing with everyday complexity and the demand it places on brain activity for rational analysis on an almost continuous basis. Cialdini writes that such automatic responses are generated by the following six principles of persuasion: (1) reciprocation to those who have already benefited an actor, (2) consistency with the prior commitments of an actor, (3) social validation, i.e. compliance with widespread societal practices, (4) readier acceptance of parties that an actor likes and feels friendly toward, (5) obedience to what an actor perceives as power and authority, and (6) preference for what appears to be scarce. If used strategically by an actor, these influence mechanisms can be quite transparent to other parties so they may well constitute formidable weapons in negotiations (Goldstein et al. 2008), especially in those taking place in the context of international environmental policy where, as mentioned previously, participants are involved in games of incomplete information (Alterman 2004; Mearsheimer 2011).

The aforementioned deviations from the assumption of perfect rationality make a player select outcomes with suboptimal payoffs; in fact, as explained by Ariely (2008), not only do players commit such systematic errors, they repeat them in a predictable way. To the important question of whether such individual judgment errors are conveyed to a collective level, Baron (1998) responded positively by employing concepts of social psychology. Baron showed that several judgmental rules, on which people tend to rely to make decisions at an individual level, have an impact on society and policy making. For example, people prefer inaction to actions that are very likely to generate significant benefits, but have an improbable chance of bringing about gravely negative effects as when, e.g. parents hesitate to have their children vaccinated against influenza. People also prefer to avoid disrupting the status quo or anything that is perceived as environmentally natural. Finally, people tend to decide and act in a way that favors the nations, the tribes or other groups to which they belong, even if that harms third parties belonging to groups that are foreign or unrelated to them. Use of these judgmental rules, affects public life and leads to suboptimal collective decisions. Baron asserts that such systematic judgmental errors play a role in gender issues, religious conflicts, resistance to change, nuclear energy policy as well as global environmental concerns such as overfishing and global climate change.

In closing this section, it is pointed out that international environmental problems such as global climate change are characterized by considerable scientific uncertainty, which surely affects their solution by bargaining. Baron (1998) stresses the uncertainty in the forecasts for global climate change expected from man-made global warming, and compares this problem to that of the ozone layer, where the international community was in possession of actual measurements of the formed hole instead of mere predictions of computer models (as in the case of greenhouse gases).

This scientific certainty may have contributed to the speed and efficiency that characterized the cooperation of the international community on the issue of the ozone hole. On the other hand, global climate change is related to carbon dioxide (CO_2) emissions, which in turn are related to the economic development of countries. Indecision characterizes the global community in its effort to combat global warming which surely has more complex economic and political dimensions than the problem of the ozone layer, despite that fact that, as Schelling (1992) warned, the world should be prepared for adverse consequences possibly even greater than those provided by the models.

3 From Game Theory to Complexity

Game theory that was discussed in the previous sections, typically analyzes games of a few players and then attempts to generalize its conclusions to systems of more actors. Yet, the world is a system of many actors, including some 206 sovereign states (http://en.wikipedia.org/wiki/List_of_sovereign_states), supranational entities such as the EU, North American Free Trade Agreement (NAFTA) and the United Nations (UN) as well as many more non-state actors and institutions. Furthermore, constructing game theoretic models of global problems, i.e. representing global affairs with the payoff matrices of game theory, is not trivial because alternative strategies are not known in advance and specific payoffs may be difficult to quantify. So it is tempting to consider that the global community constitutes a multiplayer arena that operates as a Complex Adaptive System (CAS) and examine whether such a complexity approach leads to a complementary research approach that may be of use in the study of world politics.

The Interconnected World as a Complex System
Based on his past experience at the World Economic forum and the Monthly Barometer, Malleret (2012) writes that the global community of 7 billion people and over 200 nations has become a turbulent arena of volatility, random events, uncertainties and challenges that are impossible to predict. Crises that succeeded one another and propagated beyond national borders include the US subprime lending, the EU sovereign debt, the Arab Spring and the nuclear capabilities of Iran. Malleret warns that the world is changing from a world of measurable (economic, geopolitical, environmental, societal and technological) risks to a world of discontinuous uncertainties that cannot be known in advance and cannot be measured. The world of the mid 21st century will likely bear few similarities with the present.

As noted by the naturalist John Muir, everything is linked to everything else in the universe. Indeed, the overriding characteristic of today's world is its interconnectedness. Global affairs are multidimensional, interconnected, interdependent and dominated by a multitude of cross-border relationships (Harrison 2006a). Social, political, and economic phenomena are increasingly being viewed as CASs. In fact, the global community is considered to have become not just a complex system, but a system of complex systems where everyone is interconnected to everyone else and the whole is greater than the sum of its parts (Christakis and Fowler 2011).

How easy is the study of such a complex global network? Complexity breeds more complexity because an arithmetic increase in the number of system elements leads to a geometric increase in the number of potential links and to an exponential increase in the number of possible patterns (Malleret 2012). It is not surprising then that complexity taxes the capability of analysts to understand systems and overwhelms the capabilities of politicians to analyze problems and suggest appropriate policies to solve them.

Jervis (1997) adopts a systemic approach to international politics and asserts that the social and the political life constitute complex systems with many emerging phenomena being unexpected consequences of complexity. Jervis also suggests that most social phenomena are determined by the choices and interaction of individual actors rather than at a systemic level, thus providing support to a multiplayer game-theoretic, i.e. a complexity science approach to the study of world politics. The anarchic global community may be considered a complex system in which states are agents that compete for power and security. Both regularities and non-regularities may result from such a complex global system, affected by things such as the location and the structure of system components. Furthermore, the global community constitutes not just a complex system, but a complex network, in which states exhibit memory, i.e. their behavior is influenced by their past experiences with the interaction among IR agents (memory or feedback) as well as what takes place in another part of the world (knock on effect).

Complexity, Realism and International Anarchy

In the opinion of the author of this chapter, complexity science does not (and probably cannot) constitute a paradigm shift in IR, i.e. it does not challenge, attempt to modify or even supplant established theories such as realism, constructivism or idealism. If anything, complexity may be useful as a tool that allows a better understanding of the phenomenon of international anarchy explained by realism and neorealism. In fact, an agent-centered approach (in the spirit of CAS) may suggest that, like altruism, the efforts to explain global politics with reference to an idealistic worldview are in fact confusing idealism with enlightened self-interest: unrelated agents may choose to ignore Machiavelli's advice (that a prince must learn how not to be good) and realize that they may benefit from cooperation as long as there are mechanisms in place to encourage reciprocity and punish cheating and free-riding (Harrison 2006a).

Getting into a bit more detail in IR theories, materialistic realists assume that political actors pursue power and wealth while idealists that humans should be guided by ideals. Whichever is the case, complexity science sees the actors of global politics (i.e. agents and assemblages of agents referred to as meta-agents) as closely linked and interdependent, in a position to both aid and harm one another (Harrison 2006a; Keohane and Nye 2012). The links that connect them are multiple and interwoven in a complex way, fostering relations among the actors of world politics that are nonlinear and unpredictable. One could argue that both elements of realism and idealism are required to explain developments in regimes such as environmental protection, where on the one hand Singer's deep ecology has introduced environmental assessment and Environmental Impact Statements (EIS) while, on the other hand, states play a prisoner's dilemma game as they try to talk the talk, but avoid walking the walk, i.e. adopting substantial measures that can curb anthropogenic climate change (Harrison 2006a).

The different IR theories in existence underscore the fact that IR scientists hold different opinions on the identity of the basic actors of world politics, the relevance of their interactions as well as the appropriateness of methods to study them. It is not surprising then that some eminent scholars do not consider that a complexity approach is a fruitful approach to the study of global affairs (Earnest and Rosenau 2006). Nevertheless, important scientists such as Nobel laureates Murray Gell-Mann, Kenneth J. Arrow as well as Schelling (2006) have invested efforts in complexity science, believing it to hold great promise for the analysis of IR (Harrison 2006b). As the various flavors of realism and idealism attempt to describe aspects of IR, they may benefit from attempts to provide a more comprehensive picture, a task made more difficult by the rise of nonstate actors and the interconnectedness of the world that is facilitated by the socioeconomic process of globalization and technology transfer (Binnendijk and Kugler 2006).

The author of this chapter believes that neorealism remains the most convincing conceptual model of world politics. To the state actors of this classical approach, complexity science adds an array of other agents that jointly determine the state of affairs in the world: great individuals, important groups of men as well as aggregations of nonstate actors at various levels (Harrison 2006a). So, complexity may be a fruitful methodological approach to the study of world politics with international anarchy being regarded as an emergent property of the complex global system.

Complexity and Technology

Global affairs are not optimization problems to be solved by a mathematical algorithm (Harrison 2006a). Unforeseen innovations become personal, economic, scientific, technological, cultural and political game changers and make predicting the future impossible. Reductionist approaches are suited to linear systems and are inappropriate in the nonlinear behavior typical of a highly interconnected complex system. As Malleret (2012) points out, radical shifts occurring in the fields of geopolitics, economics, society, energy and natural resources are all underscored by momentous technological advances in the form of waves changing the face of the planet and posing unprecedented challenges to world leaders. As the world becomes more connected in an accelerating fashion, the multiple intersecting links create a highly dynamic decision making context for global political leaders, policy and opinion makers as well as CEOs of multinational companies. The resulting complexity is that of a global network that is beyond hope of understanding with traditional linear thinking.

Technology has played a very important role in making the world a highly interconnected and interdependent complex system. Information and Communication Technologies (ICT) have formed a complex, dynamic global system containing billions of entities that interact over multiple spatial and temporal scales (Malleret 2012). Education and knowledge transfer has been globalized, books have become e-books and are purchased and downloaded instantaneously, product and industry life cycles have been shortened and financial markets react immediately to almost everything that happens. This has resulted in an information overload that has put the modern man in a situation not unlike that of one trying to drink from a fire hydrant, more likely to lead to info-paralysis rather than a solution, as aptly pointed out by Malleret.

Global trends are more often than not difficult to predict and, with hindsight, technology is a factor whose impact on economics and society has always been near impossible to predict, e.g. no one in the 1930s or 1940s could have foreseen how personal computers and telecommunications would change the world (Harrison 2006a). The advent of the Internet and the World Wide Web were momentous developments that opened up new possible futures that contained services like file sharing, email, search engines and social media that catalyzed the occurrence of important regional events such as the Arab Spring. The nature of world politics is such that laws, regulations and policies are mediated by socioeconomic conditions and technology and create adjacent possibilities (i.e. alternative futures) that cannot be known in advance, which means that it is very difficult to govern wisely or select those foreign policies that constitute the best responses to future events (Harrison 2006a).

Complex Systems are Interconnected and Interdependent
Complex systems are predominantly systems of interconnected and interdependent components. Herbert Simon said that a complex system is "made up by a large number of parts that interact in a nonsimple way" (Simon 1962). As Malleret (2012) writes, complexity is analogous to the number of components of a system, the interconnectedness of these components and the nonlinearity of the elements of the system. In a complex system, it is typical for changes in one component of the system to lead to unexpected and surprisingly disproportionate effects elsewhere, e.g. the Fukushima nuclear accident impacted the energy policy of European countries such as Germany. Causality is present although hidden and difficult to establish in complex system.

In the 2007 annual meeting of the World Economic Forum, Tony Blair called interdependence the defining feature of the 21st century while Thomas Freedman of the New York Times called the world hyper-connected (Malleret 2012). Professor and former diplomat Kishore Mahbubani coined an apt metaphor for what he called the greatest transformation ever, saying that the world used to be like 170 distinct ships while now it is like 193 cabins on the same boat at sea (New York Times editorial, August 18, 2011). In such a world, assessing issues in isolation (called silo thinking), a remnant of the past, is part of the reason that major crises such as the current global recession or the Arab Spring were not predicted. One of the consequences of this closely knit interdependence is that a complex system points to so many alternative possibilities that any meaningful prediction of the future is nearly impossible (Malleret 2012). Since prediction is impossible, anticipation is the optimal strategy of world leaders.

World politics take place against the backdrop of an extremely complex web of causal links, connecting a multitude of economic, geopolitical, environmental, societal and technological activities. The fate of any single actor, such as a country, is subject not only to its own choices but to the choices of other actors as well (Malleret 2012). In such a complex network, Malleret writes that risk is systemic, e.g. it is often the weakest link that brings about a system-wide crisis such as Greece (a seemingly insignificant player in terms of size) in the recent Eurozone crisis. Another similar example is provided by the United States and China that are interdependent in a very complex and deep way (Harrison 2006a). On the one hand, China has a huge population of hardworking people, but a poor resource base and possibly a diminished moral and political code. On the other hand, the United States possesses vast hard and soft power assets, but suffers from a growing deficit of wisdom and smart power. As Harrison points out, the two countries could build on their complementary need and maintain peace or enter into conflict, with neither course being predetermined nor predictable.

The web of interconnections among the 206 states of world and the thousands of non-state actors, produces a CAS. What happens at borders is very important for the functioning of a CAS (Holland 2012). Such interconnections include many types of cross-border relations, from individuals to states and the transnational level. Although scholars such as Jervis (1997) have analyzed system effects, the full explanation of cross-border complexity remains a difficult undertaking. In global politics, there are complex multilevel interactions of agents such as individuals, special interest groups, non-governmental organizations (NGOs such as Greenpeace), multinational companies (such as Toshiba or 3M), terrorist organizations (such as Al Qaeda), social classes, entire societies, political parties, governments, states, entire civilizations as well as international and transnational organizations (such as the EU or the United Nations) (Harrison 2006a). The effects of these interactions are further complicated by nonlinear responses and positive feedback loops.

Characteristics of Complex Systems

Many of the observations made in the previous paragraphs are summed up in an interesting fashion by Malleret (2012), based on his experience at the World Economic forum and the Monthly Barometer. Malleret argues that four forces constitute the key characteristics of the global community today: interdependence, velocity, transparency and immediacy. It is through the continuous and cumulative action of these forces that the global community exhibits the nonlinearity typical of complex systems, e.g. shocks propagate surprisingly fast and minor causes bring about major impacts oftentimes in a dramatic fashion. Malleret writes eloquently that velocity and immediacy, mediated by information and communication technologies (ITCs) especially the Internet, have created a "dictatorship of urgency" that runs financial markets, geopolitical upheavals, regional as well as social discontent (e.g. the Arab Spring) fast forward and allow no time to pause and ponder over developments. The effects of velocity and immediacy may be discerned in business and economics (e.g. just-in-time supply chains, high-frequency trading, news traveling around the world instantaneously, electronic books being bought and downloaded immediately) and have even spilled over into social life (e.g. fast food or speed dating). As Malleret says, as the economic value of time has increased, so has its economic scarcity. While technology and the Internet created velocity and immediacy, the social media fathered transparency which, in turn, fostered more interdependence and complexity (by exposing all dimensions and participants of current events, activating even more links). In particular, as Malleret observes, the Internet makes the younger generation (that is more adept at using it) more aware of the corruption of its leaders and facilitates the organization and orchestration of its reactions (as in the case of Tunisia and Egypt). Social media has been vested with the power to transcend occasionally the power of authorities and even the government and that is why anxiety about the social media is evident in countries such as Saudi Arabia and China. Transparency has also limited the confidentiality of many activities, forcing even Swiss bank accounts to divulge information about their clients to inquiring foreign authorities or governments. Not to mention that there can be no privacy on the Internet as long as every move of Internet users is worth something to someone and thus tracked systematically (Malleret 2012). Malleret recaps his presentation by arguing that these four characteristics of today's world have brought about confusion, a multitude of weak signals as well as asymmetry and have underscored the global government vacuum.

The behavior of a complex system emerges from the interaction of its elements (Casti 1994; Holland 1996; Levin 1999). The elements of a complex system adapt to the actions of other elements and their environment (hence such complex systems are referred to as CAS). CASs typically operate several positive feedback loops (based on memory and historical analysis), which cause small-scale disturbances to be amplified to large-scale effects. Because the interactions among the elements of a complex system are nonlinear, study of their components in isolation cannot be used to predict their outcomes (Hendrick, 2010). Harrison (2006a) essentially agrees with Hendrick, arguing that the complexity science approach considers world politics to be unpredictable in part due to chaotic dynamics, positive feedback mechanisms and the surprising power of small events. With the aid of the right tools (such as agent-based modeling, discussed in a following section) patterns may be discerned in the turbulence of world affairs and be used to understand the presence and anticipate (if not predict) the future.

There are quite a few celebrated examples of complex systems such as the weather, a beehive, or the human brain, which appear to defy a reductive approach in part because of the presence of nonlinearities. In most complex systems, outcomes are not proportional to causes and systems cannot be understood by analyzing their parts (Kauffman 2008). Nonlinearities interfere with a linear approach to aggregation and make the behavior of many complex systems more complicated than predicted by summing up the behavior of its components (Holland 1996). While linear systems are near equilibrium, complex systems typically exhibit nonlinear behavior that is far from equilibrium. Data collected from nonlinear systems typically are not normally distributed, i.e. they do not adhere to the familiar bell-shaped distribution associated with linear systems (Rihani 2014) and they tend to render quite a few outliers, making nonlinear systems less predictable. This uncertainty appears to constitute an essential part of phenomena that are now described as nonlinear.

Interestingly, although most complex systems do not exhibit linear behavior, they appear to be characterized by economies of scale, e.g. as cities expand, there are fewer gas stations per capita; and as the number of consumers increases, the length of power lines needed to serve consumers decreases (Harrison 2006a). Holland (1996) argues that another important effect of complex adaptive networks is recycling, which occurs as flows are recycled at some of the nodes of a complex network. The overall effect of many such recycles can be unexpected and contributes to the irreducibility and nonlinearity of complex systems. Furthermore, complex systems oftentimes create situations of asymmetric information, where one party is in possession of much more knowledge of the specifics of a situation than the other party and may use this asymmetry to its advantage (Malleret 2012). Cyber conflict epitomizes the power of asymmetry, which is the reason that much of what happens in the world today is beyond the control of even the most powerful states.

The maxim that complexity is the enemy of control rings true as CASs appear to be both robust and fragile and are characterized by tipping points, i.e. levels after which cascading effects are kick started by connections that used to absorb shocks, but after a certain point become shock multipliers. The behavior after a tipping point is further amplified and accelerated by positive feedback, another hallmark of complex systems. The entire process may be dynamic in a nonlinear and chaotic manner that is typified by the unexpected occurrence of rare phenomena with dramatic impacts. Taleb (2007) uses the term black swans to refer to such exceptional critical events that appear to occur with a probability higher than expected (i.e. they correspond to thick tailed distributions), making successful prediction more difficult. World affairs are often determined by such unlikely and unprecedented critical events that show up unexpectedly (like "fifty-year floods") like the First World War (Harrison 2006a). The same source describes how in a complex world dominated by chaos and uncertainty, agents and meta-agents (populated by people and institutions) have to face (and try to avail of) shocks caused by such unexpected developments.

Complexity Empowers the Individual

Another important aspect of complexity is that it empowers the individual. It is difficult to deny the influence of individual men on global politics (Harrison 2006a). In their classic analysis of the Cuban missile crisis, Allison and Zelikow (1999) focus on system-level and state-level explanations although it is not difficult to agree that other alternative possibilities might have opened up if Nikita Khrushchev, the Kennedy brothers and Robert McNamara were not present. In a similar vein, it is difficult to imagine how the 20th century would have played out without Hitler, Mussolini, and Stalin, but not difficult to agree that it would not have been the same (Harrison 2006a). A final example of how a personal relationship may have facilitated state politics is given by the evident existence of good chemistry between Papandreou and Netanyahu, Prime Ministers of Greece and

Israel correspondingly (Tziampiris 2015) that appears to have helped move the Greek-Israeli relationship forward. While these examples stress the importance of the actions of individual actors in a complex world, this argumentation is not to say that the complexity science approach to world politics accepts any monocausal model of the behavior of states and other actors (Harrison 2006a): on the contrary, complexity implies that most events are due to multiple causes and it can be very difficult to disentangle the influence of each individual cause.

Evolution and Adaptation

As the Nobel laureate Gell-Mann has reported, complexity has a tendency to grow over time. This happens with small changes that occur in the behavior of individual agents as they interact locally and adapt their rules according to the input they receive from their environment (in order to maximize their payoff, which may equal power or the possibility of their survival or that of their offsprings). As Harrison puts it (2006a), a fundamental pattern of CASs is to evolve by combining new and old building blocks and thus form unexpected emergent properties and new adjacent possibilities, i.e. alternative futures. Evolution itself is complex and even shaped by co-opetition, i.e. the simultaneous cooperation and competition of actors depending on the circumstances (Bradenburger and Nalebuff 1996; Minelli and Fusco 2008; Flannery 2011). Furthermore, the direction of evolution is impossible to predict as each adaptation opens the way to additional alternative possibilities and innovations (Harrison 2006a). As an example, consider how mainframe computers gave rise to portable computers which, in turn, opened the way to mobile telephones, the Internet, social networking and the Arab Spring. As complexity grows and deepens, it tends to breed yet more complexity and it is very difficult to predict how evolution will proceed in technological, social and political systems.

Evolution and adaptation characterizes CASs, i.e. complex systems that evolve and adapt over time. As Holland (1996) reports, the timescale of adaptive modifications varies from one system to another: while in CASs such as the central nervous system or the immune system, adaptation takes seconds to hours or days, in the field of business and economics it takes months to years while in ecology and the environment it takes days to years or even centuries. CASs may be viewed as systems composed of interacting agents whose behavior may be described with stimuli-response rules (Holland 1996). These agents can adapt to their environment, which includes other adaptive agents, by changing these rules as experience accumulates. When agents learn from their environment and modify their behavior, the system itself is modified. A large portion of an agent's time is spent adapting to other agents that are themselves adaptive (Holland 1996). All of the above imply the existence of multiple feedback loops in CASs that beget nonlinearity and the emergence of large-scale aggregate patterns that are impossible to predict by examining the behavior of an individual agent.

Holland (1996) writes that the behavior of CASs is determined more by the interactions among the agents rather than by their individual actions. Also, many CASs display an amplifier effect (related to nonlinearity): a small input can bring about major impacts. Nevertheless, researchers at the Santa Fe Institute believe that there are general principles that rule the behavior of CASs and that these principles may point to the way of answering questions and solving problems related to such systems (same source). Indeed, the world may be thought of as a CAS in which actors, i.e. agents of world politics, continuously adapt their behavior to the external signals they receive. In such a complex world, even a relatively small event may propagate like a contagious disease and end up being the main cause of consequences disproportionately big. In politics, uncertainty (amplified by complexity and fear) and nonlinear dynamics play a significant role as events unfold (Malleret 2012). Important questions must be answered such as: Is increasing complexity a sign of successful adaptation? Is greater complexity good or bad in IR? Is greater complexity a sign of stability, as in biodiverse ecosystems?

Evolution and Fitness

The complexity science approach essentially introduces concepts of adaptability, evolution and fitness to the analysis of world politics (e.g. Kauffman 1993 and other works associated with the Santa Fe Institute). Fitness results from evolution and measures the ability of states to cope with the complex global community (Harrison 2006a). As Nye points out (2011), the fitness of states and other global community agents entails the smart power that manages to amalgamate aspects of both hard and soft power into wise policies.

Harrison (2006a) mentions that fitness may be defined as the capacity of an actor to cope with complex challenges, secure its survival and advance other interests. The optimal fitness of every organism, including whole societies and the international system, is located somewhere between rigid order and chaos (i.e. anarchy in politics) but nearer to the edge of chaos (same source). Harrison also argues that fitness results from self-organization and creative responses to complexity rather than from conflictual, zero-sum approaches to problems. An agent or meta-agent that is fit, is one that thrives on the complexity of the global community.

How can fitness be measured? Harrison (2006a) argues that societal fitness may be defined as the ability to cope with complex challenges, which may be closely associated with culture, i.e. the values and way of life of each society. Some cultures may enable outstanding individuals to rise while others may suppress them; some cultures may make excellent use of their resources, while others may waste them. Harrison also suggests that quantitative measures of fitness of countries could be estimated with data such as the United Nations Human Development Index (HDI, measuring longevity, education, and income), rankings of democratization, honesty, the extent of knowledge-based economics from Freedom House, the Bertelsmann Foundation, Transparency International and the Harvard-MIT Index of Economic Complexity. Harrison further suggests that a top-down dictatorial rule with a rigid hierarchy may imply low fitness while a democratic self-organization that fosters creativity and mutual gain may be associated with high fitness. On the other hand, Malleret (2012) argues that the quantification of the fitness of state actors may not always be helpful, e.g. it is difficult to conclude that Italy (that has had about 60 changes in its governmental regime since the Second World War) is less stable than Saudi Arabia that has been run by the same family for the last 40 years.

Complexity and Future Trends of the Global Environment

What about the interplay between anthropogenic activities and the natural environment in the complex global community for the rest of the 21st Century? Malleret (2012) asserts that the following six trends will shape the future of the global community: (1) unfavorable demographics, (2) resource scarcity, (3) climate change, (4) geopolitical rebalancing, (5) indebtedness and fiscal issues as well as (6) rising inequalities.

Unfavorable demographics affect both developed and developing countries, as societies trend towards the feared 4-2-1 paradigm, i.e. one working young adult supporting two parents and four grandparents. A chronic shortage of women along with the arrival of an age wave with no pensions, no health care and no family to support the elderly, is likely to be a recipe for major social turmoil in areas such as Asia and unimaginable ramifications worldwide, considering how difficult the negotiations over retirement age in Europe and Medicare in the US have been.

Resource scarcity is likely to be caused by increased demand for water, food, energy, land and mineral resources and amplified by the complex linkages that exist among them (Malleter 2012). This increased demand will be exerted by forces such as globalization, urbanization, resource nationalism, geopolitical events that translate to negative supply shocks, the rise of emerging powers such as China and India, the prevalence of cheap technology, increased per capita consumption (attributed to the development of developing countries) as well as changes in diet and crop production for biofuels. Malleret also writes about "land grab," an interesting example that shows how resource scarcity may act together with other global mega-trends to produce surprising and serious knock-on effects in an interconnected world. Nowadays, large food importers like China, India, South Korea and Saudi Arabia tend to acquire considerable tracts of land to produce soft commodities for internal consumption. As two examples, South Korea (which imports 70 % of the grain it consumes) has acquired 1.7 million acres in Sudan to grow wheat, while a Saudi company has leased 25,000 acres in Ethiopia to grow rice (with an option to expand further). As Malleret explains, the problem is that the land grabs in Ethiopia and Sudan (which together occupy three-fourths of the Nile River Basin), have affected the flow of the Nile in the Egyptian part, which was supposed to be controlled by the Nile Waters Agreement (signed in 1959) that gave Egypt 75 % and Sudan 25 % of the river's flow. The situation has changed de facto since the wealthy foreign governments and international agribusiness companies that have snatched up large pieces of arable land along the Upper Nile, are not parties to the 1959 agreement. This is another example that shows how unexpected situations may arise in a highly interconnected world in which the power has diffused into many non-state actors.

Anthropogenic climate change, another global environmental concern, is increasingly becoming a source of geopolitical tensions and a security threat, e.g. via the intensification of water balance and water availability issues, which generate tensions and contribute to country failures in regions such as North Africa, the Middle East and South Asia (Malleret 2012). It is unfortunate that problems that are likely to be caused by anthropogenic climate change constitute a long-term threat;

as a result, policy-makers are unwilling to go into the trouble of formulating and implementing sensible climate-change policies now for fear that they will suffer immediate tangible consequences (becoming unpopular and losing the next elections).

Fiscal issues such as indebtedness are global trends that affect the richest countries, but are also likely to have knock-on effect elsewhere (Malleret 2012). Malleret points out that rich countries are likely to face lower growth and higher unemployment compared to the period that started in the mid-1980s and ended in 2007. Nevertheless, debt is likely to deepen the divide between the North, which is mainly composed of surplus countries with solid fiscal positions, and the South, which contains deficit countries with rather unsustainable fiscal positions.

A similar inequality is caused by fuel poverty, i.e. the inability of households to afford adequate warmth at home, one of the most prominent social problems of the 21st century (Boardman 1991, 2010). As pointed out by Preston et al. (2014), fuel poverty constitutes a triple injustice faced by low income households that are fuel poor: they emit the least, they pay the most yet they benefit the least (from pertinent policy interventions). So, while the poor consume and emit less, the cost of energy represents a much higher proportion of their income and their homes are usually poorly built and less energy efficient. As pointed out by the research of Paravantis and Samaras (2015), Paravantis and Santamouris (2015), Santamouris et al. (2013, 2014), fuel poor families (that are literally forced to choose among heating, eating or paying the rent) are caught in a fuel poverty trap that is self-reinforcing and very difficult to escape from, not unlike the digital divide that exists among individuals, households, businesses and geographic areas at different socio-economic levels with regard to their access and use of ICT (Iliadis and Paravantis 2011). Taking tips from complexity and the importance of weak ties (Granovetter 1973), Preston et al. (2014) proposed that energy counselling be done within the community via "green doctors" and "energy champions" that provide advice and support to fuel poor families.

Rising inequality is another global trend according to Malleret (2012) and "the single biggest threat to social stability around the world" according to economist Ken Rogoff, as people at the top have managed to avail themselves of the best educational and professional opportunities, seize most wealth and entrench themselves in privileged position, putting the prospect of a good life for most others beyond reach. On top of that, globalization expands the market and increases the demand for highly skilled individuals, but competes away the income of ordinary employees (Malleret 2012). As this inequality of opportunity and lack of transparency has exacerbated the sentiment of unfairness, Malleret asserts that a tipping point may emerge beyond which social inequality will undermine societal and global cohesion, economies will become unstable and democracy will be compromised.

Complexity and the Future

What kind of games will be played in the future complex arena of the rest of the 21st century? Gideon Rachman of the Financial Times has identified six divides that are likely to constitute fault lines and perturb the geopolitical landscape (Malleret 2012): (1) surplus countries (like Germany, China, Japan and Saudi Arabia) versus deficit countries; (2) currency manipulator versus "manipulated" countries; (3) "tightener" (like the UK) versus "splurger" countries; (4) democracies versus autocracies; (5) interventionist versus souverainist countries; and (6) big versus small countries, pitting the G20 countries against the rest of the world. Instead of one or two superpowers, Malleret thinks that the world is likely to see a number of dominant powers in the coming decades of the 21st century, i.e. China in East Asia, India in South Asia, Brazil in Latin America, Nigeria in West Africa, and Russia being a major resource power (thanks to its oil and gas exports). Malleret asks: Will the US decline? Will Europe fade into irrelevance? Will the BRICS overtake the world? Will the pauperization of the middle class continue and dominate the social arena? In a complex system that is highly interconnected, fully transparent and moving at a very high velocity, such as the global community, it is impossible to know: one can only anticipate and plan ahead.

3.1 Emergence in World Politics

CASs have many interconnected elements that interact locally and support the creation and maintenance of global emerging patterns while the system remains enigmatically coherent (Holland 1996). In fact, Holland (2014) writes that emergent behavior is an essential requirement for calling a system complex. As Rihani (2014) points out, change at the micro level takes place continuously; at the macro level, the system scrolls through many microstates as global patterns or attractors eventually emerge. Emergence occurs across the levels of a complex system (Railsback and Volker 2012) and both regularities and non-regularities may emerge, affected by things such as the location and structure of system components (Jervis 1997).

Emergent phenomena or behaviors appear at the system level. Examples include the anarchy of the international community, the bipolar structure of the Cold War and the collective memory that is oftentimes exhibited by the international community, as if it were an intelligent being (Jervis 1997). It is important to keep in mind that emerging behaviors in complex systems do not require the existence of the invisible hand of a central controller, a fact that agrees with the realist approach and a complexity science approach to the analysis of global politics (Harrison 2006a).

The Emergence of Cooperation

Cooperation is an essential trait in societies, e.g. to build and fly the International Space Station required the collaboration of thousands of people, most of whom never met in person, spoke different languages and lived very differently (Harrison 2006a). How does cooperation emerge in the CAS of world politics? As Axelrod wrote (1985), "Under what conditions will cooperation emerge in a world of egoists without central authority?"

IR and social institutions, like anything that evolves, have likely been impacted by accidental developments caused by novel combinations of ordinary features (Harrison 2006a). This makes it tougher to address several questions pertaining to emergence in world politics. Will conflict or cooperation be more prevalent in the years to come? Is unipolarity, bipolarity or multipolarity of the distribution of international power more conducive to the overall fitness and stability of the global community (Harrison 2006a)? Was the world more stable during the four and a half decades of relative calm of the Cold War and would more nuclear weapons in the hands of even states like Iran be conducive to a more stable world, as Kenneth Waltz has argued (2012)? Is the world better off now, in the era of a relatively undisputed American hegemony and would it be less stable in a likely multipolar distribution of power among the United States, China, India and other actors including smaller states with nuclear capability? Will the complexity that results from global interdependence make war unthinkable and foster peace, as the numerous intersecting interests that link countries render no single interest worth fighting for (Keohane and Nye 2012)? Or do the current troubles of the EU point to a more alarming direction, in which the more interconnected countries are, the easier it is for them to disagree, enter into disputes that are difficult to resolve and end up despising one another? Perhaps it was easier to reach agreement among the six states that formed the European Coal and Steel Community in 1951 than among the 28 EU members at the present time, as Harrison points out (2006a).

While attempting to address such questions with complexity tools may not be trivial, simulating CASs with agent-based models is a worthwhile pursuit in the context of several problems of world politics, as discussed in the next section.

4 Simulating Complexity with Agent-Based Modeling

Problems of contemporary society like inner city decay, the spread of AIDS, diffusion of energy conservation habits or diffusion of the Internet are probably best addressed by resorting to the workings of CAS elements at a micro level (Holland, 1996). CASs show coherence and persistence in the face of change while at the same time they display adaptation and learning. Such systems work based on extensive interactions among their elements. Having borrowed the term from economics, Holland uses the term agent to refer to the active components of CASs.

Agent-Based Simulation

Outside the fields of computer science and engineering, few applications of ABMs are used to solve practical problems (Railsback and Volker 2012). In a historical review and tutorial paper, Gilbert and Terna (1999) suggested that agent-based modeling is an alternative way of carrying out social science research, allowing for testing complex theories by carrying out individual-based numerical experiments on a computer and observing the phenomena that emerge. This contrasts with global politics and IR, fields in which theory is mostly formalized verbally. Axelrod (2003) described simulation as a new way of conducting research that is additional to induction and deduction. Furthermore, he asserted that although the use of simulation in social sciences remained a young field, it showed promise that was disproportionate to its actual research accomplishments at the time of his writing. Agent-based simulation is an important type of social science simulation that is in need of progress in methodology, standardization and the development of an interdisciplinary community of interested social scientists. The addition of regional and world politics to the fields that have traditionally employed decentralized adaptive system approaches (such as ecology, evolutionary biology and computer science), will be a means of advancing the state of the art.

In a colloquium paper, Bonabeau (2002) concluded that agent-based models constitute a powerful simulation technique that had seen quite a few applications at the time of writing and carry significant benefits for the analysis of human systems (although certain issues had to be addressed for their application in social, political, and economic sciences). Bonabeau described several real-world application examples of agent-based simulation in the business fields of flow simulation, organizational simulation, market simulation, and diffusion simulation. It was concluded that agent-based models are best used when (a) the interactions among the agents are discontinuous, discrete, complex and nonlinear; (b) the position of agents is not fixed and space is an important aspect of the problem at hand; (c) the population of agents is heterogeneous; (d) the topology of interactions is heterogeneous and complex; and (e) agents exhibit a complex behavior that includes learning and adaptation. The author suggested that issues that have to be taken into consideration when agent-based models are applied include: (1) a model has to be built at the right level of description (not too general, not too specific); and (2) careful consideration should be given to how human agents (i.e. actors with imperfect rationality, as discussed previously) exhibit irrational behavior, subjective choices, and complex psychology, i.e. soft factors that are difficult to quantify and calibrate. Although the author pointed out that this is a major hindrance in employing agent-based simulations in a meaningful fashion, at the same time he thought it is a major advantage of such approaches, i.e. that they are in fact the only tool that can deal with such hard to quantify applications.

How Agents Behave

As Rihani (2014) writes, a complex regime is a mix of global order and local chaos. Patterns emerge at a system level due to the chaotic interaction of agents locally. The behavior of an agent is determined by a collection of stimulus-response rules describing the strategies of the agent (Holland 1996). Oftentimes the performance of an agent is a succession of stimulus-response events. As Holland pointed out, modeling CASs is done by selecting and representing stimuli and responses. A range of possible stimuli and a set of allowed responses are determined for specific types of agents. The sequential action of these rules determines the behavior of an agent. This is where learning and adaptation enter the picture. Agents also tend to aggregate into higher level meta-agents that themselves can aggregate into meta-meta-agents, which results in the typical hierarchical organization of a CASs. Although the existence of agents and meta-agents is usual in complex systems, the components of CASs are often structured in unexpected and innovative ways (Malleret 2012). Furthermore, the formation of aggregates is facilitated by tagging as when a banner or a flag rallies members of an army or people of the same political or national persuasion (Holland 1996). Tagging is a pervasive mechanism that both aggregates and forms boundaries.

Agent-Based Networks

Many complex systems are networks with nodes and links that connect nodes (Holland 1996). Nodes represent agents and the links designate the possible interactions among the agents as well as provide transport routes for the flow of various items, e.g. goods or resources, from one node to another. In CASs the flows along the connecting links may vary over time; moreover existing links can disappear or new links may appear as agents adapt. In other words, CASs may be represented with networks and flows that are dynamic rather than fixed in time. As time elapses and agents accumulate experience, changes in adaptation are reflected in different emerging patterns, e.g. self-organization. As flows move from node to node, i.e. agent to agent, their impact is oftentimes multiplied, a phenomenon called multiplier effect that is a typical feature of CASs that may be represented by networks (same source). Holland writes that the multiplier effect is particularly evident when evolutionary changes are present, and it is one of the reasons why long-range predictions are difficult to carry out successfully in the case CASs.

Developing an Agent-Based Model

ABMs are complex systems containing a large number of autonomous agents that interact with each other and their environment (Railsback and Volker 2012). Where in mathematical models, variables are used to represent the state of the whole system, in ABMs the analyst models the individual agents of the system. Agent-based modeling can help find new, better solutions to many problems in the fields of the environment, health, economy and world politics (Railsback and Volker 2012; North and Macal 2007; Sterman 2000). Historically, the complexity of scientific models has often been limited by mathematical tractability. With computer simulation of ABMs, the limitation of mathematical tractability becomes irrelevant so one may start addressing problems that require models that are less simplified and include more characteristics of the real systems (Railsback and Volker 2012). ABMs are easier to employ while, at the same time, more sophisticated than traditional models, in that they represent a system's individual components and their local behavior rather than a mathematical representation of the macro behavior of the system (same source). Railsback and Volker also point out that, because ABMs are not defined in the language of differential equations or statistics, concepts such as emergence, self-organization, adaptive behavior, interaction and sensing are easily recognized, handled and tested when simulating with ABMs.

A model is a representation of a real system (Starfield et al. 1990), as simple as possible, but not simpler, as Einstein famously said. Usually, the real system analyzed by modeling and simulation is often too complex to be analyzed with other means such as experiments. A model is formulated by deciding on its assumptions and designing its algorithms. A decision must be made as to which aspects of the real system must be included in the model and which may be ignored. The modeling cycle is an iterative process of translating a scenario into a problem, i.e. simplifying, designing, coding, calibrating and analyzing models and then using the model to solve problems that are often intractable to solve with analytical approaches. More specifically, the modeling cycle includes the following stages (Railsback and Volker 2012):

(1) formulate clear and precise research questions that are narrow and deep;
(2) construct appropriate hypotheses for testing the research questions, possibly adopting a top down approach;
(3) formulate the model by choosing an appropriate model structure, i.e. scales, entities, state variables, processes, and parameters, starting with the simplest constructs that may test the hypotheses;
(4) implement the model using computer programming, including modern languages such as NetLogo (Wilensky, 1999) or other tools mentioned in the list of resources that may be used to implement and test simple ABMs in a short amount of time;
(5) run the model enough times to analyze, test and revise the hypotheses;
(6) communicate the final results of the simulation carried out with the model (not forgetting that the model is not identical to the real life problem, but a simplification of it).

Agents in ABMs may be individuals, small groups, businesses, institutions, nonstate actors, states and any other IR entity that pursues certain goals (Railsback and Volker 2012) and competes for resources. Agents in ABMs interact locally, meaning that they do not interact with all other agents in the system, but only with their neighbors geographically in the space of the system (same source), which in the case of IR is the highly interconnected network of the global community of states. Agents are also autonomous, implying that agents act independently of each other, pursuing their own objectives, although their behavior is adaptive as they adjust it to the current state of themselves, other agents and their environment.

ABMs may be used to look both at what happens (a) to the system because of the actions of individual agents and (b) to the individual agents because of how the system functions (Railsback and Volker 2012). Modeling and simulation via ABMs allows a researcher to reproduce and investigate emergence, i.e. dynamic phenomena at a system level that arise from how the individual components adapt by interacting with one another and their environment. The recognition of emergent phenomena, offers the only chance of prediction with the case of complex systems although it is not always easy to decide whether a pattern that emerges in an ABM is a regular pattern or a black swan event (same source).

Examples of Simple Agent-Based Systems

To give a rough idea of what may be achieved with agent-based modeling, two examples are now described: a simple educational example and one made famous by Thomas Schelling, its Nobel laureate inventor.

The author of this chapter teaches Game Theory and Complexity regularly to his undergraduate and graduate students in the Department of International and European Studies of the University of Piraeus. Both subjects are well fitted to participatory simulation in class. In one of his favorite in-class activities, the author tells his students to stand up and walk around and then he asks them to disperse so that they are homogeneously scattered in the classroom. In a first version of the activity, the author assigns the role of coordinator to one of the students, who directs each student's movement so that they all are dispersed. In a subsequent version, the author just tells the student to move as they see appropriate so that they are all dispersed more or less at an equal distance from one another and the walls.

The students quickly realize that the easiest way to achieve dispersion is when each moves on his or her own without a coordinator issuing directions. In fact, with the help of the author, they understand that all they need to do to disperse homogeneously is follow a simple rule: see who is closest to you and move away from him or her.

This activity is then demonstrated to the students with the help of a simple agent-based model programmed in the PowerBASIC Console Compiler version 5.0.5 (http://www.powerbasic.com/products/pbcc), running on a 64bit Windows 8.1 notebook computer. BASIC is a procedural language and, just as proposed by Axelrod (1997), despite its age, remains a fine way to implement quick programming tasks especially with social scientists who may not be as well versed in modern object-oriented programming techniques.

The results are shown in Figs. 1, 2 and 3 that depict a population of 160 agents. Figure 1 shows these 160 agents moving randomly, just as students do in the beginning of the in-class simulation.

Figure 2 sort of looks under the hood by having these 160 agents continue moving randomly and displaying the distance of each to its nearest agent as a solid line.

Finally, Fig. 3 shows these 160 agents moving away from the nearest agent (plus a small random jitter so that they never become immobile, for educational purposes).

The students are impressed to see how quickly (instantaneously in fact) the agents disperse in a perfectly homogeneous manner just by moving away from their closest neighbor. It is explained to the students that the homogeneous dispersion of the agents is an emergent phenomenon of this complex system and it is stressed to them that it is simple rules like this that oftentimes effect unexpected system-wide patterns such as self-organization.

Attention now turns to the second and more famous example, Schelling's segregation model (Schelling 1969, 2006). Thomas Schelling showed that even a socially fair preference for half of one's neighbors to be of the same color could lead to total segregation, placing pennies and nickels in different patterns on graph paper and then moving unhappy, ones one by one, to demonstrate his theory.

A version of Schelling's model was programmed in Netlogo (Wilensky 1999), simulating homes of two races, "black" (pictured as the darker color in Figs. 4 and 5) and "white" (pictured as the lighter color in Figs. 4 and 5.) In the version of the

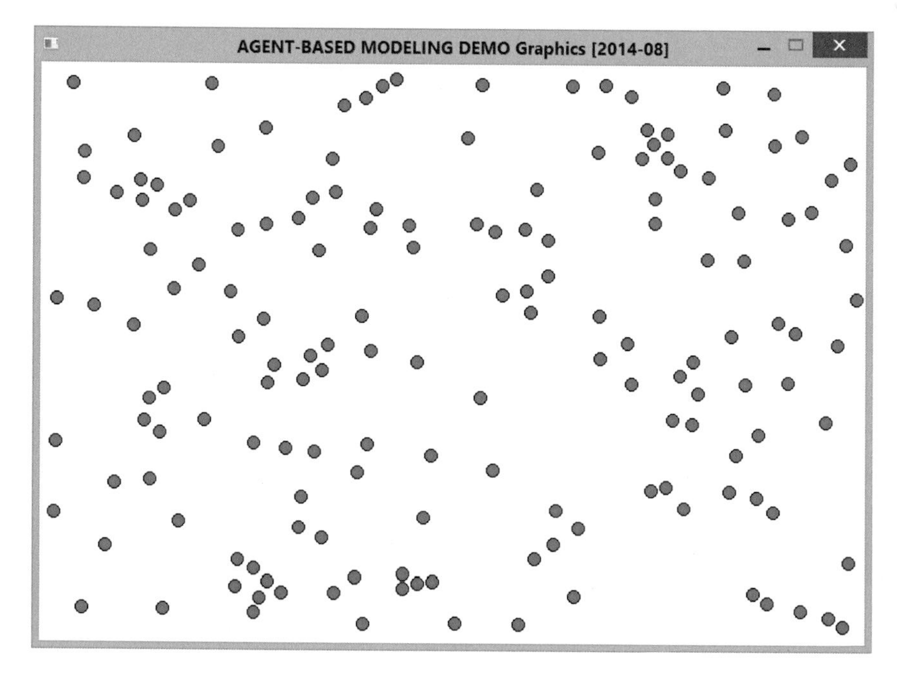

Fig. 1 160 agents moving randomly

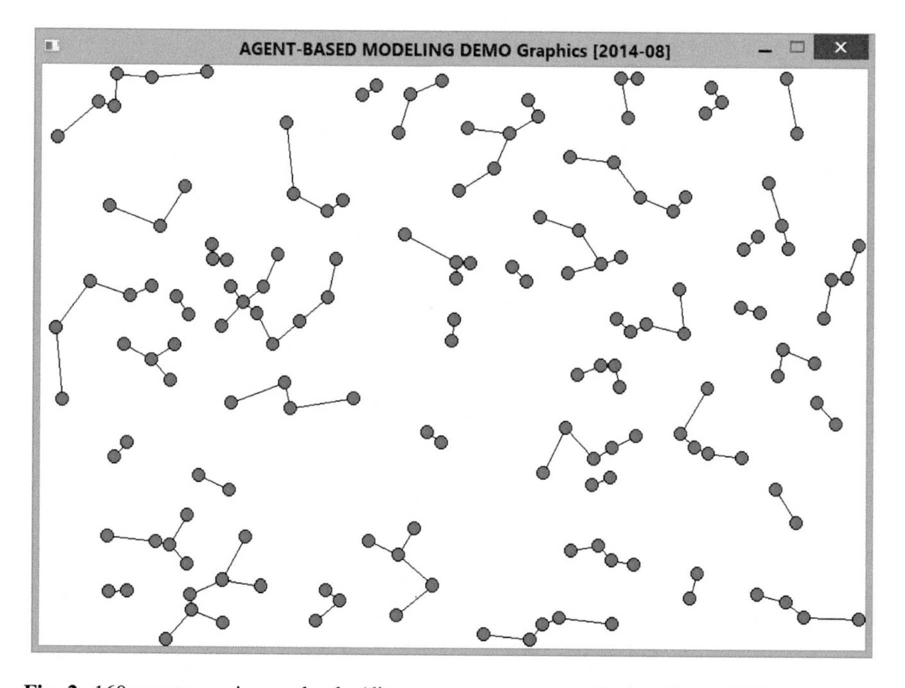

Fig. 2 160 agents moving randomly (distance to nearest agent displayed as *solid line*)

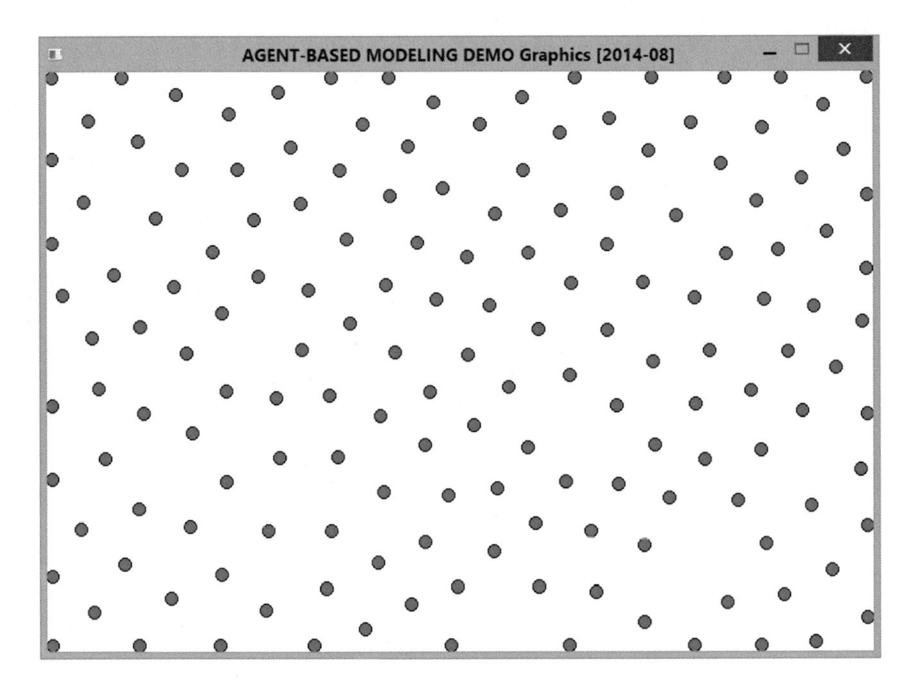

Fig. 3 160 agents moving away from nearest agent (plus a small random jitter)

segregation model simulated herein, homes are placed in a grid (employed in many cellular automata applications) so that each home has eight adjacent homes. A home would be perfectly happy if it had an evenly split number of neighbors of the two different races around it, i.e. four “black” and four “white” ones, and increasingly unhappy as the split moved further away from this socially just configuration. If unhappy, a home swaps places with the unhappiest home among its eight adjacent neighbors (if such a home existed.)

Figure 4 shows the initial random configuration of a neighborhood with one third (33 % requested by the user, 32.28 % achieved by the use of random numbers) “black” homes.

Figure 5 shows the configuration of the neighborhood after about 2500 iterations (taking about 30s at Netlogo’s fast speed on an Intel i7, 64-bit Windows 8.1 computer.) One may notice that the swapping of unhappy homes ended up in total segregation, despite the fact that individual preferences were for a perfectly balanced assortment of “black” and “white” neighbors around a home. The resulting total racial segregation is an unexpected emergent phenomenon of this Netlogo model that validates Schelling’s proposition and explains (to some extent) the racial segregation observed in many North American cities.

Validation of Agent-Based Models

While the verification and validation of any model is a critical simulation step, it is nontrivial to implement in the case of models describing social and political

Fig. 4 Random initial configuration of a simulated neighborhood with one third “black” homes

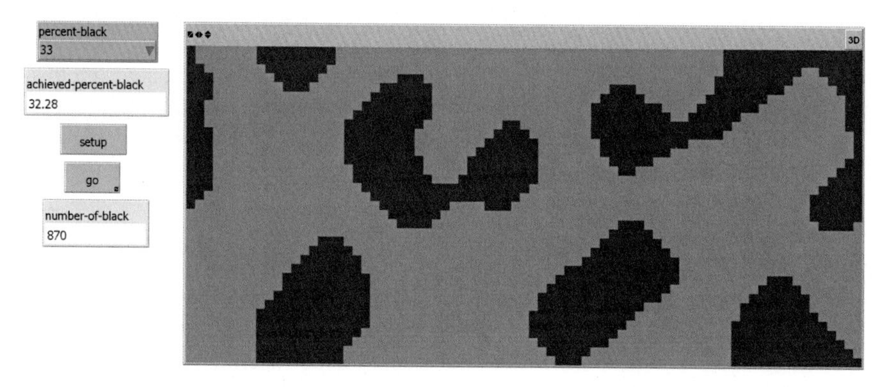

Fig. 5 The simulated neighborhood of one third “black” homes showing total segregation after many iterations

phenomena, where one is often forced to rely (to a large extent) on qualitative information (Bharathy and Silverman 2010). Global politics, in particular, with their complex path dependence and unpredictable emergence, appear to pose insurmountable verification and validation issues to researchers who attempt to analyze them with modeling.

As Schreiber (2002) explains, the way information is processed by a model, differs from how it is processed in the real world. This is also the case with agent-based models: they produce similar, but not identical output, to their targets, making ABMs paramorphic analogues of the real world phenomena they simulate. Instead of using statistical validation techniques, analysts of global political systems are oftentimes forced to resort to the believability of the output of their analyses. Certainly, it would be wrong of a researcher to expect to derive proof from the analysis of any political and social phenomenon by modeling and simulation, especially in the context of postmodernism. Nevertheless, the verification (of specifications and assumptions) and the validation (of the accuracy of the output) is a necessary endeavor in the modeling of political and social phenomena. Schreiber endorses this view by suggesting that a model-centered approach to science is appropriate for the postmodern epistemological question and discusses the following four categories of tests for the validation of ABMs of political systems:

1. *Theory-model tests* are run to confirm that theoretical concepts are properly operationalized in a model. Of these, face validity is tested by presenting the results of an ABM to scholars knowledgeable in the problem it analyzes, and asking them to confirm that the model is compatible with their informed expectations. Narrative validity is tested by comparing model results to published research accounts, essentially being a more formal version of face validity that is also more amenable to the establishment of a consensus. The Turing test (after the British pioneering computer scientist) examines the believability of the results of a model by testing whether a group of experts can tell the difference between data generated computationally by an ABM and data describing events from the real world of politics. A final form of validation is provided by the surprise test, which essentially refers to emergence, i.e. unanticipated implications arising out of an ABM; if these match some theories of global politics, an interesting form of compelling validation is achieved. All these theory-model tests should be very useful in the analysis of global politics, a knowledge area characterized by narrative theories and eminent scientists embodying key theoretical approaches.
2. *Model-model tests* compare the results obtained from an ABM to results obtained analytically or from other similar models. Of such tests, the docking test cross-validates (Axtell et al. 1996) the results of an ABM by comparing them to those obtained from similar models (some of which may be already validated); alternatively, the results of an ABM may be compared to the results of an identical model formulation, recoded from scratch by another research team (as in Rand and Wilensky 2006, see below). In analytical validity testing, the results of an ABM are compared to the results obtained from analytical methods or even formal proofs (although this is rarely the case in global politics). Fixed values testing is a particular form of such validation, where the results of an ABM are compared to hand calculations, often very easy to compute, as Schelling (1969) did with his aforementioned segregation model (that was initially developed with coins on graph paper).
3. *Model-phenomena tests* compare an ABM to the phenomena that occur in the real political world. Of such tests, historical data validity compares the results obtained from an ABM to historical data while predictive data validity compares the predictions of an ABM to actual outcomes. Out-of-sample forecast tests, mix historical with predictive data validity by calibrating an ABM on one portion of the sample and testing the predictions of the model on the rest of the sample. Experimental data validity mostly refers to the representation of the micromotives of agents in an ABM, which is linked to the validation of the macrobehavior of the entire system (Schelling 2006). Finally, event validity tests compare the occurrence of specific events in the model to their occurrence in the real world.

4. *Theory-model-phenomena tests* examine an ABM in the context of theory and phenomena simultaneously, aiming to establish the robustness of the model. Such tests may include extreme bounds analysis or extreme condition tests (i.e. testing the model under extreme values of its parameters), global sensitivity analysis (of the parameters of the model), degenerate tests (i.e. interrupting some model components in order to see the effect on the system), traces testing (i.e. examining the behavior of individual agents as they operate in the modeling environment) and animation validity (i.e. comparing the visual qualities of the model to what is seen in the real world).

To provide an example, in an ABM model of political party formation that he developed, Schreiber (2002) writes that he employed tests to confirm analytical validity (i.e. whether the model agrees with formal theory predictions), historical data validity (i.e. whether the model output reproduces historical data) and docking (i.e. how the model output compares with the predictions of similar models), at various levels of sophistication. In another ABM, Schreiber ran a model hundreds of times to ensure that the results were robust across a variety of parameter values. Finally, face validity was tested by having subject experts confirm that the results were consistent with their expectations.

An additional example is provided by the makers of Netlogo (Rand and Wilensky 2006, 2007). They mention that, while thousands of ABMs have been published over the last three decades, very few of them have been reproduced, and argue in favor of replication, as an appropriate tool for verifying and validating ABMs. They suggest that with the repeated generation of the output of an ABM, a researcher may be convinced that the original results were not an exceptional case neither a rare occurrence. Rand and Wilensky developed a distinct implementation of a well-known ABM of ethnocentrism (Axelrod and Hammond 2003) in order to study replication, essentially employing a docking test (Schreiber 2002) to validate the original model. Although they had to make numerous modifications in their ABM to match the results of the literature model, they concluded that the replication of ABMs is a necessary endeavor that must be introduced in the ABM practice, despite the fact that it may not be straightforward.

4.1 Agent-Based Modeling Research in World Politics

Having completed an overview of agent-based modeling concepts, attention how shifts to applications of ABMs in a variety of fields including the social sciences, politics, defense, IR and world politics as well as the environment.

ABM in the Social Sciences

In a theoretical work, Walby (2003) argued for the importance of insights from complexity science for sociology, asserting that complexity addresses issues that lie at the heart of classic sociological theory such as emergence, i.e. the relationship between micro and macro-levels of analysis. Walby asserted that in this era of globalization, sociology needs to expand its agenda and develop its vocabulary in order to address large scale, systemic phenomena with the aid of complexity science; such social phenomena relate to connectivity and include the coevolution of CASs in a changing fitness landscape) and path dependency. Differences within complexity (and chaos) theory were discussed, especially those between the Santa Fe Institute and Nobel laureate Ilya Prigogine's approach (Prigogine and Stengers 1984; Prigogine 1997). Finally, the example of globalization was used to illuminate the analysis, especially in relation to the changing nature of polities and how they relate to the economy and culture. This work serves to highlight the linkage between the theory of complexity and social sciences, providing further justification for the use of agent-based models in addressing pertinent problems.

A critical view of the assumptions of agent-based modeling was provided by O'Sullivan and Haklay (2000), who observed that agent-based models were an increasingly popular tool in the social sciences and thought that this trend would continue. The authors examined an overview of examples of such models in the life sciences, economics, planning, sociology, and archaeology and concluded that agent-based models tended towards an individualist view of societal systems (as does the literature), which was considered inadequate for debates in modern social theory that acknowledges the importance of the dual nature of individuals and society. It was argued that because models are closed representations of an open world it is important that institutions and other social structures be explicitly included in them (or their omission be properly justified). O'Sullivan and Hacklay based a tentative explanation for the bias of agent-based models on an examination of early research in the related fields of Artificial Intelligence (AI) and Distributed Artificial Intelligence (DAI) from which the agent-based approach was derived. Although the authors in effect asserted that institutions such as the family, the community, the capital and the state cannot be easily be accounted for in agent-based models, the author of this chapter notes that emergent phenomena are an important characteristic of complex systems and such meta-agents may well emerge from individualist modeling approaches. In closing, O'Sullivan and Hacklay noted that the underlying assumptions of agent-based models are often hidden in the implementation and concluded that such models, although powerful, must be subjected to a critical examination of their assumptions.

Saqalli et al. (2010) documented the behavior of individuals in non-pastoral villages in the Sahel and coded them in an agent-based model simulating three village archetypes that included biophysical, economic, social, agricultural and livestock modules. Social development in the Sahel, an economically deprived and environmentally challenged region with widely publicized crises in the 1970s and 1980s, depends heavily on family organization and social interaction. Simulation results showed several emerging phenomena. Villages specialized in economic activities depending on natural resource availability. Family transition and inheritance systems were implemented and contributed to the population at different sites differentiating into specialized groups according to size, assets and social status. Although validation cannot be easily done with agent-based models, sensitivity analysis was carried out to assess the robustness of parameter values.

4.1.1 Political Applications of ABM

Chakrabarti's research (2000) was motivated by the observation that most theoretical studies of corruption modeled individual acts at the micro level while empirical papers studied corruption at the country level, so the author built an agent-based model to understand the structure of corruption and provide the missing link between these two groups. An example of literature findings was provided by Treisman (2000) who asserted that countries with Protestant traditions, history of British rule, higher level of development, higher level of imports (signifying a more open economy), longer history of democracy and non-federalist structures have lower levels of corruption. Chakrabarti thought that the risk aversion of an individual towards corruption was influenced by religious traditions and cultural factors. Using the model to simulate a multi-generational economy with heterogeneous risk-averse agents showed that societies have locally stable equilibrium levels of corruption that depend upon a small number of socioeconomic determinants such as the degree of risk aversion, the proportion of income spent on anti-corruption vigilance and the level of human capital in society. However, under certain combination of the values of these parameters, there can be situations when corruption rises continuously until it stifles all economic activity. Although as the author correctly pointed out, this work opened more questions than it answered, it constitutes research that would be difficult to carry out without resorting to agent-based modeling.

Defense Applications of ABM

Hare and Goldstein (2010) extended a game-theoretic model by applying an agent-based model to an information sharing network in order to analyze investment decisions and public policies for cyber security in the defense sector. The model was used to analyze the interactions of firms from defense industry trade associations and found that the nature of these interactions (dependent on the scale of the network), driven by the topology of the network, may influence the ability of agents to influence policy makers invest in security. An important public policy implication of this research was that targeted interventions, i.e. centrally coordinated behavior, could not easily influence the investment state of the system. The fact that the characteristics of the agents, the structure of the network and the nature of the agent's interactions were perhaps more important in affecting investments in the security industry, is perhaps a testament to the power of interrelated actors and evidence of the presence and importance of complexity in that industry. The authors suggested that their work may be extended to other sectors that are characterized by intense knowledge sharing of proprietary data such as energy consortiums, the civil aerospace industry and the biotech industry.

Chaturvedi et al. (2013) maintain that the simulation of virtual worlds via agent-based modeling, emerges as an important tool in social sciences including research in the field of economics, society and politics. They present some of the technologies that underlie virtual worlds and describe Sentient World, an ultra-large-scale, multi-agent-based virtual world that has been developed as a geopolitical tool for the US military and has already been used to simulate US military operations in other countries. Demonstrating the capabilities of large agent-based models, Sentient World accounts for the political, military (including terrorist), economic, social (including citizen unrest, epidemics), information and infrastructure aspects of real systems and simulates the behavior of individuals, organizations, institutions (including religion) and geographical regions. The model displays behaviors and trends that resemble those that occur in the real world. The validation of Sentient World showed that such complex modeling systems are capable of converging to the real world and provide decision makers with a tool that helps them anticipate and evaluate potential outcomes in a realistic setting. Clearly, such models are capable of pushing the state-of-the art in global politics by providing a virtual laboratory for testing and validating theory.

ABM in International Relations and World Politics

In an international regime complexity symposium, Alter and Meunier (2008) argued that the number, detailed content and subject matter of international agreements has grown and diversified exponentially in recent decades, resulting in an "emerging density and complexity" of international governance. This phenomenon of nested, partially overlapping and parallel international regimes that are not hierarchically ordered is referred to as international regime complexity and makes it harder to locate political authority over an issue. One of the consequences of this complexity is that international governance takes place via a multitude of complexly interrelated trans-border agreements often characterized by strategic ambiguity and fragmentation. The authors noted that both feedback and competition ensue from this complex governance network, sometimes empowering and other times weakening actors of global politics. Complexity also makes spotting a true causal relationship very hard and forces bounded rationality on the actors of global politics as they have to resort to problem framing and heuristics that may vary over states, cultures and time. Alter and Meunier also argued that international regime complexity favors the generation of small group environments, making face-to-face interaction crucial and having multiple portfolios assigned to individual diplomats or experts. Small group dynamics, more likely to emerge when issues are technical and rely on expertise (opening the door to non-state actors), create depth and strong links, imprinting global politics with the touch of individual actors. The authors also mentioned that feedback effects include: competition among actors, organizations and institutions (i.e. agents and meta-agents), resulting in both positive and negative impacts; unintentional reverberations, where impacts are carried over to parallel domains; difficulty in assigning responsibility and establishing causality; loyalty in the sense that what actors, e.g. states, do in one arena of global politics may carry over into another; and facilitation of exiting, e.g. by resorting to non-compliance. Alter and Meunier concluded that while international regime complexity may empower weaker non-state actors, at the same time it may confer advantages to the most powerful states who possess the resources to sift through the maze or rules and players in order to achieve their goals.

Frej and Ramalingam (2011) examined the connection between foreign policy (which they called a field of few certainties) and CASs. The authors listed 15 global challenges facing humanity in the context of the Millennium Project (http://www.millennium-project.org) including issues related to sustainable development, climate change, clean water, ethnic conflict, population growth, resource usage, democracy, access to ICTs, new diseases, weapons of mass destruction, the role of women, organized crime, energy, science, technology and ethics. The authors discussed what they see as a "quiet revolution in complexity thinking" in foreign policy. Citing Ramalingam et al. (2008), Frej and Ramalingam suggested that foreign policy experts and analysts take into account that (a) the world is characterized by complex systems of elements that are interdependent and interconnected by multiple feedback processes; (b) system-wide behaviors emerge unpredictably from the interactions among agents; and (c) in complex systems, changes are evolutionary, dynamic, highly sensitive to initial conditions, and may exhibit non-linear tipping points. Several principles followed from these, among which that a systemic perspective should be adopted in most cases and "silver bullet" strategies should be avoided in favor of attempting several parallel experiments. The work of Frej and Ramalingam underlined the need to reevaluate foreign policy under the light of the complex interconnected world of the 21st century.

In a true IR application of agent-based modeling, Cioffi-Revilla and Rouleau (2009a) described Afriland, a moderately detailed model of the geographic region of East Africa that was programmed in MASON (Luke et al. 2005) and used to analyze inter-border socio-cultural and environmental dynamics as well as natural hazards across national frontiers that, the authors argue, are questions with scientific and policy relevance in the region. Afriland was built on Rebeland, a previous attempt by the same authors to model a single country with several provinces (Cioffi-Revilla and Rouleau 2009b, 2010), but unlike earlier models, AfriLand is capable of analyzing phenomena that transcend national boundaries. As the authors mentioned, the three basic types of research questions that were addressed by AfriLand revolved around (a) the response of a regional polity system to (anthropogenic or natural) societal stress; (b) the emergence and propagation of insurgency, domestic political instability, or even state failure across borders; and (c) the influence of the condition of the borders on regional (multi-country) dynamics. The model was used to analyze a socioeconomic system of 10 countries impacted by the distribution of resources (such as oil, diamond and gold), terrain morphology (that affected visibility) and climate. The simulation results indicated a range of human and social dynamics such as troop movements, insurgent activity, refugee flows and transnational conflict regions (shown as red dots on a map that helps establish, e.g. the creation of strongholds). Societal satisfaction was tracked and found to vary as a function of the capability of governments to manage public issues. The authors concluded that although agent-based models of international regions composed of several countries were few (at the time of writing), models of the scale of Afriland can be very useful in analyzing and understanding socio-cultural and environmental dynamics that transcend national boundaries.

Environmental Applications of ABM

In a review of agent-based modeling in coupled human-nature systems, Li (2012) explains that such systems include ecosystems that have been subjected to anthropogenic disturbances and exhibit characteristics of complexity such as heterogeneity, nonlinearity, feedback and emergence. Sites for which such empirical work has been carried out, include the Amazon, Yucatan and areas of China and Ecuador where agent-based models are employed to model human decision making and environmental consequences. Li writes that complex systems exhibit heterogeneous subsystems, autonomous entities, nonlinear relationships and multiple interactions including feedback, learning and adaptation. Complexity is manifested in many forms including path dependency, criticality, self-organization, difficulty in predictions as well as the emergence of qualities and behaviors that are not tractable to the individual system components and their attributes. No orderly and predictable relationships nor causal linkages may be found in complex systems. Li concludes that complexity science is still in its infancy, lacking a clear conceptual framework, unique techniques and an ontological and epistemological representations of complexity. Li calls agent-based modeling a major bottom-up tool for the analysis of complex systems, an approach characterized by methodological individualism rather than aggregation. Cross-pollinated by various disciplines, agent-based modeling constitutes a virtual laboratory in which numerical experiments may be done that would be impossible to carry out without the aid of computers. Nevertheless, agent-based models are difficult to validate and verify.

Angus et al. (2009) attempted to understand the complex, dynamic, spatial and nonlinear challenges facing Bangladesh, a densely populated country of around 145 million people living in a coastal area of just 145000 km^2 that is dependent on the South Asian monsoon for most of its rainfall. The authors considered that a modular agent-based model would permit the dynamic interactions of the economic, social, political, geographic, environmental and epidemiological dimensions of climate change impacts and adaptation policies to be integrated. In addition, such a model would permit the inclusion of nonlinear threshold events such as mass migrations, epidemic outbreaks, etc. The authors formulated their model in Netlogo (the most mature tool in the field) and examined (but did not fully analyze) the dynamic impacts on poverty, migration, mortality and conflict from climate change in Bangladesh for the entire 21st century. Their model combined a Geographical Information Systems (GIS) approach with a district network layer representing the spatial geography, and used level census and economic data with a methodology that allowed national scale statistics to be generated from local dynamic interactions, allowing for a more realistic treatment of distributed spatial events and heterogeneity across the country. The rest of the world was also modeled as an agent, so that exports, imports and migration could be incorporated in the model. The author's aim was not to generate precise predictions of Bangladesh's evolution, but to develop a framework that may be used for integrated scenario exploration so their work represented an initial report on progress on this project. The authors concluded that the prototype model demonstrated the desirability and feasibility of integrating the different dimensions of the CAS and hoped that the completed model (which at the time of writing was under development) could be used as the basis for a more detailed policy-oriented model.

Smith et al. (2008) addressed the issue of environmental refugees in Burkina Faso with agent-based modeling observing that even simple such models may exhibit complex emergent behavioral patterns. Burking Faso is one of the poorest countries in the world with an economy heavily based upon rain-fed agriculture and cattle-raising. Migration was considered to occur based on environmental stimuli such as land degradation. Using the longitudinal spatio-temporal data of a social survey conducted in 2000–2001, the authors developed rules of interaction between climate change and migration. Origins and destinations were considered to have push and pull factors while the influence of social structures, individuals as well as intermediate institutions were accounted for. The agent-based approach modeled the cognitive response of individuals to climate change as they appraised the situation (subject to their cognitive biases) and then considered whether to select migration as an appropriate adaptation strategy. Although the paper was limited to a conceptual description of the developed agent-based model and drawing of a few general conclusions, it may nevertheless be considered another interesting application of such models in social problems that would be quite difficult to address with any other method.

Finally, Hailegioris et al. (2010) presented an agent-based model of human-environment interaction and conflict in East Africa using the MASON agent-based simulation software system (Luke et al. 2005). Their model represented a 150 by 150 km^2 area and used a daily time step, in which it updated the regeneration of vegetation (as a function of rainfall and grazing) and activated herders randomly. Annual droughts were programmed to occur randomly with a 15 year cycle. Herders were programmed to adapt to seasonally driven changes in the grazing environment. As the carrying capacity of the landscape varied, conflict was modeled to result from trespassing incidents between herders and farmers; if it were not resolved peacefully (e.g. by cooperation between two herders of the same clan) it could escalate over time, involving more participants and ending up with the infliction of damages. All in all, the model accounted for the complex interaction of pastoral groups (herders and farmers) with their environment and other emerging external factors. Experiments with the model indicated that rainfall was an important factor that was nonlinearly related to the carrying capacity and supported the conclusion that increased seasonal rainfall variability and droughts create tremendous stress on pastoral groups and challenge their long-term resilience and adaptive response mechanisms. This example showcases how environmental complexity research may be valuable in outlining appropriate directions for environmental management policy and measures.

5 Conclusions

This chapter examined the complexity of world politics with an emphasis on global environmental issues. Concepts of game theory were reviewed and connected to the state of world politics. Game theoretic models found in IR included the prisoner's dilemma; game theoretic models encountered in global environmental negotiations included the conflict between rich North and poor South countries, the role of pollution havens and the clash of idealists versus pragmatists and optimists versus pessimists. It was suggested that the complexity of world politics, taking place on a highly interconnected global network of actors organized as agents and meta-agents, is nothing but a multiplayer extension of game theory although a complexity approach to world politics should not be regarded as a theory alternative to realism, but as a relatively novel research tool to aid with understanding and anticipating (rather than predicting) global events. Technology, interconnections, feedback and individual empowerment were discussed in the context of the complex world of global politics. Furthermore, evolution and adaptation were related to the concept of fitness and how it may be estimated for the case of actors in world politics. Furthermore, it was suggested that many events of world politics constitute emergent phenomena of the complex international community of state and non-state actors. Finally the chapter was complemented with a short overview of concepts related to ABM, arguably the most prevalent method of simulating complex systems, and a review of research problems from the fields of social science, political science, defense, world politics and the global environment that have been successfully addressed with agent-based simulation.

Thanks to the work of Nobel laureates that worked at the Santa Fe Institute such as Murray Gell-Man and other celebrated scientists such as John H. Holland and Robert Axelrod, complexity science has emerged as an umbrella science that, in the case of world politics, could be useful as a tool auxiliary to the realist worldview in modeling, understanding and perhaps anticipating the behavior of state and nonstate actors. Axelrod in particular was a key figure in linking game theory to complexity with his two seminal books (1985, 1997). The main aim of this chapter was to suggest that world politics may be considered a CAS with states being modelled as complex adaptive actors, i.e. agents and international organizations such as the United Nations or the EU being meta-agents. Understanding the system rules may be an important aspect of analyzing the international system of states as a CAS (Hoffman 2006).

Nevertheless, formulating theory for CASs is difficult because the behavior of such systems is more than the sum of the behavior of its components as it has been explained time and again in this chapter. The presence of nonlinearities means that tools such as trend analysis are destined to fail to generalize observations into theory (Holland 1996). It is, therefore, suggested that complexity and agent-based modeling may provide a framework that helps understand that entire systems exceed the sum of their parts precisely due to the interaction of their parts with each other. Moreover, it is difficult to envision any causal theoretical paradigm to be of real use in explaining today's complex, interconnected world.

The complexity science approach to world politics (including global environmental and energy policy) will help describe and explain the past and the present although it will not be able to predict future event. Nevertheless, ABMs are a formidable tool that will allow the recognition and investigation of emerging phenomena that are beyond the capabilities of classical IR theory to envision. The understanding of such phenomena though and their use in anticipating the future may best be done by resorting to theoretical and modeling tools in tandem and this is the main conclusion that is drawn from this chapter.

A list of software resources useful to those who wish to address global problems with agent-based modeling ensues after the list of references.

Acknowledgments The author thanks Dr. T. Nadasdi and Dr. S. Sinclair for their online Spell Check Plus (http://spellcheckplus.com) that was used for proofing the entire document.

References

Allan, R.: Survey of agent based modelling and simulation tools. Version 1.1, STFC Daresbury Laboratory, Daresbury, Warrington. http://www.grids.ac.uk/Complex/ABMS/ABMS.html. Last updated 10 Mar 2011

Allison, G., Zelikow, P.: Essence of Decision: Explaining the Cuban Missile Crisis, 2edn. Longman, London (1999)

Alter, K., Meunier, S.: The politics of international regime complexity. In: The Politics of International Regime Complexity Symposium, The Roberta Buffett Center for International and Comparative Studies, working paper no. 07-003, Northwestern University (2008)

Alterman, E.: When Presidents Lie: A History of Official Deception and Its Consequences. Viking, New York (2004)

Angus, S.D., Parris, B., Hassani-M, B.: Climate change impacts and adaptation in Bangladesh: an agent-based approach. In: 18th World IMACS/MODSIM Congress, Cairns, Australia, 13–17 July 2009

Ariely, D.: Predictably Irrational: The Hidden Forces that Shape Our Decisions. HarperCollins e-books, London (2008)

Axelrod, R.: The Evolution of Cooperation. Basic Books Inc.Publishers, New York (1985)

Axelrod, R.: The Complexity of Cooperation: Agent-Based Models of Competition and Collaboration. Princeton University Press, Princeton (1997)

Axelrod, R.: Advancing the art of simulation in the social sciences. Jap. J. Manage. Inf. Syst. **12**(3) (Special Issue on Agent-Based Modeling) (2003)

Axelrod, R., Hammond, R.A.: The evolution of ethnocentric behavior. In: Midwest Political Science Convention, Chicago, IL, 3–5 April 2003

Axelrod, R., Tesfatsion, L.: On-line guide for newcomers to agent-based modeling in the social sciences. http://www2.econ.iastate.edu/tesfatsi/abmread.htm. Last updated on 28 March 2015

Axtell, R., Axelrod, R., Epstein, J.M., Cohen, M.D.: Aligning simulation models: A case study and results. Comput. Math. Organ. Theory **1**(2), 123–142 (1996)

Baron, J.: Judgment Misguided: Intuition and Error in Public Decision Making. Oxford University Press, Oxford (1998)

Bennett, N., Miles, S.A.: Your Career Game: How Game Theory Can Help You Achieve Your Professional Goals. Stanford University Press, Stanford (2010)

Bharathy, G.K., Silverman, B.G.: Validating agent based social systems models. In: Proceedings of the 2010 Winter Simulation Conference, Baltimore, MD, 5–8 Dec 2010

Binnendijk, H., Kugler, R.L.: Seeing the Elephant: The US Role in Global Security. Center for Technology and National Security Policy, National Defense University Press, Potomac Press, Inc., Washington DC (2006)

Boardman, B.: Fuel Poverty: From Cold Homes to Affordable Warmth. Belhaven Press, London (1991)

Boardman, B.: Fixing Fuel Poverty: Challenges and Solutions. Earthscan, London (2010)

Boehmer-Christiansen, S., Kellow, A.: International Environmental Policy: Interests and the Failure of the Kyoto Protocol, London (2002)

Bonabeau, E.: Agent-based modeling: Methods and techniques for simulating human systems. Proc. Nat. Acad. Sci. USA (PNAS) **99**(3) (2002)

Bradenburger, A.M., Nalebuff, B.J.: Co-opetition. Currency Doubleday, New York (1996)

Brams, S.J.: Negotiation Games: Applying Game Theory to Bargaining and Arbitration. Revised Edition, Routledge, London (2005)

Brams, S.J., Kilgour, D.M.: Game Theory and National Security. Basil Blackwell, Oxford (1988)

Casti, J.L.: Complexification: Explaining a Paradoxical World Through the Science of Surprise. HarperCollins, New York (1994)

CBC News.: Canada pulls out of Kyoto Protocol. http://www.cbc.ca/news/politics/story/2011/12/12/pol-kent-kyoto-pullout.html, 12 Dec 2011

Chakrabarti, R.: An agent based model of corruption. In: 2000 Computing in Economics and Finance (CEF) conference, Barcelona, Spain (2000)

Chaturvedi, R., Armstrong, B., Chaturvedi, A., Dolk, D., Drnevich, P.: Got a problem? Agent-based modeling becomes mainstream. Global Econ. Manage. Rev. **18**, 33–39 (2013)

Christakis, N.A., Fowler, J.H.: Connected: The Surprising Power of Our Social Networks and How They Shape Our Lives—How Your Friends' Friends' Friends Affect Everything You Feel, Think, and Do. Back Bay Books, Reprint edition, New York (2011)

Cialdini, R.B.: Influence: Science and Practice, 5th edn. Pearson, London (2009)

Cioffi-Revilla, C., Rouleau, M.: MASON AfriLand: a regional multi-country agent-based model with cultural and environmental dynamics. In: Proceedings of the Human Behavior-Computational Modeling and Interoperability Conference 2009 (HB-CMI-09), Joint Institute for Computational Science, Oak Ridge National Laboratory, Oak Ridge, Tennessee, USA, 23–24 June 2009

Cioffi-Revilla, C., Rouleau, M.: MASON RebeLand: An agent-based model of politics, environment, and insurgency. In: Proceedings of the Annual Convention of the International Studies Association, New York, 15–18 Feb 2009

Cioffi-Revilla, C., Rouleau, M.: MASON RebeLand: an agent-based model of politics, environment, and insurgency. Int. Stud. Rev. **12**, 31–52 (2010)

Cohen, M.D., March, J.G., Olsen, J.P.: A garbage can model of organizational choice. Adm. Sci. Q. **17**(1), 1–25 (1972)

Damasio, A.R.: Descartes' Error: Emotion, Reason and the Human Brain. Avon Books, New York (1994)

Dixit, A., Nalebuff, B.: Thinking Strategically: The Competitive Edge in Business, Politics and Everyday Life. Norton, New York (1991)

Dixit, A., Nalebuff, B.: The Art of Strategy: A Game Theorist's Guide to Success in Business and Life. Norton, New York (2008)

Dixit, A., Skeath, S.: Games of Strategy, 2nd edn. W.W. Norton and Company, New York (2004)

Donnelly, J.: Realism and International Relations. Cambridge University Press, Cambridge (2004)

Earnest, D.C., Rosenau, J.N.: Signifying nothing? What complex systems theory can and cannot tell us about global politics. In: Harrison, N.E., Rosenau. J.N. (ed.) Complexity in World Politics: Concepts and Methods of a New Paradigm, State University of New York (SUNY) Series in Global Politics (2006)

Finus, M.: Game Theory and International Environmental Cooperation. New Horizons in Environmental Economics. Edward Elgar, London (2001)

Fiorino, D.J.: Making Environmental Policy. University of California Press, Berkeley (1995)

Fisher, R., Ury, W.: Getting to Yes: Negotiating Agreement Without Giving in. Random House Business Books, New York (1999)
Flannery, T.: Here on Earth: A Natural History of the Planet. HarperCollins, Toronto (2011)
Frej, W., Ramalingam, B.: Foreign policy and complex adaptive systems: Exploring new paradigms for analysis and action. Santa Fe Institute (SFI) Working Paper 11-06-022, 16 June 2011
Gilbert, N., Terna, P.: How to build and use agent-based models in social sci*ence*. Iowa State University, http://www2.econ.iastate.edu/tesfatsi/howtousebuildabmss.gilbertterna1999.pdf (1999)
Gladwell, M.: Blink: The Power of Thinking Without Thinking. Back Bay Books, Little, Brown and Company, New York (2005)
Goldstein, N.J., Martin, S.J., Cialdini, R.B.: Yes! 50 Scientifically Proven Ways to be Persuasive. Free Press, New York (2008)
Granovetter, M.S.: The strength of weak ties. Am. J. Sociol. **78**(6), 1360 (1973)
Grossman, G.M., Krueger, A.B.: Environmental impacts of a North American free trade agreement. National Bureau of Economic Research, working paper 3914. NBER, Cambridge, MA (1991)
Hailegiorgis, A.B., Kennedy, W.G., Rouleau, M., Bassett, J.K., Coletti, M., Balan, G.C., Gulden, T.: An agent based model of climate change and conflict among pastoralists in East Africa. International Environmental Modelling and Software Society (iEMSs). In: 2010 International Congress on Environmental Modelling and Software Modelling for Environment's Sake, Fifth Biennial Meeting, Ottawa, Canada (2010)
Hardin, G.: The tragedy of the commons. Science **162**, 1243–1248 (1968)
Hare, F., Goldstein, J.: The interdependent security problem in the defense industrial base: an agent-based model on a social network. Int. J. Crit. Infrastruct. Prot. **3**(3–4), 128–139 (2010)
Harrison, N.E.: Thinking about the world we make. In: Harrison, N.E., Rosenau, J.N. (ed.) Complexity in World Politics: Concepts and Methods of a New Paradigm. State University of New York (SUNY), Series in Global Politics (2006a)
Harrison, N.E.: Complexity systems and the practice of world politics. In: Harrison, N.E., Rosenau, J.N. (eds.) Complexity in World Politics: Concepts and Methods of a New Paradigm. State University of New York (SUNY), Series in Global Politics (2006b)
Hendrick, D.: Complexity theory and conflict transformation. In: Young, N.J. (ed.) Oxford International Encyclopedia of Peace. Oxford University Press, New York (2010)
Hoffman, M.J.: Beyond regime theory: complex adaptation and the ozone depletion regime. In: Harrison, N.E. (ed.) Complexity in World Politics: Concepts and Methods of a New Paradigm. State University of New York Press, New York (2006)
Holland, J.H.: Hidden Order: How Adaptation Builds Complexity. Helix Books (1996)
Holland, J.H.: Signals and Boundaries: Building Blocks for Complex Adaptive Systems. The MIT Press, Cambridge (2012)
Holland, J.H.: Complexity: A Very Short Introduction. Oxford University Press, Oxford (2014)
Iliadis, M., Paravantis, J. A.: A multivariate cross-country empirical analysis of the digital divide. In: 2011 IEEE Symposium on Computers and Communications (ISCC), Corfu, Greece, pp. 2785–28788, 28 June–1 July 2011
Jervis, R.: System Effects: Complexity in Political and Social Life. Princeton University Press, Princeton (1997)
Kauffman, S.A.: The Origins of Order: Self-organization and Selection in Evolution. Oxford University Press, New York (1993)
Kauffman, S.A.: Reinventing the Sacred: A New View of Science, Reason, and Religion. Basic Books, New York (2008)
Kennedy, R.: Cubans sound off on détente efforts with the US. Al Jazeera, 11 April 11 2015. http://www.aljazeera.com/indepth/features/2015/04/cubans-sound-detente-efforts-150411150513419.html
Keohane, R.O., Nye Jr, J.S.: Power and Interdependence, 4th edn. Longman Classics in Political Science, Pearson, London (2012)

Lehrer, J.: How We Decide. Houghton Mifflin Harcourt, Boston (2009)
Levin, S.A.: Fragile Dominion: Complexity and the Commons. Perseus, Reading (1999)
Li, A.: Modeling human decisions in coupled human and natural systems: review of agent-based models. Ecol. Model. **229**, 25–36 (2012)
Lindblom, C.: The science of muddling through. Public Adm. Rev. **19**, 79–88 (1959)
Lindblom, C.: Still muddling, not yet through. Public Adm. Rev. **39**, 517–526 (1970)
Lomborg, B.: Cool It: The Skeptical Environmentalist's Guide to Global Warming. Cyan Marshall Cavendish Editions, London (2007)
Luke, S., Cioffi-Revilla, C., Sullivan, K., Balan, G.: MASON: a multiagent simulation environment. Simulation **81**(7), 517–527 (2005)
Malleret, T.: Disequilibrium: A World Out of Kilter. Amazon Kindle Edition (2012)
McCain, R.A.: Game Theory: A Non-Technical Introduction to the Analysis of Strategy. Thomson South-Western, Mason (2004)
McMillan, J.: Games, Strategies and Managers. Oxford University Press, Oxford (1992)
Mearsheimer, J.J.: Why Leaders Lie: The Truth About Lying in International Politics. Oxford University Press, Oxford (2011)
Miller, J.D.: Game Theory at Work: How to Use Game Theory to Outthink and Outmaneuver Your Competition. McGraw-Hill, New York (2003)
Minelli, A., Fusco, G. (eds.): Evolving pathways: Key Themes in Evolutionary Developmental Biology. Cambridge University Press, Cambridge (2008)
Morgenthau, H.: Politics Among Nations. McGraw-Hill, New York (2005)
Navarro, J., Karlins, M.: What Every Body is Saying: An ex-FBI Agent's Guide to Speed Reading People. HarperCollings e-books (2008)
Neumayer, E.: Pollution havens: an analysis of policy options for dealing with an elusive phenomenon. J. Environ. Develop. **10**(2), 147–177 (2001)
Noll, D.E.: Elusive Peace: How Modern Diplomatic Strategies Could Better Resolve World Conflicts. Prometheus Books, New York (2011)
North, M.J., Macal, C.M.: Managing Business Complexity: Discovering Strategic Solutions with Agent-Based Modeling and Simulation. Oxford University Press, Oxford (2007)
Nye Jr, J.S.: The Future of Power. Public Affairs, New York (2011)
O'Sullivan, D., Haklay, M.: Agent-based models and individualism: Is the world agent-based? Environ. Plan. A **32**(8), 1409–1425 (2000)
Paravantis, J.A., Samaras, A.: Fuel poverty in Greece and the European Union. In: Society, Borders and Security Symposium, , University of Siegen, Germany, 24–25 (2015)
Paravantis, J.A., Santamouris, M.: An analysis of indoor temperature measurements in low and very low income housing in Athens, Greece. Special Issue on Indoor Environmental Quality in Low Income Housing in Europe of the Journal: Advances in Building Energy Research, Taylor and Francis, in print (2015)
Patterson, K., Grenny, J., McMillan, R., Switzler, A.: Crucial Conversations: Tools for Talking When Stakes are High. McGraw-Hill, New York (2002)
Pirages, D.C., Manley DeGeest, T.: Ecological Security: An Evolutionary Perspective on Globalization. Rowman and Littlefield Publishers, Lanham (2004)
Preston, I., Payne, G., Bevan, I., Hunt, M.: Fuel poverty: tackling the triple injustice. The green housing forum, Sustainable futures for social housing. http://www.daikin.co.uk/binaries/GHF14_Fuel%20Poverty-tackling%20the%20triple%20injustice_Whitepaper_tcm511-342469.pdf. Accessed on Feb 2015 (2014)
Prigogine, I.: The End of Certainty: Time, Chaos, and the New Laws of Nature. Free Press, New York (1997)
Prigogine, I., Stengers, I.: Order Out of Chaos: Man's New Dialogue with Nature. Heinemann, London (1984)
Railsback, S.F., Volker, V.: Agent-based and Individual-Based Modeling: A Practical Introduction. Princeton University Press, Princeton (2012)
Railsback, S.F., Lytinen, S.L., Jackson, S.K.: Agent-based simulation platforms: Review and development recommendations. Simulation **82**(9), 609–623 (2006)

Ramalingam, B., Jones, H., Reba, T., Young, J.: Exploring the science of complexity: Ideas and implications for development and humanitarian efforts, 2nd edition. Working Paper 285, Overseas Development Institute, London (2008)

Rand, W., Wilensky, U.: Verification and validation through replication: a case study using Axelrod and Hammond's ethnocentrism model. In: North American Association for Computational Social and Organization Sciences (NAACSOS), South Bend, IN, June 2006

Rand, W., Wilensky, U.: Making models match: Replicating an Agent-Based Model. J. Artif. Soc. Soc. Simul. **10**(4), 2. http://jasss.soc.surrey.ac.uk/10/4/2.html. Accessed May 2015 (2007)

Rapoport, A., Chammah, A.M.: Prisoners' dilemma. University of Michigan Press, Ann Arbor (1965)

Rihani, S.: Nonlinear systems. http://www.globalcomplexity.org/nonlinear-systems. Accessed April 2015 (2014)

Rubino, J.: Clean Money: Picking Winners in the Green-tech Boom. Wiley, London (2014)

Sandler, T., Arce, M.D.G.: Terrorism and game theory. Simul. Gam. **34**(3), 319–337 (2003)

Santamouris, M., Alevizos, S.M., Aslanoglou, L., Mantzios, D., Milonas, P., Sarelli, I., Karatasou, S., Cartalis, K., Paravantis, J.A.: Freezing the poo—indoor environmental quality in low and very low income households during the winter period in Athens. Energy Build. **70**, 61–70 (2014)

Santamouris, M., Paravantis, J.A., Founda, D., Kolokotsa, D., Michalakakou, P., Papadopoulos, A.M., Kontoulis, N., Tzavali, A., Stigka, E.K., Ioannidis, Z., Mehilli, A., Matthiessen, A., Servou, E.: Financial crisis and energy consumption: a household survey in Greece. Energy Build. **65**, 477–487 (2013)

Saqalli, M., Gérard, B., Bielders, C., Defourny, P.: Testing the impact of social forces on the evolution of Sahelian farming systems: a combined agent-based modeling and anthropological approach. Ecol. Model. **221**, 2714–2727 (2010)

Schelling, T.C.: Models of segregation. In: The American Economic Review, Papers and Proceedings of the Eighty-first Annual Meeting of the American Economic Association, vol. 59(2), pp. 488–493, May 1969

Schelling, T.: The Strategy of Conflict. Harvard University, Cambridge (1980)

Schelling, T.: Some economics of global warming. Am. Econ. Rev. **82**, 1–14 (1992)

Schelling, T.C.: Micromotives and Macrobehavior. Revised edition, W. W. Norton and Company, New York (2006)

Schreiber, D.M.: Validating agent-based models: From metaphysics to applications. In: Midwestern Political Science Association's 60th Annual Conference, Chicago, IL, 25–28 April 2002

Simon, H.: A behavioral model of rational choice. In: Models of Man, Social and Rational: Mathematical Essays on Rational Human Behavior in a Social Setting. Wiley, New York (1957)

Simon, H.: Bounded rationality and organizational learning. Organ. Sci. **2**(1), 125–134 (1991)

Simon, H.: The architecture of complexity. Proc. Am. Philos. Soc. **106**(6), 467–482 (1962)

Smith, C., Kniveton, D., Black, R., Wood, S.: Climate change, migration and agent-based modelling: Modelling the impact of climate change on forced migration in Burkina Faso. University of Sussex, http://users.sussex.ac.uk/~cds21/infweb/christopher%20smith1.pdf (2008)

Starfield, A.M., Smith, K.A., Bleloch, A.L.: How to model it: Problem solving for the computer age. McGraw-Hill, New York (1990)

Sterman, J.D.: Business Dynamics: Systems Thinking and Modeling for a Complex World. Irwin McGraw-Hill, New York (2000)

Stevens, S.P.: Games People Play: Game Theory in Life, Business and Beyond. Parts I and II. The Teaching Company (2008)

Susskind, L.E.: Environmental Diplomacy: Negotiating More Effective Global Agreements. Oxford University Press, Oxford (1994)

Taleb, N.N.: The Black Swan: The Impact of the Highly Improbable. Random House, New York (2007)

Tannen, D.: You Just Don't Understand. Ballantine Books, New York (1990)

Tesfatsion, L.: Repast: a software toolkit for agent-based social science modeling. Self-Study Guide for Java-Based Repast. http://www2.econ.iastate.edu/tesfatsi/repastsg.htm. Last updated on 28 Mar 2015

Treisman, D.: The causes of corruption: a cross-national study. J. Public Econ. **76**, 399–457 (2000)

Tziampiris, A.: The Emergence of Israeli-Greek cooperation. Springer, Berlin (2015)

Von Neumann, J., Morgenstern, O.: Theory of Games and Economic Behavior. Princeton University Press, Princeton (1944)

Walby, S.: Complexity theory, globalisation and diversity. In: Conference of the British Sociological Association, University of York, April 2003

Waltz, K.: Man, the State, and War: A Theoretical Analysis. Columbia University Press, New York (1954)

Waltz, K.: Realism and International Politics. Routledge, London (2008)

Waltz, K.: Theory of International Politics. Waveland Press Incorporated, Long Grove (2010)

Waltz, K.N.: Why Iran should get the bomb. Foreign Affairs, July/August 2012

Weber, C.: International Relations Theory: A Critical Introduction, 3rd edn. Taylor and Francis, London (2009)

Wilensky, U.: NetLogo. Center for Connected Learning and Computer-Based Modeling, Northwestern University, Evanston, IL, USA, http://ccl.northwestern.edu/netlogo (1999)

Zagare, F.C.: The Geneva conference of 1954: a case of tacit deception. Int. Stud. Quart. **23**(3), 390–411 (1979)

List of Software Resources

Lists of ABM software resources may be found at

1. Axelrod's dated but still excellent "Resources for Agent-Based Modeling" may be found as Appendix B in his "Complexity of Cooperation" book (1997). In particular, I strongly concur with his recommendation that a beginner use a procedural language (like BASIC or Pascal) to start working on developing an agent-based model
2. Axelrod and Tesfatsion's outstanding online guide, http://www2.econ.iastate.edu/tesfatsi/abmread.htm. Last updated on 28 March 2015
3. An excellent introductory to the Repast software as well as agent-based models is maintained by Tesfatsion at http://www2.econ.iastate.edu/tesfatsi/repastsg.htm. Last updated on 28 Mar 2015
4. A thorough list of software resources is presented and commented online by Allan (2011)
5. Another list of software resources is given by Railsback et al. (2006)
6. A list of ABM platforms in the CoMSES Network, https://www.openabm.org/page/modeling-platforms
7. The following ABM systems (with an emphasis on open source and free packages) are suggested for social science simulation by the author of this chapter (in order of personal preference)
8. Netlogo, https://ccl.northwestern.edu/netlogo, that has inspired a few other tools that are based on it, such as AgentScript (http://agentscript.org) and Modelling4All (http://m.modelling4all.org)
9. Starlogo at http://education.mit.edu/starlogo, Starlogo TNG at http://education.mit.edu/projects/starlogo-tng, and Starlogo NOVA at http://www.slnova.org
10. Anylogic, which has a free edition for academic and personal use (Anylogic PLE), http://www.anylogic.com/blog?page=postandid=237
11. Repast, which offers the capability of being programmed in Java, Relogo or Python http://repast.sourceforge.net
12. MASON, http://cs.gmu.edu/~eclab/projects/mason (Luke et al., 2005)

13. Swarm, presently at http://savannah.nongnu.org/projects/swarm and MAML, http://www.maml.hu/maml/introduction/introduction.html, that uses easier code that translates to Swarm
14. FLAME, http://www.flame.ac.uk
15. Agent Modeling Platform (AMP), http://eclipse.org/amp
16. Breve, http://www.spiderland.org
17. Cormas, a "natural resources and agent-based simulation" system, http://cormas.cirad.fr/en/outil/outil.htm
18. Agentbase, http://agentbase.org, a tool aimed at educational uses, with the capability of coding models in Coffeescript (a simplified version of Java, http://coffeescript.org) and running them on the browser

A Semantic Approach for Representing and Querying Business Processes

Eleni-Maria Kalogeraki, Dimitris Apostolou, Themis Panayiotopoulos, George Tsihrintzis and Stamatios Theocharis

Abstract As business information is getting more and more difficult to manage, business entities are increasingly adopting business process management practices and tools, providing new models and techniques for process-related information management. In parallel, semantic web technologies are emerging, which are capable of providing methodologies and tools to represent and process the semantic content of business processes. Applying semantic web technologies in business process management can enhance process understanding and can facilitate querying for pertinent process information. This chapter discusses the challenges and benefits associated with the coupling of semantic technologies with business process management and describes a methodology for representing the semantic content of the BPMN specification in the form of ontology. Moreover, it explores query mechanisms for conventional and semantics-based querying of business models.

Keywords BPMN · Business process model · Business process management · Knowledge management · Semantic process model · BPMN ontology · Information retrieval · Query language

E.-M. Kalogeraki (✉) · D. Apostolou · T. Panayiotopoulos · G. Tsihrintzis
S. Theocharis
Department of Informatics, University of Piraeus, 80,
Karaoli & Dimitriou str., Piraeus, Greece
e-mail: elmaklg@unipi.gr

D. Apostolou
e-mail: dapost@unipi.gr

T. Panayiotopoulos
e-mail: themisp@unipi.gr

G. Tsihrintzis
e-mail: geoatsi@unipi.gr

S. Theocharis
e-mail: stheohar@unipi.gr

G.A. Tsihrintzis et al. (eds.), *Intelligent Computing Systems*,
Studies in Computational Intelligence 627, DOI 10.1007/978-3-662-49179-9_4

1 Introduction

The modern business world is characterized by an increasing international integration of markets and technologies where different business cultures are able to interact with each other, causing a rapid technological development in Business Information Systems. In the highly competitive business environment, enterprises are seeking for new techniques that leverage their values to improve their productivity and business effectiveness. Knowledge Management reflects this scope, as it is the act of identifying, creating, capturing, extracting and sharing organizational knowledge resources (Taghieh and Taghieh 2013). Knowledge Management aims to promote relationships between experts and individuals lacking of specialized knowledge (Taghieh and Taghieh 2013).

Companies have to deal with hundreds of business processes in their day-to-day operation. Handling them manually is often a difficult, time-consuming and costly process which can easily undermine effective decision making and distract business entities from their goals. Business process modeling refers to the act of representing all aspects of the tasks taking place in order to fulfill a business process, aiming to better understand, analyze and restructure the business process to improve the efficiency and facilitate the growth strategies of an enterprise. Business Process Management is serving these tasks, being the technique for organizing, automating, monitoring, controlling business processes in organizations. Therefore, we argue that knowledge management and business processes management have synergies the investigation and analysis of which is a growing research area (Taylor 2012).

Business process modeling is hitherto directly expressed with process modeling languages such as UML Activity Diagrams, Business Process Modeling Notation (BPMN), Event-driven Process Chains (EPCs), RosettaNet, Activity Decision Flow Diagram (ADF), the family modeling language of Integration Definition (IDEF), the Electronic Business using the Extensible Markup Language Business Process Specification Schema (ebXML BPSS) (Hoang et al. 2010). Business process modeling notations intend to increase the abstraction level in software deployment and enhance the processes with semantic information (Fernández et al. 2010). However, there exist some generic weaknesses in business process modeling which current approached do not address adequately. Since several flowchart notations are using natural language, the complexity of homonyms, synonyms and further naming and structural conflicts makes it difficult for machines to retrieve information from these implicit representations during the execution process (Bergener et al. 2015). As information of business process models is getting deeper and more complex, business process management toolkits are required to support efficient information retrieval (Möller et al. 1998). Several query languages have been proposed for querying business process models. Nevertheless, the complexity of business process information poses challenges to existing languages while there is lack of a generic approach for representing dynamically process semantics in machine processible manner. This fosters a growing gap between the business world and that of information management and retrieval.

Another challenge in business process information management is the "*change of medium*" (Bergener et al. 2015): When a radical change to the structure of a business process is performed or its primary information is enriched or updated, applications hinder to correspond and restructure the model due to the insufficient level of automation in the process execution. Moreover, another crucial point is that developing graphical models under formal specifications lacks simplicity. Therefore, it can mostly be done by business analysts as expert knowledge is required and training business users would take a considerable time, generating a much more noticeable cost to an enterprise. Business analysts primarily focus on the application-domain when depicting business processes rather than concerning the business-domain. The fact that they cannot be aware of all the particular aspects of a business process causes a deficiency in business process representation (Fernández et al. 2010). Last but not least, the complexity of different terminologies adopted by business entities in process modeling leads to semantically inconsistent interpretations lacking of interoperability in the information exchange between organizations.

Defining the appropriate semantics in business process models can make them fully machine-readable (Taylor 2012). Many researchers have considered the combination of semantic technologies with Management Information Systems as a promising approach to confront the aforementioned problems. This growing technological trend, has led to the establishment of a new research field, known as Semantic Business Process Management. Significant work has been made in the last decade applying semantic web methodologies to business information systems. The potential of a worldwide standardization unifying these topics is no longer an unreachable expectation. Attempting to investigate the aforementioned issues, in our previous work we have modeled business processes using BPMN, evaluated them with a business process management toolkit and defined their semantic concepts by developing a semantic process model in a knowledge-based framework (Kalogeraki et al. 2014). Continuously, we set the same queries both in a conventional database query language of a business process management tool and in a semantic query language mechanism of a knowledge acquisition system (Kalogeraki et al. 2015). In this work, we aim to generalize the semantic concepts of BPMN 2.0 and propose a BPMN 2.0 ontology for modeling business processes in a semantic web environment. Our work is organized as follows: Sect. 2 summarizes the generic aspects of semantic business process management, conceptual modeling and information retrieval, underlying some related work. In Sect. 3 we describe our research methodology and the structure of the BPMN ontology. In Sect. 4 we argue about conventional database querying versus semantic querying, according to our work results. In Sect. 5 we draw conclusions and suggest methods for improving the current project and set issues for future work.

2 Semantic Web Techniques in Management Information Systems

2.1 What's Worth in Combining Management Information Systems with Semantic Web Technologies?

Kő and Ternai (2011) state that "*The main challenge in Business Process Management is the continuous translation between the business requirements view and the IT systems and resources*". Business analysts' perception of business process models is formulated in a high level application environment. The contextual information of the presented business processes must be translated into a low-level computational representation in order to be further analyzed and executed by the machine's programming language. Traditionally, application developments tend to be programming driven rather than business process designed. Besides, the absence of an ad hoc notation when representing business processes into business models results to an ambiguous description of organizational aspects and its needs, unable for a computing machine to interpret information to a low-level machine language. This is well-known in literature as the fostering semantic gap between the administrative view and the IT's view.

Applying Semantic Web technologies in Management Information Systems can be a solution to bridge this gap. The new research area that studies this issue is called Semantic Business Process Management considered as the fusion of Semantic Web Services and Business Process Management that aims to identify and standardize the semantic patterns of business process models to create an integrated solution between business' and IT's world and to augment the automation level in the Business Process Management lifecycle (Hoang et al. 2010).

Semantic web techniques are able to provide methodologies and tools to support the semantic content of a business model dynamically and deal with the B2B integration challenge (Hoang et al. 2010). Applying semantic web technologies in Management Information Systems can benefit business environments generally for the following reasons:

- A semantic representation development adopts methods of structured conceptualization and produces ontology-based interpretations of natural language. Ontologies considered as formal representations of a domain providing not only the explicit statements of a development, but also the additional, implicit information which can be extracted by inference and reasoning mechanisms, helping an implementation to be machine-readable (Thomas and Fellmann 2009).
- Semantic web programming languages and specifications such as OWL programming language (Ontology Web Language), RDF (Resource Description Framework) and SPARQL standard are open source fast and flexible software. OWL Language and RDF standard can be expressed in other formats such as OWL/XML as well as RDF/XML which are both human and machine readable. Converting semantic models into XML formats enables extra-organizational

data sharing and use. Hence, it simplifies the information exchange and increasing interoperability between organizations.
- As stated previously, ontology models, instead of traditional business process models, have a semantic dynamic content, supporting run-time knowledge exploitation (Atkinson 2007) (A further analysis on traditional and ontology models is discussed in Sect. 2.2).
- Ontologies have a formal-structure character, missing from the conventional modeling representation (Atkinson 2007).
- Semantic models are able to share and retrieve information on the Web.
- Adopting semantic techniques in business process models enables business entities to have an explicit presentation of the involving participants' relationships in a process and helps them improve their business performance.
- Enhancing business modeling with semantic annotation can provide much more relevant information to the user who sets queries, due to automatic reasoning mechanisms. This is a considerable advantage for business owners because adjusting the deployed reference models offered from third-parties consultants becomes an additional cost for the enterprise (Thomas and Fellmann 2009).

2.2 Process Models, Conceptual Models and Ontologies

While a process model tends to be an abstract depiction of a business process, a meta-model is considered as a copy of a model which reflects all the types of elements that can be found within the model and identifies their relationships between them (such as a relational model) (Sánchez et al. 2009). The prefix "meta"—expresses a recursive function described as "a model for models" (Sánchez et al. 2009).

Conceptual modeling is "*a knowledge structure of a subject domain*" (Kogalovsky and Kalinichenko 2009), characterized by three main principles; *Modeling language constructs*, *Application domain gathering* and *Engineering* as in (Thalheim 2012). The conceptual model of the subject domain can be exploited along with ontology (ontological model) and analyzed in detail, enhancing the semantic representation of the subject domain (Kogalovsky and Kalinichenko 2009). Conceptual models differ from ontologies as they are high level rather abstract representations and do not use any specific language in presenting aspects of the physical world. High level data models are also known as semantic models.

According to Gruber (1993) ontology is "an explicit specification of a conceptualization". In terms of Philosophy and Artificial Intelligence theories, ontology is a systematic account of existence where everything that exists can be represented (Gruber 1993). A considerable advantage of an ontology development in software engineering is its high-level structure, including all possible variations of a concept able to mediate and balance variables between different systems and embrace interoperability (Gruber 1993).

The main difference between database models and ontologies relies on the different perception an agent is considered to approach information. Ontologies are built under the Open World Assumption in which an agent or an observer is thought to have incomplete knowledge (Atkinson 2007). In Open World Assumption what has not been stated is assumed to be unknown whether is true or false. Database models follow the principles of the Close World Assumption where an agent is thought to have complete knowledge; therefore what has not been stated is set as false. As for example, supposing adding a statement in each different system such as "citizen *A* is an owner of a sport car and can afford *only one* car". If we then add the state "citizen *A* is an owner of a red car" the Close World Assumption system will set the second statement as a false statement because the red car is considered as an additional car for citizen *A* which is not allowed in this case. This is because database applications follow the Unique Name Assumption, where every single name is presented by a different entity. Unlikely, the Open World Assumption system would consider the statement of citizen *A* having a red car as true, making the logical inference that the sport car of citizen *A* is also red. Hence, Open World Assumption systems are able to infer information that is not explicitly stated in graph representation using reasoning mechanisms. The Owl Semantic Web language is a typical example of applying the principles of the Open World Assumption. In our work, we have used Protégé Knowledge Management tool of Stanford University to develop the semantic ontology of the business process models.

2.3 Querying Business Process Models

Querying data considered as a precise request for information. Queries can be stated either on a Database Management System or on an Information Retrieval System.

Business Process Models has a lot of information to share and retrieve. Query Languages can be divided into table-based and graph-based. The representative standard of table-based queries is the SQL standard. SQL is the most famous declarative language, supported by a majority of tools. Graph-based query languages follow the Object Query Language (OQL) standard of object-oriented database management systems. Several traditional query languages have been made to extract information from business models. An example of a BPM query language is the BPMN-Q for BPMN Models, first defining patterns by drawing them and then translating them to SQL statements (Bergener et al. 2015). Another typical example is the APQL, concentrated on temporal logic and set restrictions on the processing control flow (Bergener et al. 2015). VMQL is a query language, although, intended to extract information for UML diagrams succeeded to be extended and used also for BPMN modeling specification (Bergener et al. 2015). It works with drawing patterns mechanisms like BPMN-Q. The BPQL query language although it functions by making patterns, it is limited to work only for a particular BPQL modeling language. There is, also, the Adonis Query Language

(AQL) of the Adonis Business Process Management Toolkit. The AQL language of Adonis Tool is based on object-driven OQL Language. In current work we have used the AQL Language mechanism of Adonis software to set queries to a conventional database of a business process model.

A traditional Information Retrieval search engine is able to use content-based methods to extract information from business process models. In keyword search methods, the query is treated as a list of independent keywords (Ma and Tian 2015). As a result, it can only succeed in explicit knowledge search (Ma and Tian 2015). In most cases, such queries end up losing its primary intention (Ma and Tian 2015). A web user has to deal with a great surfeit of data, unable to manage them properly, according to his needs. Semantic querying aims to depict tacit knowledge to extract implicit information from data sources (Ma and Tian 2015). The potential solution for querying efficiently and effectively information in business process models is to enrich them with semantic search techniques. Semantic search is focused on the meaning and importance of logical expressions instead of literal strings. This can be done by using Inference and Reasoning mechanisms, whereby an application can discover additional information that is not explicitly stated in the initial data. Inference and reasoning procedures in knowledge-based environments run by Inference Engines, allow both static and dynamic reasoning, such as using either a fixed set of rules or creating their rules dynamically according to data input.

Ontologies, as stated previously, can represent complex relationships between objects and include rules and axioms to describe the knowledge of a particular field. Evaluation, checking the logical consistency and exporting implicit information during a logical process of an ontology business process model, are functions provided through *Reasoners* (Sirin et al. 2007).

Among various theories for the representation of knowledge and its further management the theory of Description Logic has particular resonance. Key advantages include subjects such as the affinity with the ontology developing languages, the development of related algorithms that support the overall expressiveness of OWL and the development of corresponding tools of inference. According to this approach, knowledge is represented in terms of object oriented programming. Such business modeling uses concepts/classes, roles/properties, instances/individual and literals (Sirin et al. 2007).

Protégé, the knowledge-based Management tool of Stanford University, can provide information retrieval process through simple and complex semantic querying methods using the SPARQL standard. SPARQL is the predominant query language to derive information from the triple statement pattern of RDF (Resource Description Framework), a W3C specification. In this work, semantic queries were set in the developed BPMN ontology using SPARQL in order to compare the results with those of the conventional database.

2.4 Related Work

Several suggestions have been made for integrating the two scientific fields of Business Process Management and Semantic Web technologies. These approaches were mostly based on Ontology-driven Architectures, Ontology-based software engineering, Ontology Definition Metamodel, Model Driven Semantic Web (Atkinson 2007). The main purpose is to discover the know-how for building semantic bridges to eliminate the gap between business process management applications and IT environments.

There are many works developing methods for increasing the interoperability between different business technologies. IDEF family projects, the horizontal type of ebXML BPSS, the vertical standard of RosettaNet PIPs and UBL are some business process standards examples trying to represent organizational business information (Hoang et al. 2010), although during the informational exchange among business partners the semantic content was degraded.

Bringing together semantic web services with business process management by developing ontologies under the BPMN standard is a promising perspective for improving knowledge representation in Management Information Systems and providing intelligent skills of querying.

The Simple Business Process Management Notation (SBPMN) approach aims to represent semantic content of BPMN notation efficiently and make a comparative study between BPMN notation and SBPMN proposal (Fernández et al. 2010). An innovative approach which aims to enhance business applications with semantics and integrates them has been developed in the European-funded project FUSION (Hoang et al. 2010). It provides an integrated process modeling BPMN approach (Hoang et al. 2010), that transforms processes to XPDL and BPEL formats. A similar effort focusing on defining virtual partners during the exchange among actual and virtual trading partners and representing semantic information of integrated procedures has been put forward by the project SUPER, a Semantic Business Process Management (sBPMN) meta-ontology, based on semantic transactions applying Web Services Modeling Ontology (WSMO) to sBPEL executable model (Hoang et al. 2010). SemBiz, FIT and OntoGov are some other projects using ontologies and automated reasoning in support of business processes (Hoang et al. 2010; Sánchez et al. 2009). Another approach analyzes the semantics of business process models using the BPM standards of EPC, BPMN and OWL. The suggested formalization of the semantics of the individual model elements in conjunction with the usage of inference engines targets to facilitate query functionalities in modeling tools as in (Fill 2011). Semantic extensions in BPMN modeling adopting the use of web services have been also described applying the Maestro tool. A class hierarchy ontology of BPMN v2.0 has already been proposed by Natschläger (2011) and the presented ontology has been validated by different reasoners (Natschläger 2011). Even though, the first steps to this practice have been made, the ontology still stands incomplete. The project eBEST (Empowering Business Ecosystems of Small Service Enterprises to Face the Economic Crisis, 2011) maps conceptual models to ontology models. It uses the ADONIS suite to structure business processes, which they are exported in an xml ADONIS format and transferred to the OWL/XML format of Protégé application. Fill et al. (2013) have also worked on a generic service oriented architecture to support the semantic annotation of conceptual models using the Adoxx platform and Protégé format expanding BPMN elements with semantic extensions using WSDL and OWL ontology. An ontological analysis of the particular BPMN elements of events and activities using DOLCE framework (A Descriptive Ontology for Linguistic and Cognitive Engineering), an account of perdurant entities described in (Sanfilippo et al. 2014).

M. Rospocher, C. Ghidini and L. Serafini have currently worked on building a BPMN ontology, using the latex2owl tool,5 with its textual editor. The ontology created, a text file containing axioms encoded in latex-like syntax and then converted with the tool in a standard OWL RDF/XML serialization (Rospocher et al. 2014). The Business Process Diagram was also represented in an A-box to allow the exploitation of ontological reasoning services as in (Rospocher et al. 2014). There is also the pattern-based proposal for automatically detecting potential process weaknesses in semantic process models, aiming to improve business process performance. It provides a direct identification of potential weaknesses in business processes supporting a critical phase of BPI, showing how the referred pattern matching could be realized through the application of GMQL as in (Bergener et al. 2015).

3 A BPMN Semantic Process Model

3.1 The Research Methodology

Our research methodology focuses on upgrading our BPMN semantic description according to BPMN version 2.0.2 (Object Management Group 2013) and formalizing the abstract results by proposing a BPMN ontology model. To develop the business process models we have used the object-oriented architecture of ADONIS Business Process and Knowledge Management Toolkit provided by BOC Group. To create the BPMN ontology model we have used the ontology-based architecture of Protégé, a Business Process and Knowledge Management Toolkit created by Stanford University. Our method consists of the following steps:

- *Choose a real-life scenario of a particular aspect of the business world to analyze*

  The real-life scenario must be capable of giving a clear and brief description of the business aspect in question. A brief statement expressing of the activity description can facilitate the business analyst to divide the business scenario into distinguishable units of work. The selected aspect in question should have a considerable number of alternative instantiations for better estimation of results.

In order to efficiently represent business information and expand the fields of knowledge, it is a good practice to experiment on more than one real-life scenarios. In our work, the selected real-life scenarios describe different matters of higher education studies.
The first real-life scenario refers to *managing the preparation works for organizing a practical training seminar for students of the Hellenic National School of Public Administration and Local Government*, considered probationers (Kalogeraki et al. 2014). In order to represent precisely the business process three different graphical models were developed; the business process diagram, the choreography diagram and the working environment, each one illustrating different aspects of the entire process (Kalogeraki et al. 2014).
The second real-life scenario aims to *describe all the appropriate procedures taken place for the students' enrollment in the E-learning Program offered by the Department of Informatics of the University of Piraeus* (Kalogeraki et al. 2015). The analyzed business process encapsulates a sub-process of students paying the course tuition fee for attending the E-learning Program (Kalogeraki et al. 2015).

- *Define the underlying activities that are able to describe accurately the entire business process*

In this step we organize all the relevant information into groups. As a consequence, we identify special key-words, generic terms and headings to represent with the BPMN elements in our graphical model. For example, in the second real-life scenario all the e-payments for tuition fees to attend an e-learning course can be characterized as "web-banking payments".

- *Identify hierarchy in the activities and record them in individual steps*

Set a data hierarchy of the knowledge-base and distinguish the most important procedures of the business process to describe them step-by-step.

- *State clearly the involving participants and their roles in the process*

Analyzing the processes can help the business analyst to define business roles, the process stakeholders and participants, both from within the company and any external partners and customers. This will help to adequately represent Business-to-Business (B2B), Business-to-Consumer (B2C) and Business-to-Government (B2G) interactions.

- *Develop the appropriate Business Models with a Management toolkit, supporting the BPMN 2.0.x standard, to represent the business process*

Business Process modeling is supported by a variety of graphical representations each of them describing a certain point-of-view, depending on the business process perspective which the analyst wants to explore. BPMN v.2.0.2 specification consists of the following set of diagrams; the business process diagram (including private and public processes), the choreography diagram, the collaboration diagram (including the informal depiction of the conversation diagram) (Object Management Group 2013). In order to represent precisely and effectively the business processes of these two real-life scenarios, we use two different types of BPMN graphical models for each case:

(i) the business process diagram, to describe the private business processes involving the organization's internal performers and the public business processes to denote all the interactions between organization's internal and external performers and
(ii) the choreography diagram, to show explicitly the information exchange between two different agents in a process
(iii) Additionally, we create a working environment graph for each of two business scenarios to illustrate the business roles of participants within a process and the organizational structure.

- *Define semantics concepts of the business process model elements*

  Delving into implicit relations and common attributes of BPMN elements
- *Express their semantic relations in a form of an ontology using a knowledge management toolset and defining logical restrictions whenever needed according to BPMN v.2.0.2 specification*

  Using Protégé platform we develop a BPMN ontology by creating all the BPMN elements in the form of an ontology, defining hierarchies between them, while creating classes and subclasses and set their attributes by adding properties to classes.
- *Evaluate the ontology by checking the accuracy of taxonomy, its logical consistency, satisfiability, classification, and realization using reasoning techniques*

  Checking the ontology's consistency and evaluated the accuracy of the taxonomy using the inference mechanisms of HermiT and FaCT++ reasoners of Protégé. Assure no logical errors or conflicts are found.
- *Setting up queries both in the business process model and the semantic process model*

  Setting the same queries both in the business process model and in the BPMN ontology. Queries in the business process model can be set using expressions of the AQL (Adonis Query Language), which is based on an object-driven database architecture. Queries in the BPMN ontology can be structured using the SPARQL Query Language (Fig. 1).
- *Comparing the querying results of traditional and semantic information retrieval*

  Comparing the search results and drawing conclusions about these two querying methods.

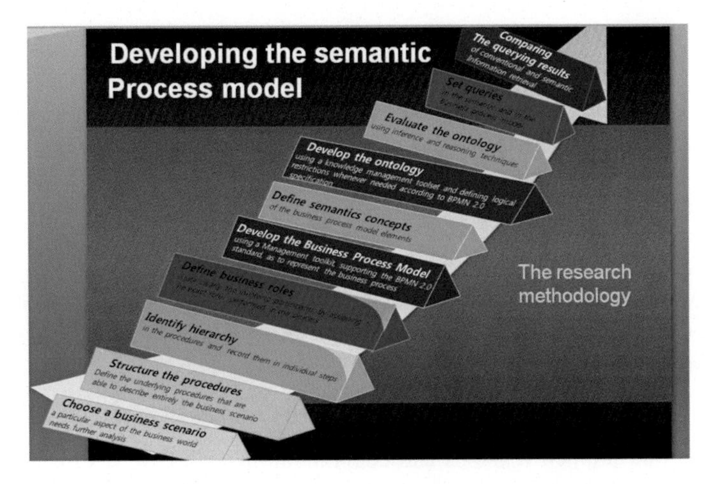

Fig. 1 The research methodology for developing the semantic process model

3.2 Developing Business Process Models

As it has already been stated, a business process model was developed for each real-life scenario, described previously in Sect. 2.1, under the upgraded version 2 of BPMN specification using the ADONIS platform, an integrated Business Process and Knowledge Management Toolkit of object-oriented architecture.

The *business process diagram* (as shown in Figs. 2 and 4) corresponds to the form of BPMN 2.0 standard presenting all the internal and external procedures taken place in the process, the organization's involved units and its shareholders, even though the external performers of the business process, such as customers.

The *choreography diagram*, shown in Fig. 3 reflects the idea of enriching business process models with process choreography principles, expressing a two-dimensional status of the executing procedures. Choreography is the state of incorporating the interactions of independent agents occurred while they exchange information in order to perform their tasks and achieve particular goals (Robak et al. 2012) (Fig. 4).

The *working environment* (seen in Fig. 5) is a graph of the Adonis Management Tool (which follows the organizational chart principles; it describes three object types; the organizational units, the performers and their roles either in the process or within the organization.

3.3 Developing the Ontology

According to Guarino's (1998) consideration "*every (symbolic) information system (IS) has its own ontology, since it ascribes meaning to the symbols used according to a particular view of the world*". In this section we describe the transaction of business

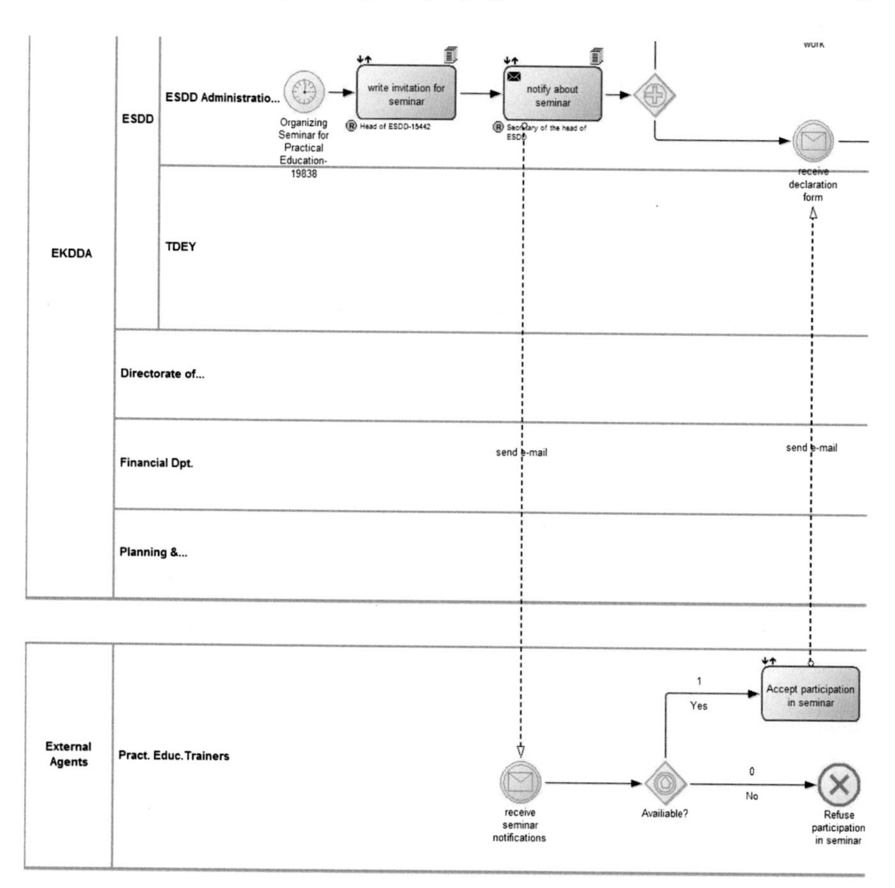

Fig. 2 An abstract from the first real-life scenario of the business process diagram, using BPMN 2.0 specification referring to the procedures of inviting practical educational trainers to the seminar and getting information about their participation

process models to conceptual models. This can be done by defining the semantics concepts of the BPMN elements and asserting these concepts in the form of ontology.

The semantic representation of the business process model in our work is characterized as "*semantic process model*". Protégé is an ontology Editor and Knowledge-based Framework of Stanford University. The older versions of Protégé (such as v.3.x.x) contain two editors; the frames editor for constructing frame-based domain ontologies under and the Owl editor supporting files of various formats such as OWL/RDF, HTML, XML. Frames editor makes the Closed World Assumption in which everything that has not been explicitly asserted in the knowledge base will not return true value (Atkinson 2007). On the contrary, Owl editor makes the Open World Assumption that anything can be entered into an OWL Knowledge Base in spite of violating one of the constraints (Atkinson 2007). Owl editor follows the subject-predicate-object triple theory of the Resource

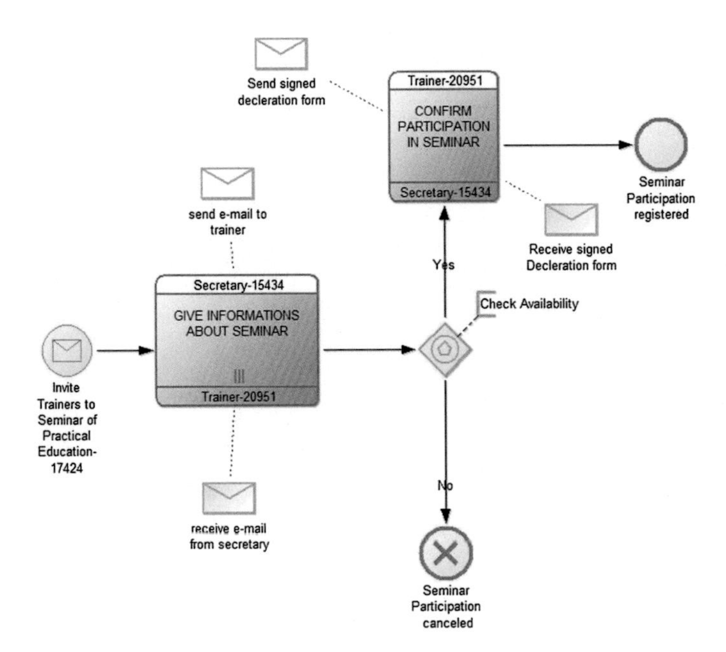

Fig. 3 Depicting the same abstract of the first real-life scenario in the form of BPMN's 2.0 choreography diagram, giving a two-dimensional representation objective of the procedures

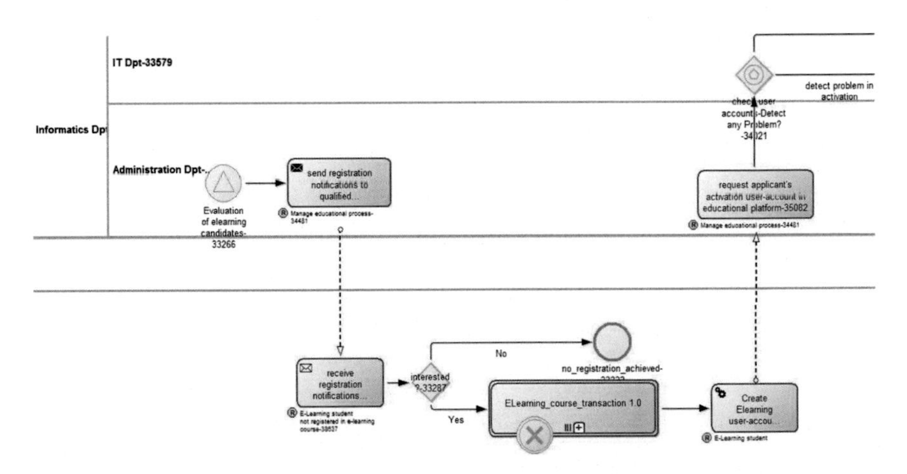

Fig. 4 A part from the business process diagram of the second real-life scenario analyzes all the appropriate procedures for registering in the E-learning course, encapsulating a sub-process about paying the tuition fee

Description Framework (RDF) standard. In Owl editor predicates of RDF standard are called properties, divided into data/type, annotation and object properties, while RDF objects expressed as individuals. A deeper analysis of two protégé editors is out of the current discussion topic.

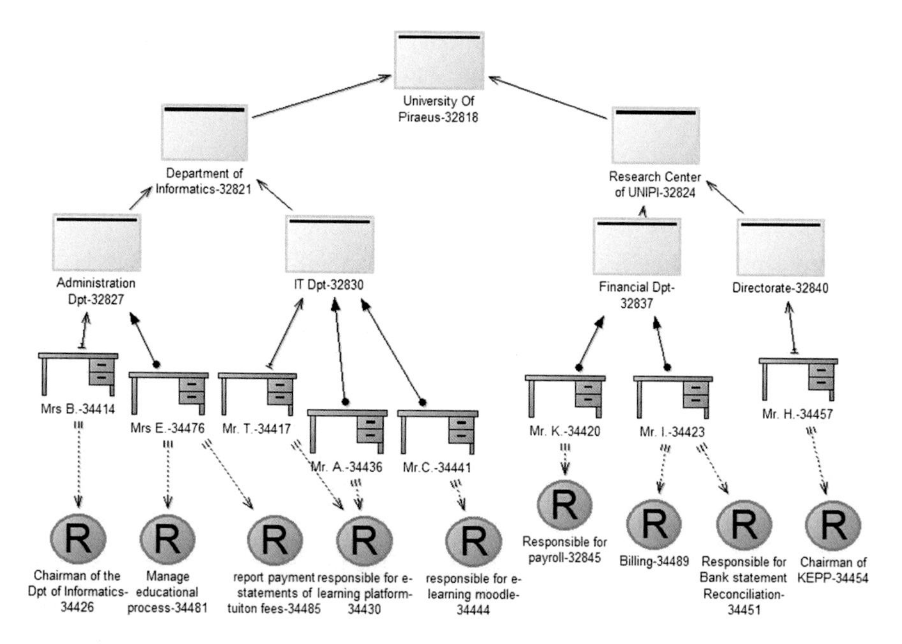

Fig. 5 Defining business roles and participants of the second real-life scenario using the working environment diagram of ADONIS management toolkit

Next, we describe the semantic process model develop in the Owl Editor of Protégé. The definition of classes, class hierarchies and variables in our semantic process model follows the structure of BPMN v.2.0.2 suite (Object Management Group 2013). According to BPMN 2.0.2 specification, we consider our Knowledge Base divided into three different aspects which are the three major superclasses in the ontology:

- *The scope of the BPMN v.2.0.2 elements, defined by the superclass "BpmnElements"*
- *The scope of the generic BPMN 2.0 alternative models, defined by the superclass "BpmnModel"*
- *The scope of the agent or actor participating in the process, defined by the superclass "Person"* (Fig. 6).

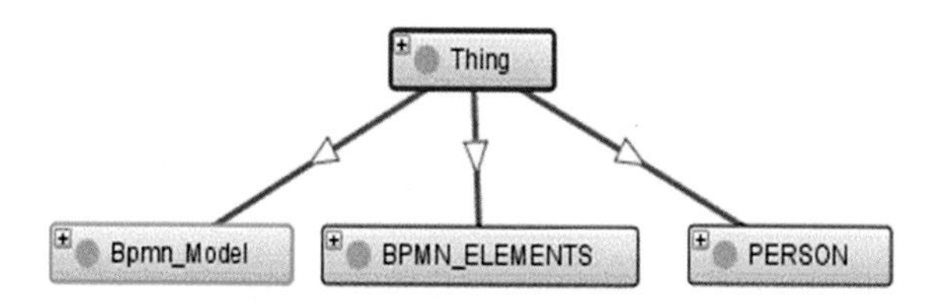

Fig. 6 The three main superclasses of the BPMN ontology from the OntoGraf of Protégé Owl Editor

Classes are linked with individuals, through object and data properties, shown in Fig. 11, by structuring RDF statements and setting semantic assertions as the example in Fig. 12.

3.3.1 The Scope of the BPMN Elements

We consider the "BpmnElements" superclass, which is divided into five subclasses, presenting the main categories of BPMN v.2.0.2 elements, named as "*Swimlanes*", "*FlowObjects*", "*Connecting Objects*", "*Artifacts*" and "*Data*" (Object Management Group 2013). Each of the five sub-class is further analyzed in the sub-categories of BPMN v.2.0.2 standard as in Fig. 7. Additional division of the

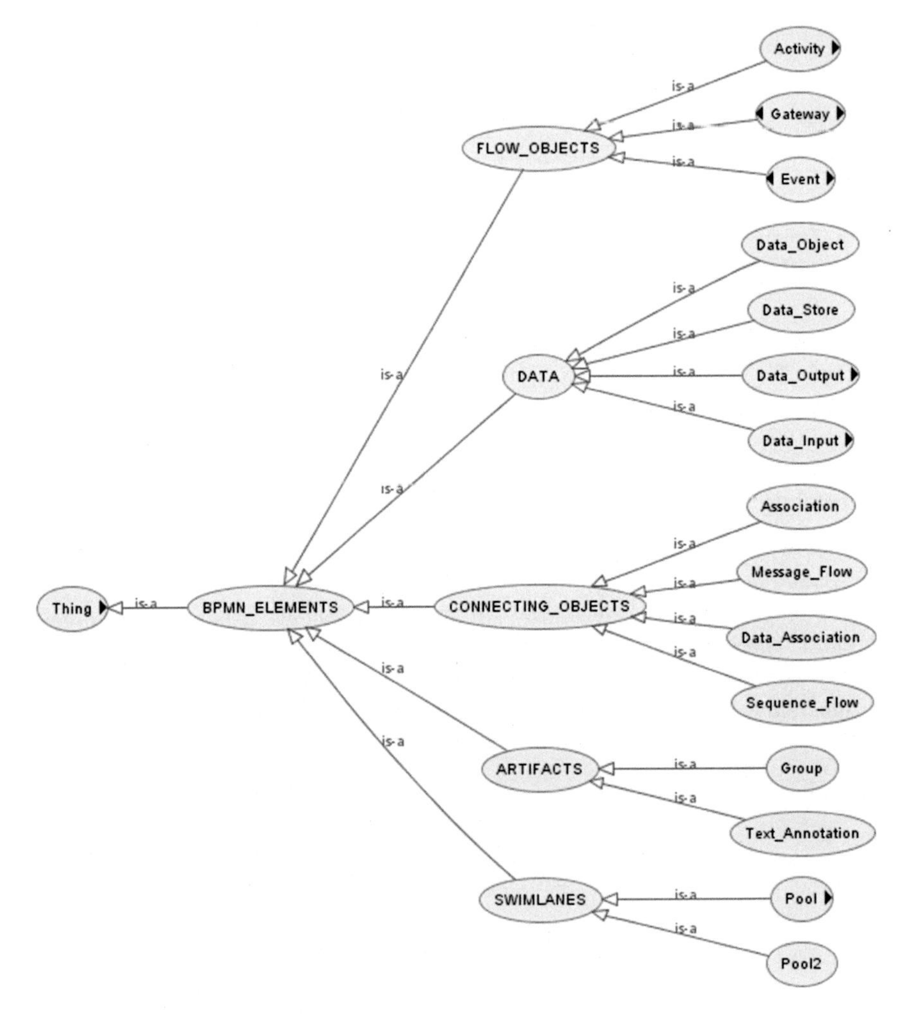

Fig. 7 The main subclasses of "BpmnElements" superclass as presented from the Owl Viz Graph of Protégé Owl Editor

above sub-classes refers to the extended BPMN modeling elements (Object Management Group 2013). In our ontology the extended BPMN analysis basically focuses on the *Flow Object* elements: *Event, Activity and Gateway*.

Depending on the *flow dimension*, the sub-class of "*Event*" is separated into three further sub-classes "*start*", "*intermediate*" and "*end*". Events are caused by a particular trigger. The triggers that cause events are: *message, timer, error, escalation, cancel, compensation, conditional, link, signal, terminate, multiple, parallel multiple*. Every type of event corresponds to a different set of possible triggers.

According to the type of process in which *a start event* occurs its class is divided into three sub-classes: "*Event_Sub_Process_Trigger*", "*Sub_Process_Start_Event*" and "*Top_Level_ Process_Event*". Start Events of sub-processes can either interrupt the processing or not affect it. To better describe a start event we consider 3 different data properties: the "*Top_Level_Event_Definition*", the "*Type_of_ Interrupting_StartEvent*" and the "*Type_of _non_interrupting_Event*". Each data property has a particular set of possible triggers to choose; although one of them should be linked with the appropriate start event class, as in p. 259 of (Object Management Group 2013).

Intermediate events can be divided into the sub-classes of boundary and non boundary events, catching or throwing. Boundary events can also categorized into the sub-classes of interrupting and non interrupting. We consider the following data properties for intermediate events each one referring to a different set of possible triggers:
"*catching_Intermediate_Event*","*throwing_Intermediate_Event*","*Type_of_Interrupting_IntermediateEvent*" and "*Type_of _non_interrupting_Event*", which is the same previously mentioned data property for start events. Finally, end events can be linked with a particular type from the data property set called "*End_ Event_definition*".

The "*FlowObject*" class has the sub-class of "*Activity*" which is separated into five further sub-classes regarding the BPMN v.2.0.2 conformance (Object Management Group 2013): "*Call_Activity*", "*Event_Sub_Process*", "*Sub_Process*", "*Task*" and "*Transaction_Sub_Process*". Depending on the type of behavior, tasks can have one of the following characteristics: abstract, business rule, manual, receive, script, send, service, user. These characteristics are settled into a data property called "type_of_task".

The "*FlowObject*" class also consists of the sub-class of "*Gateway*", which can be analyzed into further sub-classes such as *"AND_Parallel_Fork_Join", "Complex_Decision_Merge", "OR_Inclusive_Decision_Merge" and "XOR_ Exclusive_Decision". Gateways might be either event-based or event-based instantiate or data based. This distinction is asserted by the data property "type_of_Gateway".*

Figure 8 shows the extended modeling elements of the developed BPMN ontology.

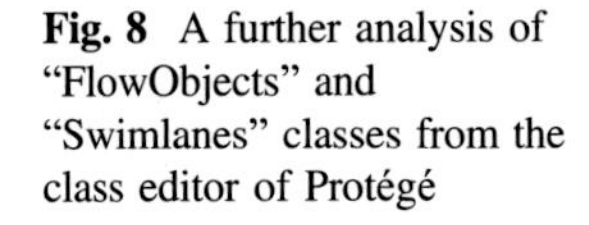

Fig. 8 A further analysis of "FlowObjects" and "Swimlanes" classes from the class editor of Protégé

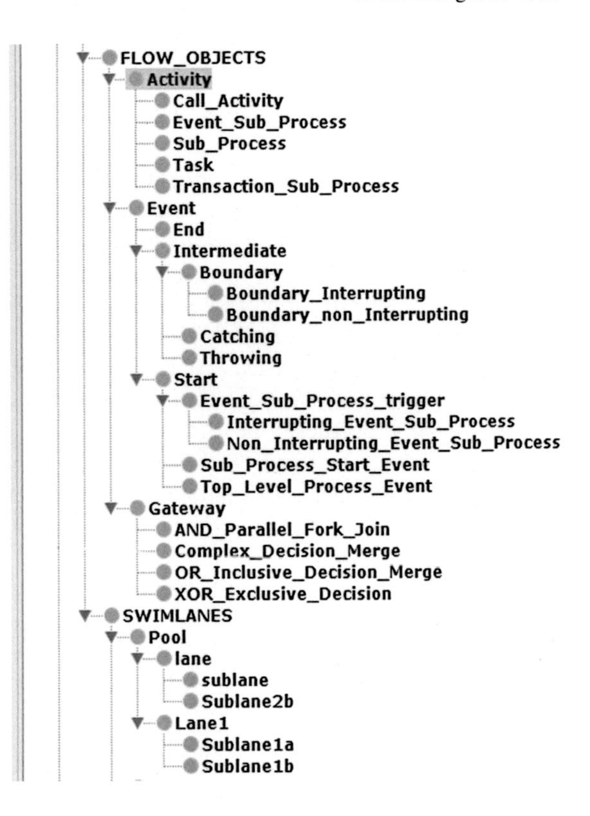

3.3.2 The Scope of the Generic BPMN Alternative Models

The "BpmnModel" superclass consists of four different models of BPMN 2.0 specification subclasses; the *"BusinessProcessModel"*, the *"Choreographies"*, the *"CollaborationModel"* and the *"ConversationModel"* as in Fig. 9. Our two real-life scenario implementations are presented with a Business Process Model and Choreography Model.

As shown in Fig. 9, the subclasses "Gateway" and 'Events" of the "Choreographies" superclass are the same subclasses depicted in the "FlowObjects" superclass having the same properties and individuals as seen in Fig. 10.

3.3.3 The Scope of the Agent or Actor Participating in the Process

We consider the superclass "Person" as the actor or agent of the business process, regarding the kind of business role in the process and the relationship between the actor and the business entities. The superclass "Person" consists of the two subclasses of "*InternalPerformer*" and "*ExternalPerformer*" showing in Fig. 11. We have defined these two generic classes two declare a distinction between the internal and external Performers of an organization participating in the business

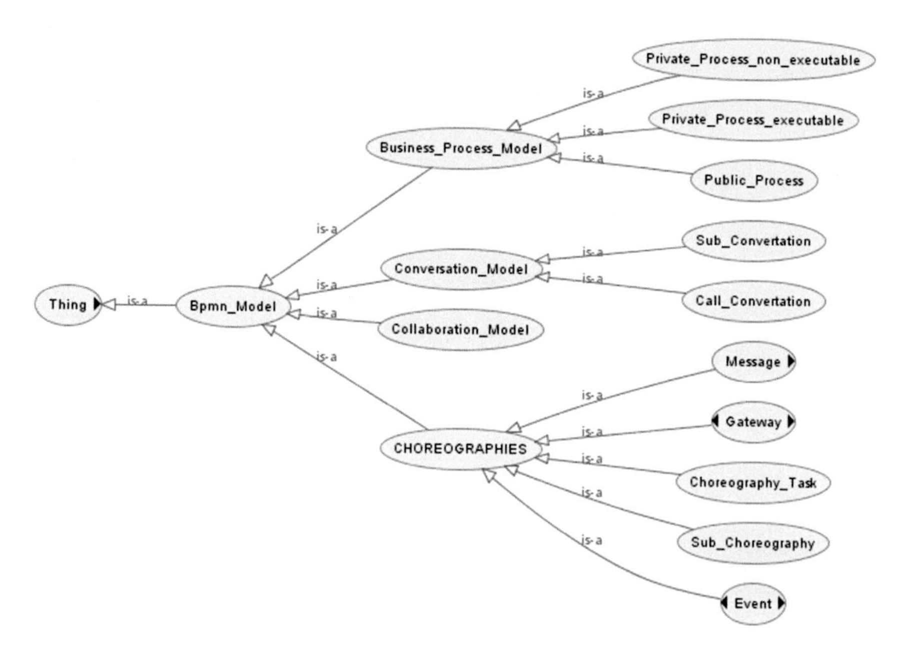

Fig. 9 The subclasses of "BpmnModel" superclass as presented from the Owl Viz Graph of Protégé Owl Editor

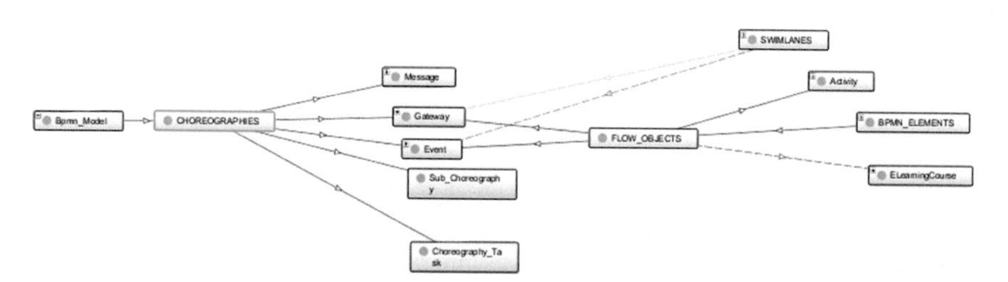

Fig. 10 A graphical analysis of the class "Choreographies" as presented from the OntoGraph of Protégé Owl Editor showing that "FlowObjects" is a common class for "Choreographies" and "FlowObjects"

process. Figure 11 is a snapshot of the implemented business process "Enrollment in the E-Learning Program", shows that the class "*InternalPerformer*" is linked with the organization class which is the individual "*University_of_Piraeus*". These are the general classes analysis of BPMN 2.0.2 that are likely to fit in any alternative real-life scenario as they are expressed in general. Instead of these defined classes, we may occasionally have to add few others in the ontology, depending on the real-life scenario we choose to examine. For example, in the first-real life scenario about organizing practical education seminar, the proposed ontology has all the appropriate classes to represent the business process so none definition class needed to be added. On the other hand, for representing the second real-life

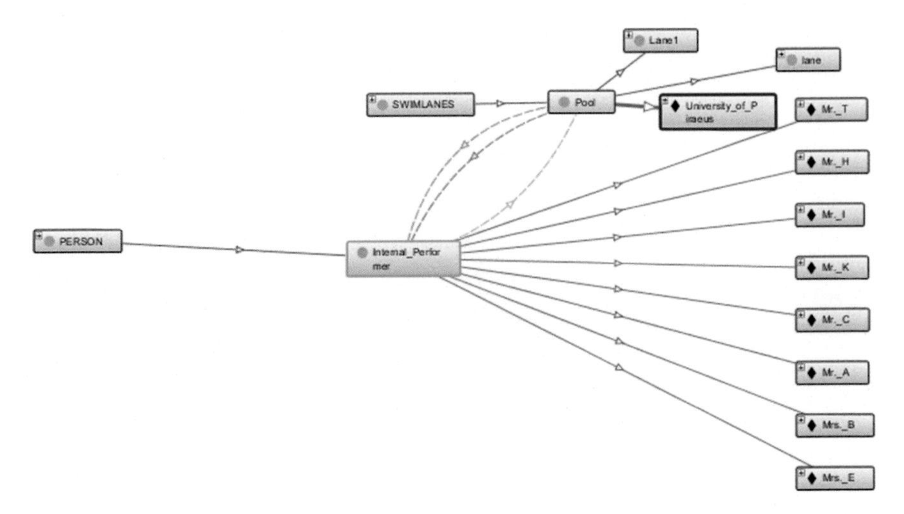

Fig. 11 A graphical analysis of the class “InternalPerformer” as presented from the OntoGraph of Protégé Owl Editor, showing that “FlowObjects” is a common class for “Choreographies” and “FlowObjects”

scenario, about the enrollment of students to e-learning program and the sub-process of their registration to e-learning courses, a new superclass required to be defined, considered as “ELearningCourse” superclass having children called “ELesson” as subclasses.

We have already mentioned that Owl makes use of the Open World Assumption when everything can be stated unless is declared not to be true. As a consequence, in previous example for the class “InternalPerformer” must somehow been stated that its subclasses are University of Piraeus employees linked only with the organizations swimlane. These can be expressed by setting disjoints between siblings classes “Pool” (which has the “University_of_Piraeus” individual) and “Pool2” (which has the “Elearning_Applicant” individual). Figure 12 shows this restriction. The semantic process model classes are linked either with classes or individuals through properties following the triple logical statement of subject-predicate-object of RDF standard (Fig. 13).

3.4 Validating the Ontology

We use three different reasoners to provide reasoning services for OWL documents: FaCT++, Pellet and HermiT in order to have an evaluation for the ontology we developed and to retrieve and fix any logical error. The logical inference process generates some implicit information such as the following statements (Fig. 14).

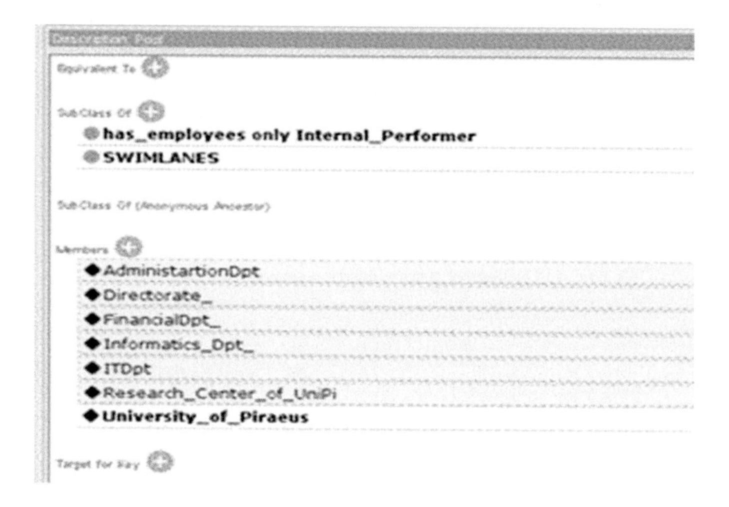

Fig. 12 The restriction of the University of Piraeus having only Internal Performers employees

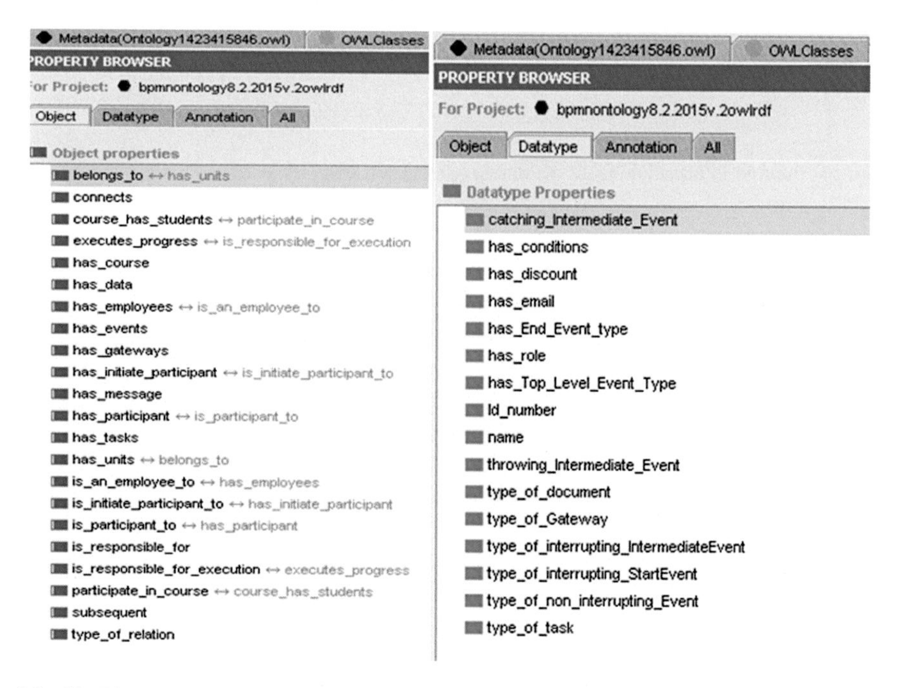

Fig. 13 The data properties matrix shows all the properties of the ontology *that have literals* as a range

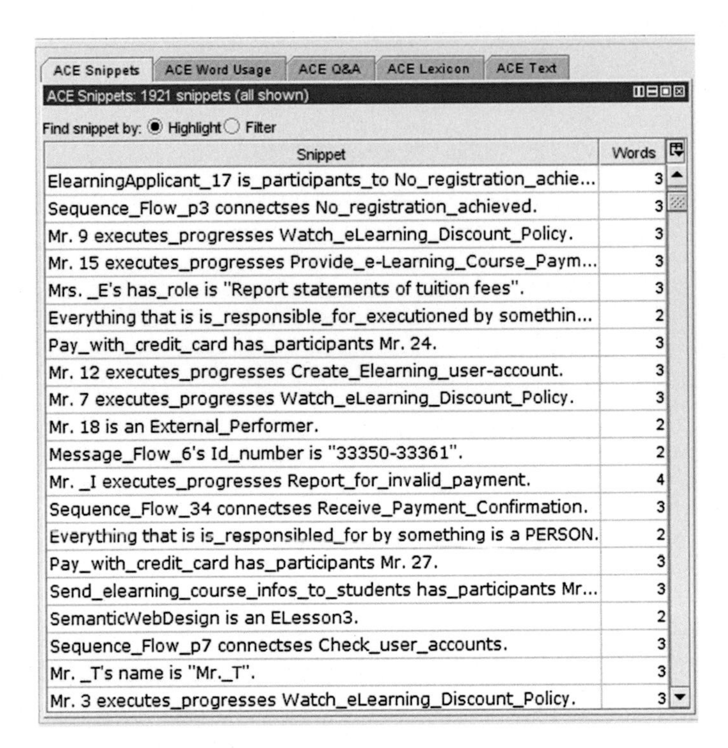

Fig. 14 Inferred information about the ontology extracted from reasoners

4 Querying Conventional Databases and Semantic Models

Once we have developed the two kinds of models implying the business proccss model and the semantic process model, we set the same queries in both models in order to compare the two alternative methods of business information retrieval. In the following sections, we describe the querying methods and comment on the results.

Adonis platform includes the Adonis Query Language (AQL), an object-driven architecture for running queries on business models (BOC Group 2012). Object-oriented databases (OODB) are databases, depicting and storing information in the form of objects. They provide a combination of database characteristics and object-oriented programming capabilities. Queries in the business process models were set using the expressions of Adonis Query Language (AQL). AQL syntax has adopted the basic rules of the Extended Backus Naur Form (EBNF) metasyntax notation. Moreover, Adonis Management Toolkit Community Edition (ADONIS: CE) supports the following methods of querying business models for evaluation purposes: (i) ad hoc queries; user-defined and standardized queries. User-defined queries give users the ability to structure their own queries using the AQL expressions (ii) setting queries from a library of predefined queries (BOC Group 2012).

The following 12 queries were composed for the business process models of BPMN notation so as for the working environment diagrams of Adonis suite, giving the results presented below:

1. Give all the business entities and its Departments/sub-Departments
2. Give the University of Piraeus employees
3. Find "Task" BPMN element used in processes having the type of "send"
4. Give the University of Piraeus employees in every Department
5. Find e-learning students with large family discount paying e-learning fee with credit card
6. Find unemployed students pay in cash that have chosen Web Design E-Learning Course
7. Find students pay in cash that have chosen Web Design E-Learning Course and are not necessary unemployed
8. Which progresses are executed by the IT Department? And who executes them?
9. In what actions is the IT Department involved?
10. Find e-learning Applicants that they did not register in the E-Learning Course
11. Give E-learning students with a large family member discount
12. Give E-learning students that pay with credit card

Queries 1, 2, 4, 5, 6, 7, 10, 11, 12 listed above, were made on the developed working environment diagrams, while queries 3, 8, 9, were run on the developed business process models. The standardized queries format of Adonis Query Language is performed with simple natural language syntax. The user-defined query structure is translated into an AQL expression. Standardized queries can alternatively be described in an AQL expression. Predefined queries are also given in a natural language format. Table 1 shows further details for the business model querying method; queries 1–4 are standardized questions, queries 4–9 are user-defined and queries 10–12 have the predefined format of Adonis Query Language. By characterizing this query method as "conventional", we mean retrieving information from a non-semantic database.

Queries in the BPMN ontology were structured with the SPARQL Query Language, using the Protégé Knowledge Management Tool. SPARQL is a semantic query language, capable of querying data in RDF triple statements subject—predicate—object. We have set in SPARQL the same queries listed in the previous section. In order to retrieve the requested information from SPARQL engine, we have used the general form SELECT-FROM-WHERE and the OPTIONAL query structure. Table 1 shows the semantic queries written in SPARQL.

Next, we present some query results (shown in Figs. 15, 16, 17), retrieved using AQL and SPARQL query languages (queries from Table 1):

1. Give all the business entities and its Departments/sub-Departments
3. Find "Task" bpmn element used in processes having the type of "send"
7. Find students pay in cash that have chosen Web Design E-Learning Course and are not necessary unemployed

Table 1 Conventional versus semantic query method

Query Serial number	*Description in Natural Language*	Conventional Query Method *OBJECT-ORIENTED DATABASE (AQL)*	Semantic Query Method *SEMANTIC-BASED SEARCH (SPARQL)*
1	Give all the business entities and its Dpts /sub-Dpts	query: Give all objects of class "Organizational Unit"	PREFIX rdf: <http://www.w3.org/1999/02/22-rdf-syntax-ns# PREFIX owl: <http://www.w3.org/2002/07/owl#> PREFIX xsd: <http://www.w3.org/2001/XMLSchema#> PREFIX rdfs: <http://www.w3.org/2000/01/rdf-schema #> PREFIX search:< http://www.owl-ontologies.com/Ontology1423415846.owl#> SELECT ?subject ?object WHERE { ?subject search:belongs_to ?object }
2	UniPi employees	query: Get all objects of class "performer"	.....SELECT ?subject ?object WHERE { search:University_of_Piraeus search:has_employees ?object }
3	Give "Task" bpmn element used in processes having the type of "send"	query: Get all objects of class "Task" with attribute "Task type" = "send"	....SELECT ?subject ?object WHERE { ?subject search:type_of_task "send"^^http://www.w3.org/2001/XMLSchema#string> }
4	UniPi employees in every Dpt.	AQL expression: (<"Is manager">) OR (<"Belongs to">)	......SELECT ?subject ?object WHERE { ?subject search:has_employees ?object }
5	E-learning students with large family discount paying e-learning fee with credit card	AQL expression: ({"pay with credit card":"Role"}<-"Has role") OR ({"pay with credit card":"Role"}->"Has role") DIFF({"unemployed":"Role"}<-"Has role") OR ({"unemployed":"Role"}->"Has role") DIFF ({"no discount":"Role"}<-"Has role") OR ({"no discount":"Role"}->"Has role") DIFF ({"special needs":"Role"}<-"Has role") OR ({"special needs":"Role"}->"Has role") DIFF ({"unipi graduated/postgraduated":"Role"}<-"Has role") OR ({"unipi graduated/postgraduated":"Role"}->"Has role") DIFF ({"company employee benefit":"Role"}<-"Has role") OR ({"company employee benefit":"Role"}->"Has role")	.......SELECT ?subject ?object WHERE { search:Large_family_benefits search:has_participant ?object . search:Pay_with_credit_card search:has_participant ?object }
6	Find unemployed students pay in cash have chosen Web Design E-Learning Course	AQL expression: (({"Course ELI01: WEB DESIGN":"Organizational unit"}<-"Belongs to") OR ({"Course ELI01: WEB DESIGN":"Organizational unit"}->"Belongs to")) AND (({"unemployed":"Role"}<-"Has role") OR ({"unemployed":"Role"}->"Has role")) AND (({"pay in cash":"Role"}<-"Has role") OR ({"pay in cash":"Role"}->"Has role"))	.........SELECT ?subject ?object WHERE { search:Pay_in_cash search:has_participant ?object . search:Unemployement_benefits search:has_participant?object . search:WebDesign search:course_has_students ?object }
7	Find students pay in cash that have chosen Web Design E-Learning Course and are not necessary unemployed	AQL expression: (({"Course ELI01: WEB DESIGN":"Organizational unit"}<-"Belongs to") OR ({"Course ELI01: WEB DESIGN":"Organizational unit"}->"Belongs to")) AND (({"pay in cash":"Role"}<-"Has role") OR ({"pay in cash":"Role"}->"Has role"))	...SELECT ?subject ?object WHERE { search:Pay_in_cash search:has_participant ?object . search:WebDesign search:course_has_students ?object .OPTIONAL {search:Unemployement_benefits search:has_participant ?object }
8	Which progresses are executed by the IT Dpt.? And who executes them?	AQL expression: (<"Task">[?"Done by" = "Mr.C.-34441"])	...SELECT ?subject ?object WHERE { ?subject search:executes_progress ?object . ?subject search:is_an_employee_to search:ITDpt }
9	In what actions is the IT Dpt. involved?	AQL expression: ({"IT Dpt-33579":"Lane"}<-"Is inside") OR ({"IT Dpt-33579":"Lane"}->"Is inside") DIFF ({"ELearning Applicant-33227":"Pool"}<-"Is inside") OR ({"ELearning Applicant-33227":"Pool"}->"Is inside") DIFF ({"Directorate-33410":"Lane"}<-"Is inside") OR ({"Directorate-33410":"Lane"}->"Is inside") DIFF ({"Financial Dpt-33405":"Lane"}<-"Is inside") OR ({"Financial Dpt-33405":"Lane"}->"Is inside") DIFF ({"Administration Dpt-33230":"Lane"}<-"Is inside") OR ({"Administration Dpt-33230":"Lane"}->"Is inside")	.....SELECT ?subject ?object ?x ?y WHERE { search:ITDpt search:has_gateways ?object . search: ITDpt search:has_events ?x . search: ITDpt search:has_tasks ?y .}
10	E-Learning Applicants that they did not register in E-Learning Course"	query: All performers with the role "*not registered in e-learning course-38637*"	...SELECT ?subject ?object WHERE { search: No_registration_achieved search:has_participant ?object }
11	E-learning students with a large family member discount	query: All performers with the role "*large family member*"	...SELECT ?subject ?object WHERE { ?subject search: has_discount"Large_Family_Members"^^<http://www.w3.org/2001/XMLSchema#string> }
12	E-learning students that pay with credit card	query: All performers with the role "*pay with credit card*"	......SELECT ?subject ?object WHERE { search: Pay_with_credit_card search:has_participant ?object }

Query Method Expression	Search Result

AQL

query: Give all objects of class "Organizational Unit"

Name	Class
Administration Dpt-32827	Organizational unit
Department of Informatics-32821	Organizational unit
Directorate-32840	Organizational unit
Financial Dpt-32837	Organizational unit
IT Dpt-32830	Organizational unit
Research Center of UNIPI-32824	Organizational unit
University Of Piraeus-32818	Organizational unit

SPARQL

```
PREFIX rdf:
<http://www.w3.org/1999/02/22-rdf-
syntax-ns#
PREFIX owl:
<http://www.w3.org/2002/07/owl#>
PREFIX xsd:
<http://www.w3.org/2001/XMLSchema#>
PREFIX rdfs:
<http://www.w3.org/2000/01/rdf-schema #
PREFIX search:< http://www.owl-
ontologies.com/Ontology1423415846.owl#>
SELECT ?subject ?object
                WHERE { ?subject
search:belongs_to ?object }
```

Subject	Object
Research_Center_of_UniPi	University_Of_Piraeus
Directorate_	Research_Center_of_UniPi
FinancialDpt_	Research_Center_of_UniPi
InformaticsDpt_	University_of_Piraeus
IT Dpt	InformaticsDpt_
AdministationDpt	InformaticsDpt_

Fig. 15 1st query results

Query Method Expression	Search Result

AQL

query: Get all objects of class "Task" with attribute "Task type" = "send"

	Model	Name	Class	Attribute	Value
1.	ELearning_registration 1.0				
		send elearning course infos to students - 34104	Task	Task type	Send
		send registration notifications to qualified candidates-33274	Task	Task type	Send
1.1.	ELearning_transaction 1.0				
		Report for invalid payment-39164	Task	Task type	Send
		Send Supporting Documentation-33350	Task	Task type	Send

SPARQL

```
PREFIX rdf: <http://www.w3.org/1999/02/22-rdf-
syntax-ns#
PREFIX owl: <http://www.w3.org/2002/07/owl#>
PREFIX xsd: <http://www.w3.org/2001/XMLSchema#>
PREFIX rdfs: <http://www.w3.org/2000/01/rdf-schema
#
PREFIX search:< http://www.owl-
ontologies.com/Ontology1423415846.owl#>
SELECT ?subject ?object
              WHERE { ?subject
search:type_of_task
"send"^^http://www.w3.org/2001/XMLSchema#string> }
```

Subject	Object
Send_registration_notifications_to_qualified_candidates	
Send_Supporting_Documentation	
Send_elearning_course_infos_to_students	
Report_for_invalid_payment	

Fig. 16 3rd query results

Query Method Expression	Search Result

AQL

```
query: Find  students pay in
cash that have chosen Web Design
E-Learning Course and are not
necessary unemployed

(has AQL expression as table I)
```

Model	Name	Class	Attribute	Value
E-learning Transaction 1.0				
	Mr.1	Perfo rmer	-	-
	Mr.22	Perfo rmer	-	-
	Mr.23	Perfo rmer	-	-

SPARQL

```
PREFIX rdf:
<http://www.w3.org/1999/02/22-rdf-
syntax-ns#
PREFIX owl:
<http://www.w3.org/2002/07/owl#>
PREFIX xsd:
<http://www.w3.org/2001/XMLSchema#>
PREFIX rdfs:
<http://www.w3.org/2000/01/rdf-schema #
PREFIX search:< http://www.owl-
ontologies.com/Ontology1423415846.owl#>
SELECT ?subject ?object
    WHERE { search:Pay_in_cash
search:has_participant ?object .
search:WebDesign
search:course_has_students ?object
.OPTIONAL
{search:Unemployement_benefits
search:has_participant ?object }
```

Subject	Object
	Mr.1
	Mr.22
	Mr.23

Fig. 17 7th query results

Comparing the two retrieved query results, we observe that both methods of queries are able to extract the right information, but only the semantic-based method can give further analysis about the requested subject. For example, when setting the 1st query "Give all the business entities and its Dpts/sub-Dpts" the information retrieved from the object-oriented database is all the recorded departments (Dpts) and sub-departments (sub-Dpts) of the Organization in an abstract list, On the contrary, when the same query performed in SPARQL, the retrieved information shows not only the requested organizational units but also the relations between them, as depicted in Fig. 15. Additionally, the object-oriented method of AQL query language represents the results in the form of classes (shown in Figs. 15, 16, 17) while semantics-based method of SPARQL query language is more efficient because it defines the implicit relationships between classes in the triple form of subject-predicate-object. Comparing the syntax of the AQL and SPARQL query languages we can see that AQL expression encapsulated complex procedures for data processing unlike SPARQL's which is based on plain triple patterns, conjunctions and disjunctions (seen in Table 1).

5 Conclusions

Coupling semantic web technologies with business process modeling techniques can facilitate bridging the gap between the business world and IT experts. In this chapter, we presented our work on enhancing the semantic content of the widely adopted BPMN business process modeling specification. Specifically, we have provided a research methodology for representing semantically BPMN business process models and have proposed an ontology for BPMN. This representation enables a deeper analysis and understanding of the BPMN models, helping comprehend not only the explicit asserted information but also inferred information which is derived from implicit, hard to observe BPMN information. The implicit information is generated from the logical consequences and the asserted axioms, performed by semantic reasoners' inference operations. Another advantage of the proposed BPMN ontology is that it provides an open access database, which can facilitate information exchange and improve interoperability within and between organizations.

We have experimented on querying the BPMN models using both conventional and semantic search engines. By using the word 'conventional' we refer to a 'non semantic' search engine. For the conventional method of information retrieval, we have used the object-oriented AQL query language. To support the semantic query method, we have used the SPARQL semantic query language. Object-based search engines are graph-based, operated on objects and graphs schemes, while semantic search engines follow the triple RDF logical statements of Description Logic theory. Comparing the results of these two query methods, we conclude that both object-oriented and semantic-based search can retrieve information from complex and heterogeneous database systems. Nevertheless, the semantic query method can generate logical conclusions on the requested issues (shown in Figs. 14, 15).

Two real-life scenarios, taken from the world of education, have been used to demonstrate the applicability of our ontology. Future work will include enriching the BPMN ontology and adjusting it to a more generic structure, so that it can be used also in other business domains. Additionally, further investigation is planned on applying semantic search techniques for querying business process models.

References

Atkinson, C.: Models versus ontologies—What's the difference and where does it Matter?. Presentation for University of Birmingham, 19 April 2007

BOC Group.: Volume 2: Users Manual. Adonis Version 5 (2012)

Bergener, P., Delfmann, P., Weiss, B., Winkelmann, A.: Detecting potential weaknesses in business processes: an exploration of semantic pattern matching in process models. Bus. Process Manag. J. **21**(1), 25–54 (2015)

Fernández, H.F., Palacios-González, E., García-Díaz, V., Sanjuán Martínez, O., Lovelle, C.J.M.: SBPMN—an easier business process modeling notation for business users. Comput. Stand. Interf. **32**(1–2), 18–28 (2010)

Fill, H.-G.: On the Conceptualization of a Modeling Language for Semantic Model Annotations. Lecture Notes in Business Information Processing, vol. 83, pp. 134–148. Springer, Berlin (2011)

Fill, H.G., Schremser, D., Karagiannis, D.: A generic approach for the semantic annotation of conceptual models using a service-oriented architecture. Int. J. Knowl. Manag. **9**(1), 76–88 (2013)

Gruber, T.R.: A translation approach to portable ontology specifications. Int. J. Hum. Comput. Stud. Knowl. Acquis. **5**(2), 199–220 (1993)

Guarino, N.: Formal ontology and information systems. In: Proceedings of the First International Conference (FOIS'98), Trento, Italy, 6–8 June 1998

Hoang, H.H., Tran, P.-C.T., Le, T.M.: State of the Art of Semantic Business Process Management: An Investigation on Approaches for Business-to-Business Integration. Lecture Notes in Computer Science, vol. 5991, pp. 154–165. Springer, Berlin (2010)

Kalogeraki, E.-M., Panayiotopoulos, T., Apostolou, D.: Semantic queries in BPMN 2.0: a contemporary method for information retrieval. In: Proceedings of the 6th International Conference on Information, Intelligence, Systems and Applications (IISA 2015), Corfu 6–8 July 2015, (in press)

Kalogeraki, E.-M., Theocharis, S., Apostolou, D., Tsihrintzis, G., Panayiotopoulos, T.: Semantic concepts in BPMN 2.0. In: IISA 2014—5th International Conference on Information, Intelligence, Systems and Applications, vol. 6878791, pp. 204–209 (2014)

Kő, A., Ternai, K.: A development method for ontology based business processes. In: Cunningham, P., Cunningham, M. (eds.) eChallenges e-2011 Conference Proceedings of IIMC International Information Management Corporation Ltd 2011. ISBN 978-1-905824-27-4

Kogalovsky, M.R., Kalinichenko, L.A.: Conceptual and ontological modeling in information systems. Program. Comput. Soft. **35**(5), 241–256 (2009)

Ma, S., Tian, L.: Ontology-based semantic retrieval for mechanical design knowledge. Int. J. Comput. Integr. Manuf. **28**(2), 226–238 (2015)

Möller, R., Haarslev, V., Neumann B.: Semantics-based information retrieval. Available from Ralf Möller at http://www.researchgate.net/ since 01 Dec 2012. (1998)

Natschläger, C.: Towards a BPMN 2.0 ontology. Lect. Notes Bus. Inf. Process. **95**, 1–15 (2011)

Object Management Group.: Business process model and notation (BPMN), version 2.0.2, 2013. http://www.omg.org/spec/BPMN/2.0.2/ as it is 25 May 2015

Robak, S., Franczyk, B., Robak, M.: Applying linked data concepts in BPM. In: IEEE, 2012 Federated Conference on Computer Science and Information Systems, FedCSIS 2012, vol. 6354386, pp. 1105–1110 (2012)

Rospocher, M., Ghidini, C., Serafini, L.: An ontology for the business process modelling notation. In: Frontiers in Artificial Intelligence and Applications, vol. 267, pp. 133–146 (2014)

Sánchez, J.M., Cavero, J.M., Marcos, E.: The concepts of model in information systems engineering: a proposal for an ontology of models. Knowl. Eng. Rev. **24**(1), 5–21 (2009)

Sanfilippo, E.M., Borgo, S., Masolo, C.: Events and activities: is there an ontology behind BPMN?. In: Frontiers in Artificial Intelligence and Applications, vol. 267, pp. 147–156 (2014)

Sirin, E., Parsia, B., Grau, B.C., Kalyanpur, A., Katz, Y.: Pellet: a practical OWL-DL reasoner. Web Seman. **5**(2), 51–53 (2007)

Taghieh, S., Taghieh, M.B.: Knowledge management, with the emphasis on the key success factors and the cycle of active knowledge management system. Adv. Environ. Biol. **7**(10), 2789–2794 (2013)

Taylor, C.: Reunifying Knowledge and Business Process Management. BPTrends, Feb 2012

Thalheim, B.: The science and art of conceptual modeling. Lecture Notes in Computer Science (including subseries Lecture Notes in Artificial Intelligence and Lecture Notes in Bioinformatics), 7600 LNCS, pp. 76–105 (2012)

Thomas, O., Fellmann, M.: Semantic process modeling—design and implementation of an ontology-based representation of business processes. Bus. Inf. Syst. Eng. **51**(6), 506–518 (2009)

Using Conversational Knowledge Management as a Lens for Virtual Collaboration in the Course of Small Group Activities

Demosthenes Akoumianakis and Dionysia Mavraki

Abstract For several decades scholars from different disciplines explore knowledge management using different theoretical and engineering lenses. This paper focuses on a relatively recent approach to knowledge management and collaborative learning, namely Conversational Knowledge (CK) management. Using empirical data from two different case studies, we identify conversational patterns and assess intrinsic properties of their prominent media types. This helps anchor these patterns as 'Dialogue for Action', thus informing the design of portlets (in the Liferay portal) that foster creation of new knowledge through collaborative conversational engagements of peers.

Keywords Conversational knowledge management · Collaborative technologies · Virtual work · Small groups

1 Introduction

Conversational Knowledge (CK) coins knowledge that is embedded in informal dialoguing patterns of collaborating peers (Lee and Lan 2007). Contrasting traditional models that focus largely on the explicit and formal representation of knowledge in computer-based systems, CK emphasizes the tacit and social aspects of knowledge management (Honga et al. 2011; Davison et al. 2013; Holste and Fields 2010). In this respect, CK favors collective intelligence rather than mere accumulation of knowledge as codified in repositories (Lee and Lan 2007) by embracing a more interactive and collaborative approach which relies heavily on an

D. Akoumianakis (✉) · D. Mavraki
Department of Informatics Engineering, Technological Education Institute of Crete, Heraklion, Crete, Greece
e-mail: da@ie.teicrete.gr

D. Mavraki
e-mail: mdionysia@hotmail.com

G.A. Tsihrintzis et al. (eds.), *Intelligent Computing Systems*,
Studies in Computational Intelligence 627, DOI 10.1007/978-3-662-49179-9_5

emergent process of co-creation and sharing amongst peers (Davison et al. 2013). This focus is justified by the weakness of conventional approaches to cope with sharing tacit knowledge (Davison et al. 2013; Holste and Fields 2010; Leea et al. 2009) as well as the barriers (both individual and organizational) preventing firms from effective knowledge sharing (Honga et al. 2011). To remedy for these problems, CK management relies on informal mechanisms to capture the tacit, the situated and the circumstantial aspects of knowledge (Honga et al. 2011; Davison et al. 2013; Majchrzak et al. 2013). Arguably, this is valid both for small group activities as well as organization-wide settings and cross-functional teams (Davison et al. 2013).

Due to its nature, CK is typically embedded in "conversational" patterns such as narratives, storytelling and word-of-mouth where peers can raise issues and engage in argumentation. As such, CK is profoundly evident and proliferates in "Communities of Practice" (CoP) (Etienne 1991). Nonetheless, the advent of web 2.0 and the variety of social technologies have brought about novel linguistic vocabularies that serve the purposes of CK management just as well as the more traditional means (Honga et al. 2011). Such vocabularies provide novel means for social contracting (i.e., extending invitations, sharing resources, content co-creation, etc.), expressing opinion, intentions or state of mind using social widgets such as like-buttons, externalizing cognitive anchors through bookmarks, or even making online discourse searchable by using cultural markers such as the hashtag. Thus, conversations need no longer be conceived as language exchanges between co-present peers. Rather, they may be thought of as co-engagements in a linguistic domain or an artificial system of learned communicative behaviors that arise between actors as the result of their particular history of co-existence.

In this context, the present work concentrates on a particular thread of research which explores CK management in small group collaborative settings where the outcome of the collaboration is determined by the reliability of both tacit and explicit knowledge shared amongst credible peers. A representative case is when remote experts co-engage to code data sets embedded in interview transcripts and other supplementary resources. The coding scheme may be either suggested to peers (based on an existing theory) or developed from scratch to suite the refinement or development of new theory. Clearly such small group endeavors offer a useful baseline for assessing not only the value of CK management but also its practical implications for non-trivial virtual work. To address the challenges involved, we critically appraise prominent media types that host CK and then recruit the theoretical lens of the Language-Action perspective (Winograd 1988) to inform the design of a system for CK management.

The rest of the paper is structured as follows. The next section reviews CK in terms of its constituents and popular media types. This anchors the rationale for and motivates the present work. Then, we present the methodology guiding an empirical inquiry aimed at analyzing conversational knowledge across two different settings with the objective to identify core conversational patterns and model them. Based on these models, we then discuss the implementation of a system for CK management. The paper is concluded with a discussion and suggestions for future work.

2 Related Work and Motivation

CK management draws upon various scholarships. This section sets out to review basic concepts and highlight related works in Information Systems research and Computer-supported collaboration with a focus on conversational technologies.

2.1 *Conversational Patterns*

A conversation may be broadly defined as a series of related message exchanges between groups of peers which can last for seconds, hours or days. In case of multiple conversation instances active at the same time, then messages belonging to each instance of conversation are correlated typically through an appropriate identifier or code. On the other hand, a pattern coins a chunk of information that describes an intended solution to a common problem within a specific context. Patterns are typically observed from actual experience and may be useful in declaring what is known about a design problem or domain. Conversational patterns can therefore be conceived as structured contexts of related message exchanges. Depending on the messages and the medium through which messages are exchanged, it is possible to envision multiple (and frequently non-comparable) conversational patterns. For instance, one of the earlier patterns for exchanging opinion is Word-of-Mouth (WoM). In Goyette et al. (2010), WoM is defined as an exchange that facilitates flow of information, communication and conversation between individuals. Telling stories (about life, experience and work situations) is another pattern particularly suited for people to pass on their knowledge to next generations. Stories are especially useful for sharing tacit knowledge (Smith 2001; Hansen et al. 1999) and proliferate in communities of practice (CoP). Frequently, conversational patterns such as story telling invoke additional acts such as turn taking, question and answer, argumentation, etc. All of them may occur either in face-to-face, mass-mediated, computer-mediated contexts, or in oral and written discussion. Of particular interest to the present work is argumentative conversation, where people try to get to the core of the topic by putting forward different points of view, claims, perspectives, arguments and counterarguments (Laurinen and Marttunen 2007). Due to these features, there have been attempts to create frameworks and systems for organizing and structuring argumentation, thus aiding the process of decision making among the group's members (Stewart et al. 2007).

Another qualification of conversational patterns may be obtained by examining the sequence of message exchanges. Thus, for instance one common pattern is the 'Request-Reply' pattern that entails sending a request and waiting for a reply. This is the simplest form of conversation with a single conversation state (assuming no error conditions). Other more complex conversational patterns may also be devised as shown in Fig. 1. 'Request-Reply with Retry' anchors the case where the sender can repeat the request a number of times before giving up. Although, this pattern appears

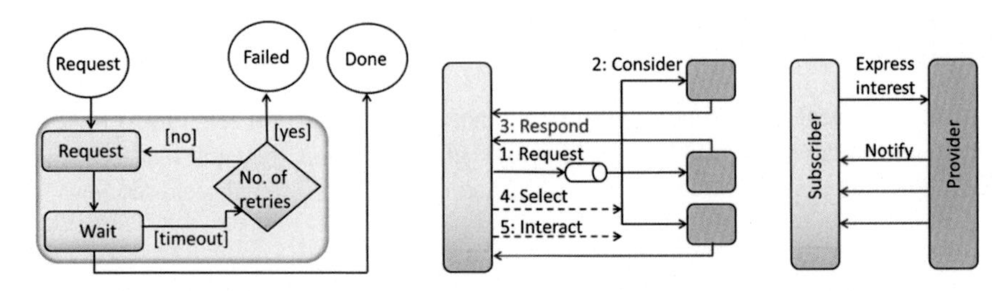

Fig. 1 Layered conversation (*left*); Dynamic discovery (*middle*); Subscribe-Notify (*right*)

to be more flexible, it cannot avoid situations where the sender might receive responses after it gave up. 'Dynamic Discovery' relies on broadcasting requests for providers. Provider(s) consider whether to respond and those interested send their responses. Once the 'best' option is chosen, conversation proceeds between the requestor and the chosen provider. In the 'Subscribe-Notify' pattern, the subscriber expresses interest in receiving notifications until a condition is reached. Notifiers send their offers until they are notified otherwise by the subscriber.

The advent of web 2.0 and the supporting technologies (i.e., bookmarking, social networks, blogs, etc.) have popularized new conversational patterns through 'artificial' constructs that progressively remediate conversations. As an example, social tagging systems allow users to freely select keywords to externalize cognition, thus declaring an association of users with this resource. The aggregation of these tags constitutes part of the resource metadata and provides useful insights into how the resource is used and appropriated. For instance, tags may also serve the purpose of exploring collections by inspiring users to search for new information (Kimmerle et al. 2010), thus creating and sharing new knowledge. Indeed, in Robu et al. (2009) it is argued that tagging enables users to order and share data more efficiently, and as a result, it may be used to enhance both informal and formal knowledge sharing by allowing users to augment resources with own concepts and points of view. Similarly, social widgets, 'like' buttons, tweets and the hashtag can be conceived as linguistic conventions for sharing content, expressing state of mind or opinion and rendering online discourse searchable.

2.2 Design Frames and Technologies for CK Management

A variety of technologies have been developed to facilitate conversational exchanges of various sorts and types. Early works were inspired by the Language-Action perspective (LAP) on Information Systems as developed in the joint authorship of 'Understanding Computers and Cognition' by Fernando Flores and Terry Winograd (Winograd 1988). They presented a new foundation for designing information systems by conceptualizing actions performed through communications as recurrent communicative patterns (Winograd 1988). The key concept in LAP is that language is not only used for exchanging information but also to perform actions like promises, orders, declarations, etc. Accordingly, it is claimed that information systems should not be viewed as repositories for storing information and real facts, but instead as tools supporting communication among people who perform certain actions together (Schoop 2001). For LAP, people are members of the community that coordinates their actions. Moreover, LAP emphasizes the communication between people, describing how language is used in order to create a common reality as well as how people coordinate their actions through communication. Hence, the LAP approach is based on the linguistic and social rules that control the use of language.

Interestingly LAP, in its various forms and versions, has been very influential for a number of technology genres widely available today. Amongst the most common are discussion forums, chatrooms, Weblogs or blogs and Wikis. In recent years, these technologies have been used to anchor tacit knowledge management in both intra- and inter-organizational settings. Discussion forums offer a technology that fosters question-answering as a means for tacit knowledge sharing (Bibbo et al. 2012; Standing and Kiniti 2011). Micro-blogging platforms such as the Twitter also resemble WoM, thus acting as news media (Kwak et al. 2010). Web logs (blogs) are also commonly referred to as story telling media (Bibbo et al. 2012; Wagner and Bolloju 2005). However, blogs may also convey bias or limit the forms of exchanging opinion to a comments section which is often controlled by the author (Bibbo et al. 2012). Wikis improve on blogs, by exploiting collaborative writing and linkage of Web pages created by a group, to offer an alternative story telling medium (Leuf and Cunningham 2001). Authors write pages in plain text or with a simplified markup language by asking and answering to questions, creating hyperlinks to non-existing page or narrating their stories thus sharing their knowledge (Bibbo et al. 2012). Wikis also have drawbacks. Their open nature may distract people from trusting its content, while in certain cases Wikis may also necessitate a support team to address any potential problems (Bibbo et al. 2012). Another disadvantage is that wikis tend to represent the collective notion of their user group (Bibbo et al. 2012) rather than the tacit and implicit details behind it.

Conversations have also been of interest to the information and knowledge visualization community (McDonald 2007). For instance, Collaborative Computer-Supported Argument Visualization (CCSAV) (Iandoli et al. 2014) coins techniques that support online collective discussion over complex dilemmas. Compared with more traditional conversational technologies, such as forums, wikis or blogs, CCSAV foster critical thinking and evidence-based reasoning and grounding, by recruiting representations that highlight notions, ideas and conceptual relationships between contributions (Iandoli et al. 2014). Collaborative tagging systems offer another genre of technology for CK management. This time the conversation is anchored by or exploits meta-data that users attach to shared digital resources such as photos, videos, websites, e-mails or any other piece of digital information (Kimmerle et al. 2010; Golder and Huberman 2006). Illustrative cases of highly recognized collaborative tagging systems include delicious, Flickr, Technorati, and Amazon that allow users to "tag" objects with keywords to facilitate retrieval. Nonetheless, tagging has its drawbacks as it is prone to ambiguity in the meaning of tags or the use of synonyms which creates informational redundancy (Robu et al. 2009).

2.3 Consolidation and Research Focus

In spite of the variety of technologies available for CK management, there are aspects of conversations that are still 'hidden' in tacit details or constrained by preconceptions inscribed in technology. Moreover, as new social technologies are deployed conversational practices become enriched and extended, allowing for novel linguistic encounters that expand the scope and bandwidth of conversational exchanges. Thus, it is argued that conversation (as linguistic domain) entails certain practices (story-telling, word of mouth, argumentation, etc.) which subsume activities (declaring presence, taking positions, expression of opinion, argumentation, etc.) that obtain their meaning relative to designated artifacts (text, photos, video, etc.) and referent objects. Such a lens sets the focus on certain affordances of the software and/or the medium that enable or constrain the type and range of computer-mediated message exchanges. These affordances may determine:

(a) How users are represented in the digital setting—for instance in email conversations users may be represented with text on screen or their digital identifier, while in synchronous conferencing systems users may have multiple representations (i.e., voice, digital identifier and highlighter);
(b) What is possible to be expressed—for example pressing a button may be sufficient for conveying state of mind or opinion, while tags may be used to externalize cognition, etc.;
(c) What may not be easily expressed using virtual referents (i.e., civic inattention, metaphoric referents, etc.) or what was unnoticed and is now brought to the forefront as a result of new linguistic constructs (i.e., hashtags make 'talk' searchable).

In light of the above, CK management systems are therefore compelled to make provisions for these more challenging patterns of expression so as to allow collaborators to recruit different linguistic codes to convey situated and local (rather than global and general) conditions in online discourse. Motivated by the above, the present research revisits the premises of conversational acts in an effort to create tools that capture the social and collaborative aspects of knowledge management in small group activities.

3 Methodology

To address the research challenges set above, we studied conversational patterns in two distinct and radically different settings. The first setting entails moderated team learning where peers become acquainted with new topics. The second setting is the analysis of interview transcripts codifying conversations between one interviewer and five interviewees. In both cases, the analysis followed in three steps. Firstly, we screened raw data to identify conversational patterns in relation to human intentionality such as creating arguments, raising issues, making claims, inviting participation, etc. Secondly, the identified patterns were modeled using state diagrams in the tradition of the Language-Action perspective (Winograd 1988; Schoop 2001). Finally, the resulting insight was used to inform the design of portlets of the Liferay portal (https://www.liferay.com/) to provide basic support for CK management.

3.1 Data Samples and Analysis

The first set of materials used in this analysis comprises the moderated exchanges between a group of postgraduate students and one instructor during the 2012–2013 summer semester class of the course 'Computer-Supported Collaboration' in the postgraduate program 'Informatics & Multimedia'. The sample comprised email exchanges, discussion threads in a forum and informal exchanges amongst group members. All data items coined the group's effort to prepare a class assignment which is a pre-requisite for the course. The second set of data comprised transcripts of five semi-structured interviews compiled by the first author in the context of a research inquiry into current practices of a cooperative active in organic farming in the southern part of the region of Crete. Each interview transcript was 4–6 pages long and represented work undertaken by actors within the cooperative.

Analysis of both sets of transcripts was iterative, aiming to identify user roles and affiliation, human intentionality, choice of technology, computer-mediated tasks, as well as communicative patterns. Of particular interest for the purposes of the present work, are the conversational patterns revealed through this exercise. Specifically, we coded patterns designating role assignment and responsibility (moderator vs. participant), social acts such as extending invitations and tracking invited contributions, but also linguistic patterns such as adjacent and non-adjacent turns, cross-turn quoting, turn-entry and turn-exit devices, indexing, qualifying and tagging, and taking a decision. The information gathered was coded using excel and provided the common reference for subsequent analytical inquiries.

3.2 Language-Action Models

Once the key conversational patterns were identified and decided, LAP was recruited to model them as parts of a coherent online discourse. In the tradition of LAP, actions and argumentation can be depicted through the use of state diagrams anchoring a coordinated sequence of acts that can be interpreted as having linguistic meaning without the need of a spoken dialogue, or even involve the use of ordinary language. An initial model was developed with a minimal set of clauses which was subsequently aligned with the requirements of a document-based system. The final set of acts compiled is as follows:

- Create (a, s) Actor a creates a new document.
- Invite (a, [p1, … pn]) Owner invites peers.
- Propose (a, s) Actor proposes issue s.
- Agree (a, s) Actor a accepts statement s as a valid statement. A statement can be accepted once proposed.
- Disagree (a, s) Actor a rejects statement s, because a finds s unacceptable. Reject is the counterpart of accept.
- Ask (a, q) Actor a asks question q, to be answered by some actor. Queries can be withdrawn or answered.
- Answer (a, q, s) Actor a answers question q with statement s; an answer functions as a special Propose.
- Withdraw (a, s) Actor a withdraws document s.
- Accept (a, s) User a accepts to contribute to document s.
- Reject (a, s) User a does not contribute to document.
- Take position (a, I, p) User a takes position p on issue I.
- Tag (a, T, E) A user a attaches a tag t to resource E.
- Comment (a, C, E) Actor a comments on a resource E.
- Rate (a, R, s) Actor a rates a document s.
- Vote(a, V, p) Actor a votes a position p.

4 Implementation

Using the above as guiding lens, we have implemented the required extensions in two portlets to support the needed functionality. The chosen platform is the Liferay enterprise portal, which in addition to basic functions of a portal platform, provides numerous applications and tools in the form of portlets and third-party plugins. At core, our system provides for two roles, namely administrators (owners) and peer users. Administrators create 'documents' for conversing and invite peers to collaborate and contribute. Any conversation is therefore bound to a specific document which affords designated conversational patterns.

4.1 Transformable Document Templates

In order for 'documents' to invoke the required human intentionality they are assigned to an appropriate transformable document template (Akoumianakis et al. 2012). This is an abstraction that designates the structure of a document in terms of elements and the affordances of each element. Elements are components hosting designated pieces of content. Simple documents may contain just one element. For the purposes of the present work, the transformable document template (CKM-Doc) is depicted in Fig. 2. Thus, each document assigned to this template comprises elements that support specific elements and means for assembling structured content. Once a document is defined and populated, the author/owner can invite other peers to engage using the second portlet and the DfA introduced earlier.

4.2 *The Portlets*

Figure 3 depicts a typical CKM document while being created by its owner. In this portlet, the user can start creating issues and coding the document. While working with a document, issues are elaborated by defining properties (i.e., a brief description, attaching user positions, supporting materials and tags) as shown in Fig. 4. Once document is ready, author invites peers to contribute.

In the second portlet (Fig. 5), the Document List mode, which facilitates summative lists of documents, provides a summative account of the argumentative process and mechanisms for exploring and consolidating the issues. More specifically, in case of a new issue, the user should provide the necessary details to define the issue. For issues already contributed (by others), it is possible to contribute with new position statements and codes/tags. An interesting feature is the way in which semantically equivalent tags are negotiated and mapped to metatags or top-level categories (see Fig. 4, bottom right). Detailed analysis of these linguistic

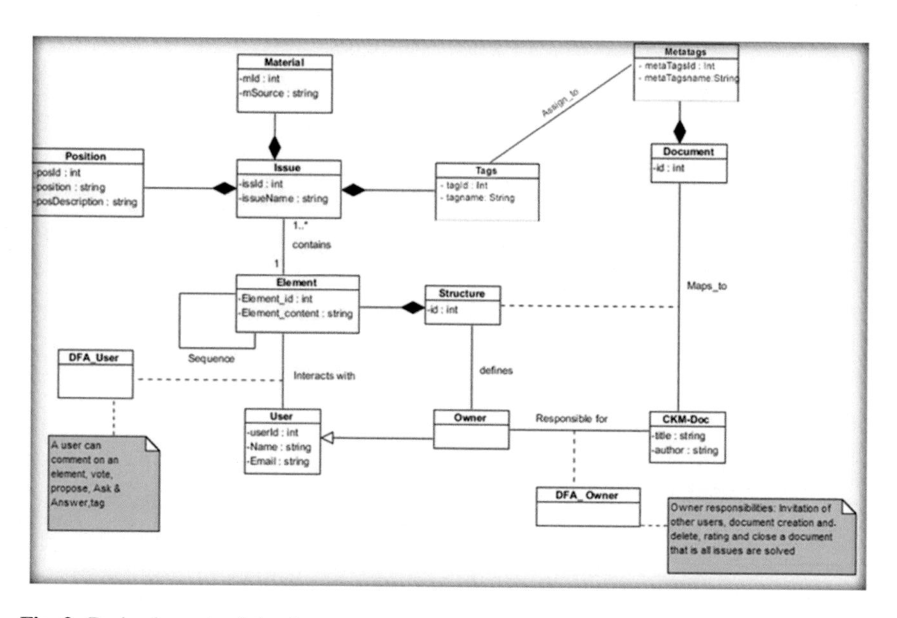

Fig. 2 Basic elements of the document structure

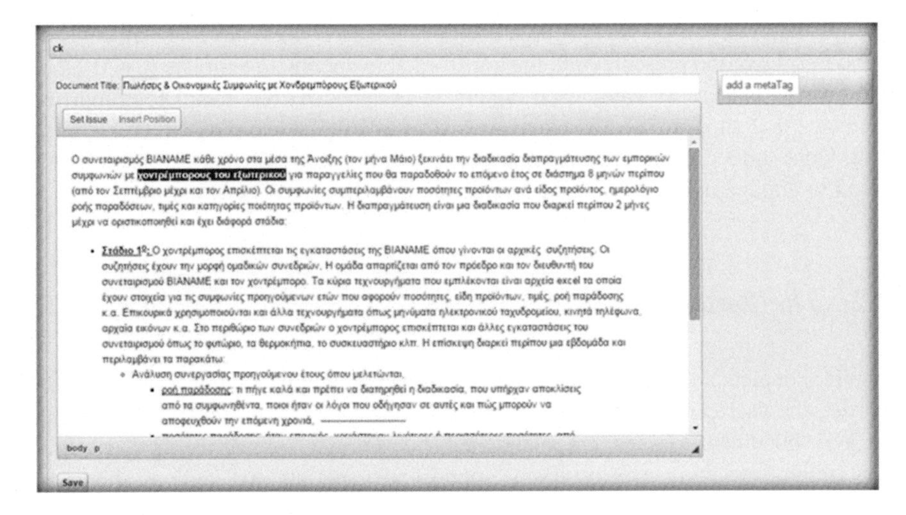

Fig. 3 Overview of CKM document

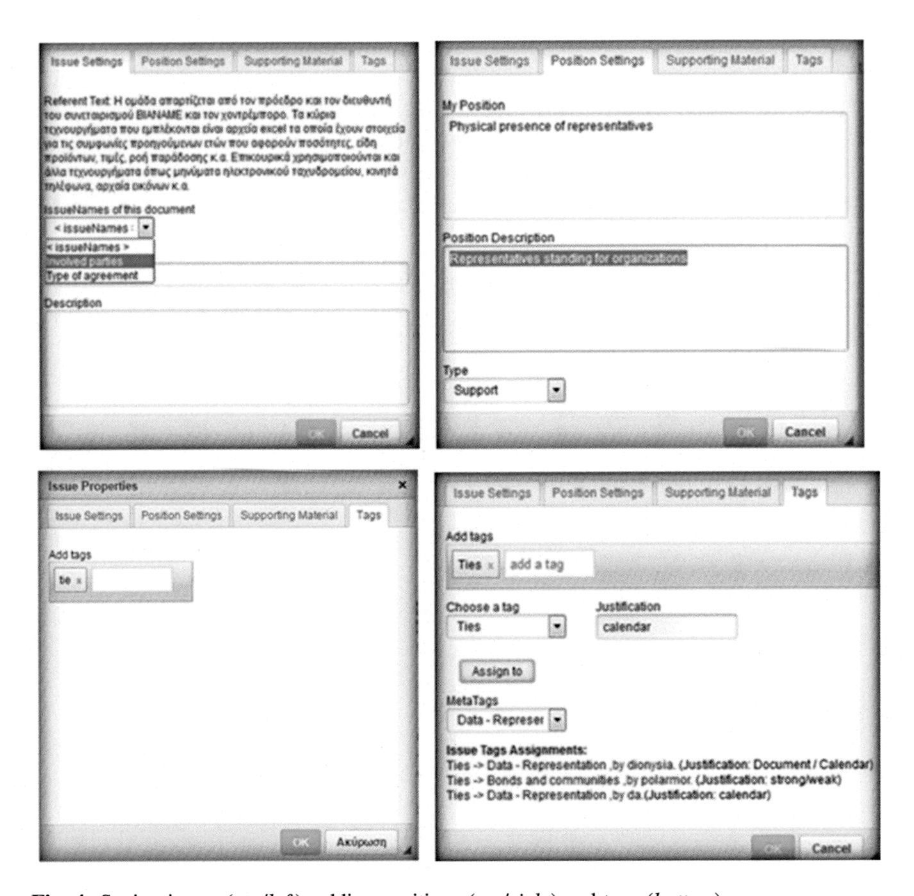

Fig. 4 Setting issues (*top/left*), adding positions (*top/right*) and tags (*bottom*)

conventions can provide useful insight to the collaboration that takes place, agreements and disagreements between group members as well as who teams up with whom on specific issues.

Besides, as shown, the user can select a specific document from the list (Fig. 5 top) and review the wisdom available (bottom). Such wisdom comprises issues, sub-codes (or tags) and top level codes (or metatags). The word cloud provides an overview of the current set of issues, thus allowing users to filter documents that contain a specific issue. Once a document is selected it reveals the accumulated wisdom (see Fig. 5 bottom). From this portlet the author can also extend invitations to other peers to collaborate on a specific document. With respect to the marking scheme used, highlighted text in an element depicts issues designated by the collaborators. Peer issues are colored blue so as to be distinguished from the issues contributed by the owner which are marked red. The two components on the right hand side summarize the collective consensus of the peer group regarding the specific document. As shown, tags and metatags are scented to depict assignment frequency of each entry amongst the peer group.

Invited peers at their own pace and asynchronously may assess the document's online discourse and accordingly contribute by extending established issues or

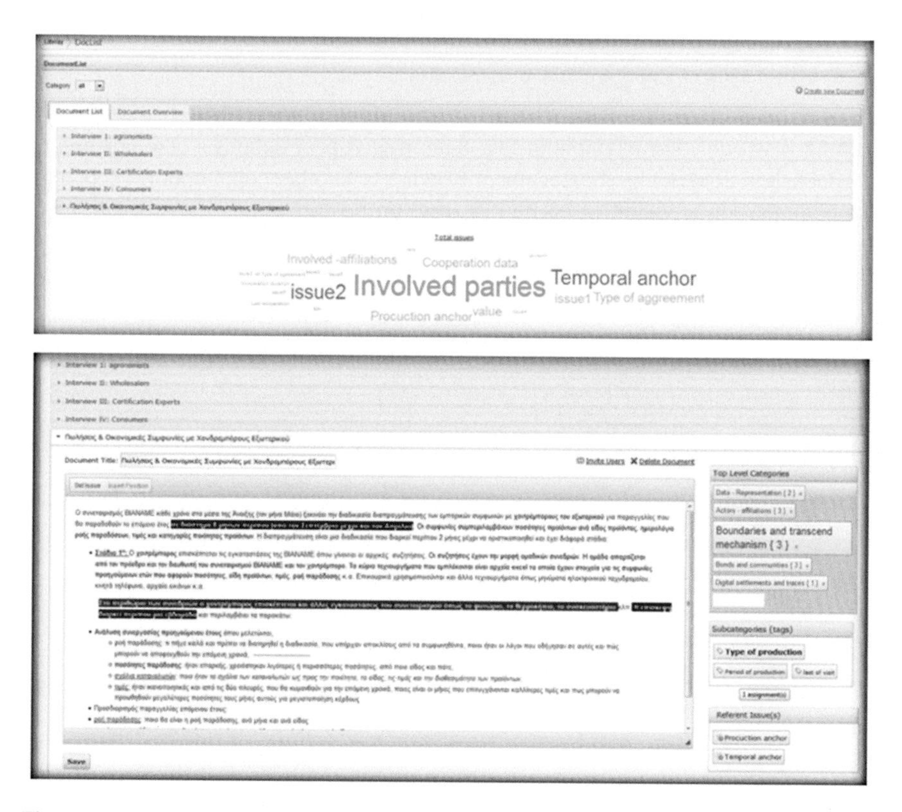

Fig. 5 Collective wisdom related to a document

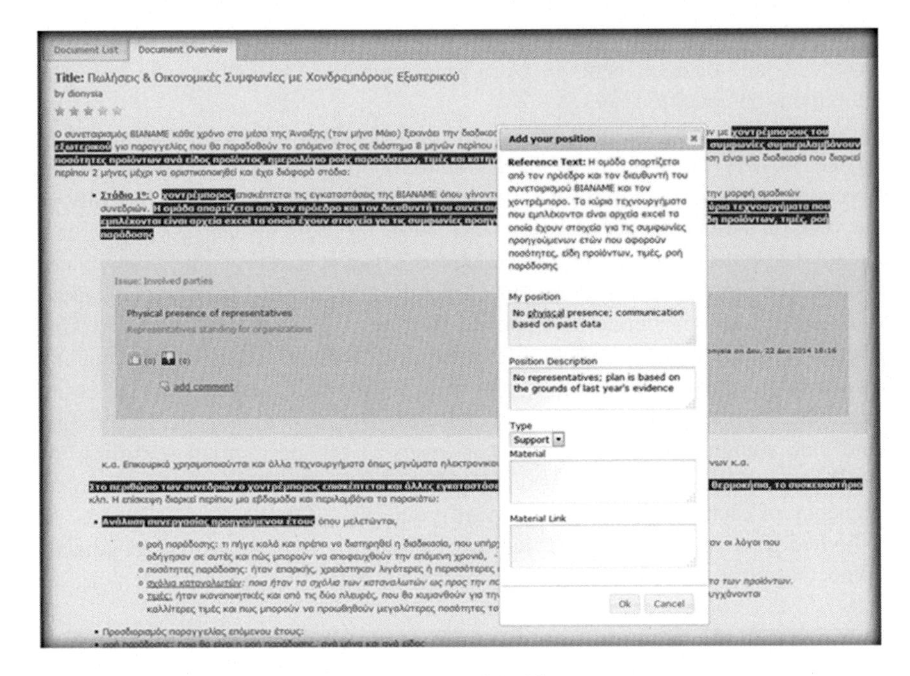

Fig. 6 Contribution of a new position to an established issue

defining new ones. For example in Fig. 8 an invited peer defines a new position to an established issue set by the document's owner or another collaborator. Peers may also comment on a position (see Fig. 7), vote on positions (i.e., like/dislike), raise and respond to questions. Nonetheless, they are not allowed to create metatags as

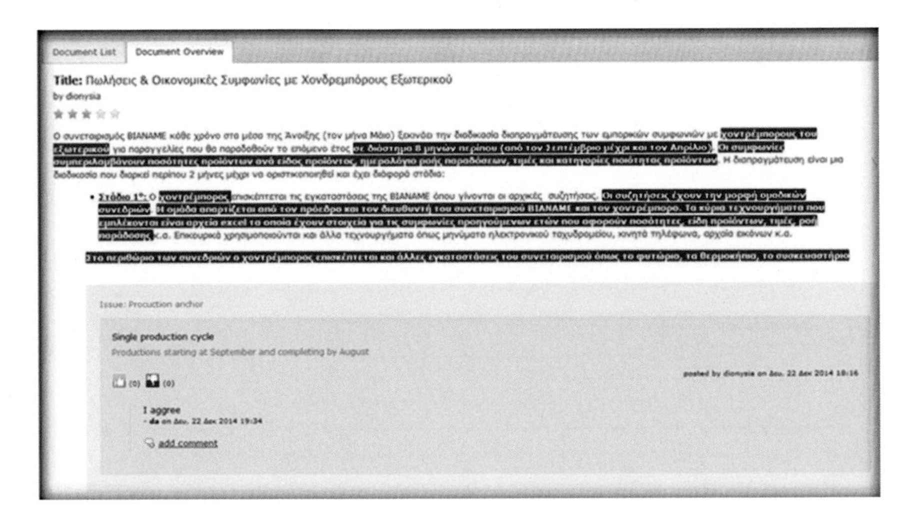

Fig. 7 Commenting on the issue's online discourse

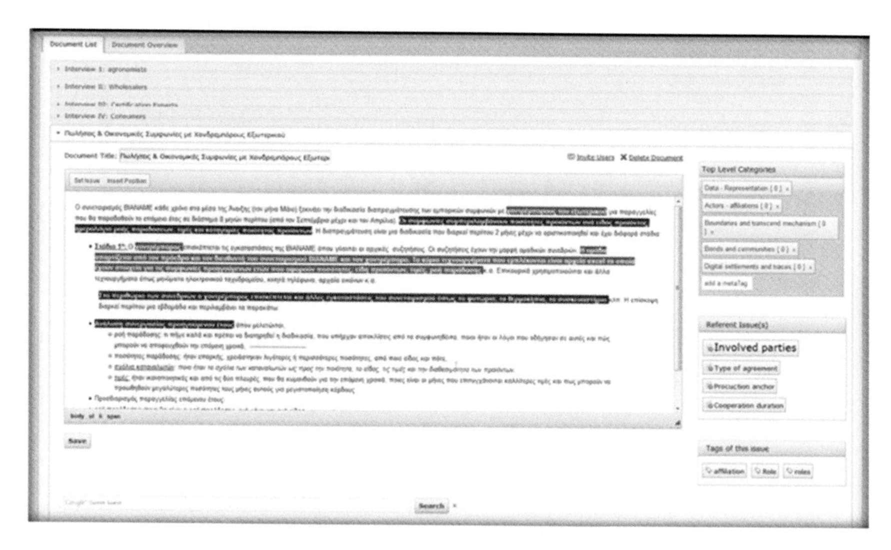

Fig. 8 Exploring the document using tags and metatags

this is a responsibility of the document's owner. Metatags are set by the author and presented to peers as a word cloud with scent indicating frequency of assignment (Fig. 6).

Another interesting point is that users can use tags, metatags and issues to explore a specific document or the entire collection of documents. For instance in Fig. 8 the user has selected the issue 'Involved parties' and filters the document to highlight all referents assigned to this issue, including the tags. In this manner, it is possible to assess consistency of referents but also the rationale behind each occurrence. In the

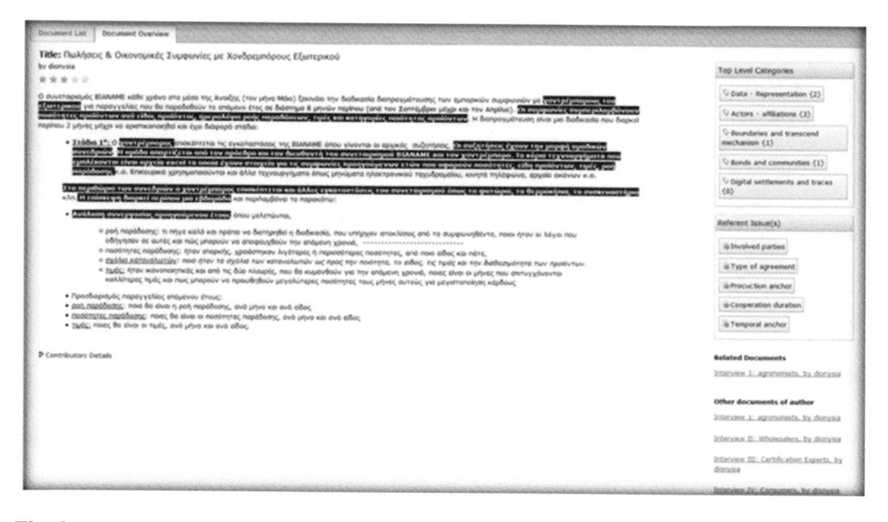

Fig. 9 The document overview mode

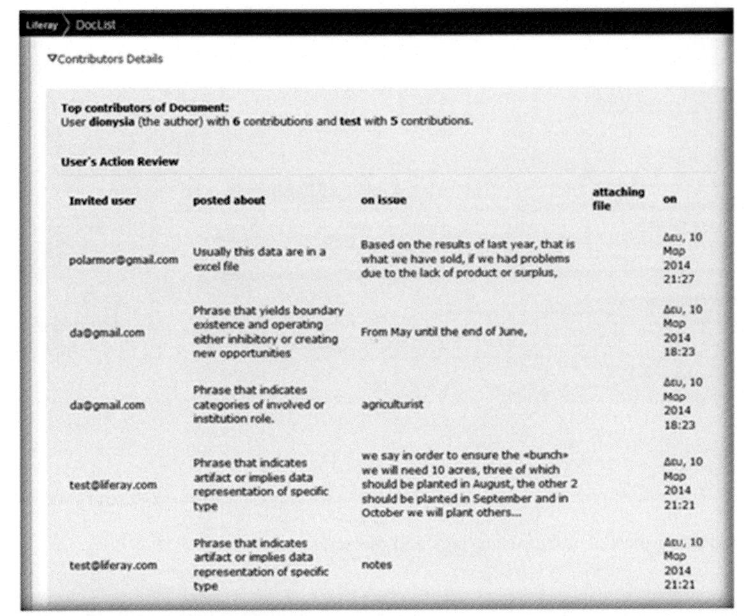

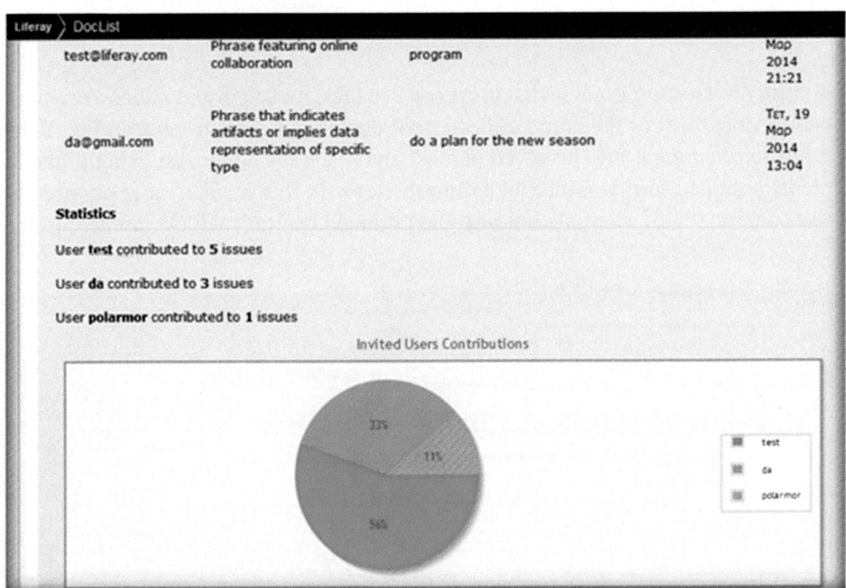

Fig. 10 Statistics and summative reflections

'Document overview' mode (see Fig. 9), users can spot other documents that may be of interest or somehow related to the current document, thus linking to documents of the same author or to documents with the same issues, tags or metatags (top-level categories). When all issues have been resolved, the document is closed.

Finally, owners of documents may request graphical representations and statics of online discourse, including contributions per participant, issues resolved/unresolved, positions per issue but also cross-document comparisons highlighting documents of the same type or scope, as in the case of the five interviews (see Fig. 10).

From the above brief description, it becomes evident that our ‘conversations’ are bound to documents, just as face to face discussions are anchored by topics or threads. A conversation starts with a proposition by a user (i.e., a document and its metadata prepared by an author) and proceeds as a series of moderated exchanges between the members of the invited peer group. Such exchanges rely on linguistic patterns intended for raising or responding to issues, commenting, voting and tagging. Moderation entails two prime functions exclusively undertaken by authors/owners of documents. These entail the initial proposal and consolidation of metatags and the resolution of issues during the conversation. Issue resolution is achieved by designating preference to a proposal, either because the proposal is deemed appropriate or because it is the most popular.

5 Concluding Remarks

This chapter presents the design rationale and implementation tactics of a system intended to facilitate CK management in the course of small group activities. One useful feature of the system is that it supports an incremental dialoguing pattern for negotiating contents and reaching collective consensus. Such a pattern is built around documents that adhere to transformable information templates to invoke a kind of conversing using linguistic anchors such as issues, positions, votes, comments, etc. The resulting online discourse can be interactively explored and negotiated so as to enable creation of new knowledge creation. Typically, such new knowledge manifests itself through coding schemes (i.e., tags and metatags) and can be further accessed through issue tracking provisions and assessment of the document corpus.

Thus far, the system has been tested in two applications with very promising feedback in both cases. The case of qualitative data analysis brought to the surface intrinsic properties of tacit knowledge management and computer-mediated discourse, which is the reason why it was chosen for elaboration in this paper. Specifically, the system can support equally well cases where codes are developed as well as cases where established codes are validated by independent coders.

Acknowledgment The research has been co-financed by the EU European Social Fund (ESF) and Greek national funds through the Operational Program “Education and Lifelong Learning” of the National Strategic Reference Framework—Research Funding Program: ARCHIMEDES III “Investing in knowledge society through the ESF”.

References

Akoumianakis, D., Milolidakis, G., Akrivos, A., Panteris, Z., Ktistakis, G.: Clinical practice guideline management information systems: cancer guidelines as boundary spanning transformable objects of practice. In: Proceedings of the 2nd INCoS, IEEE, pp. 230–237, (2012)

Bibbo, D., Michelich, J., Sprehe, E., Lee, Y.E.: Employing wiki for knowledge management as a collaborative information repository: an NBC universal case. J. Inf. Technol. Teach. Cases **2**, 17–28 (2012)

Davison, R.M., Ou, C.X.J., Martinsons, M.G.: Information technology to support informal knowledge sharing. Inf. Syst. J. **23**(1), 89–109 (2013)

Etienne, W.: Situated Learning: Legitimate Peripheral Participation. Cambridge University Press, Cambridge (1991)

Golder, S.A., Huberman, B.A.: Usage patterns of collaborative tagging systems. J. Inf. Sci. **32**(2), 198–208 (2006)

Goyette, I., Ricard, L., Bergeron, J., Marticotte, F.: e-WOM Scale: word-of-mouth measurement scale for e-services context. Can. J. Adm. Sci. **27**, 5–23 (2010)

Hansen, T.M., Nohria, N., Tierney, T.: What's your strategy for managing knowledge? Harvard Business Rev. **77**, 106–116 (1999)

Holste, J.S., Fields, D.: Trust and tacit knowledge sharing and use. J. Knowl. Manag. **14**(1), 128–140 (2010)

Honga, D., Suha, E., Koob, C.: Developing strategies for overcoming barriers to knowledge sharing based on conversational knowledge management: a case study of a financial company. Expert Syst. Appl. **38**(2), 14417–14427 (2011)

Iandoli, L., Quinto, I., De Liddo, A., Shum, S.B.: Socially augmented argumentation tools: rationale, design and evaluation of a debate dashboard. Int. J. Hum Comput. Stud **72**, 298–319 (2014)

Kimmerle, J., Cress, U., Held, C.: The interplay between individual and collective knowledge: technologies for organisational learning and knowledge building. Knowl. Manag. Res. Pract. **8**, 33–44 (2010)

Kwak, H., Lee, C., Park, H.: What is twitter, a social network or a news media? In: Proceedings of 18th WWW'2010, pp. 591–600 (2010)

Laurinen, L.I., Marttunen, M.J.: Written arguments and collaborative speech acts in practising the argumentative power of language through chat debates. Comput. Compos. **24**, 230–246 (2007)

Lee, M.R., Lan, Y.-C.: From Web 2.0 to conversational knowledge management: towards collaborative intelligence. J. Entrepreneurship Res. **2**(2), 47–62 (2007)

Leea, H.J., Kim, J.W., Kohc, J.: A contingent approach on knowledge portal design for R&D teams: relative importance of knowledge portal functionalities. Exp. Syst. Appl. **36**(2), 3662–3670 (2009)

Leuf, B., Cunningham, W.: The Wiki Way: Collaboration and Sharing on the Internet. Addison-Wesley Professional, Boston (2001)

Majchrzak, A., Wagner, C., Yates, D.: The impact of shaping on knowledge reuse for organizational improvement with wikis. MISQ **37**(2), 455–469 (2013)

McDonald, D.W.: Visual conversation styles in web communities. In: HICSS'2007, vol. 40, pp. 1238 (2007)

Robu, V., Halpin, H., Shepherd, H.: Emergence of consensus and shared vocabularies in collaborative tagging systems. ACM Trans. Web. **3**(4), Article 14, (2009)

Schoop, M.: An introduction to language action perspective. SIGGROUP Bull. **22**(2), 3–8 (2001)

Smith, E.A.: The role of tacit and explicit knowledge in the workplace. J. Knowl. Manag. **5**(4), 311–321 (2001)

Standing, C., Kiniti, S.: How can organizations use wikis for innovation? Technovation **31**, 287–295 (2011)

Stewart, C.O., Setlock, L.D., Fussell, S.R.: Conversational argumentation in decision making: Chinese and U.S. participants in face-to-face and IM interactions. Discourse Process (2007)
Wagner, C., Bolloju, N.: Supporting knowledge management in organizations with conversational technologies: discussion forums, weblogs, and wikis. J. Database Manag. **16**(2), 1–8 (2005)
Winograd, T.: A language/action perspective on the design of cooperative work. Hum Comput. Interact. **3**, 3–30 (1988)

Spatial Environments for m-Learning: Review and Potentials

Georgios Styliaras

Abstract Nowadays, the technological advancement in mobile user interfaces has made possible the development of hypermedia applications that exploit space. These applications can represent knowledge modules and relations among them both explicitly and implicitly. Such modules may be shared, commented and further reused under other circumstances. Hypermedia spatial applications can benefit educational environments executed on mobile devices. The paper intends to review existing spatial hypermedia interfaces as well as related environments and their potential use in educational platforms for mobile devices. The review includes issues such as content representation, relations, interface operations, sharing and annotation properties, exemplary usage and the exploitation of spatial structures for educational needs.

1 Introduction

Smartphone sales have reached outstanding numbers. According to Gartner (www.gartner.com), one billion smartphones were sold in 2013. Furthermore, smartphones outsold feature phones in 2013 for first time ever, whereas smartphone growth is driven by developing markets. Therefore, it is obvious that smartphone penetration is a global phenomenon and applications developed for them have potentials to reach billion users worldwide. The characteristics of smartphones, such as spatial interface, touch interactivity, wireless connectivity and voice commands make them ideal for a lot of categories of applications.

G. Styliaras (✉)
Department of Cultural Heritage Management and New Technologies,
University of Patras, Agrinio, Greece
e-mail: gstyl@upatras.gr

G.A. Tsihrintzis et al. (eds.), *Intelligent Computing Systems*,
Studies in Computational Intelligence 627, DOI 10.1007/978-3-662-49179-9_6

On the other hand, spatial hypermedia (Shipman and Marshall 1999) extends classic hypertext and hypermedia in the following ways: Firstly, it allows new ways of explicit or implicit linking of multimedia nodes. More specifically, in classic hypermedia, an explicit directed hyperlink is employed in order to join two nodes, one of those being the anchor or target of the link. In spatial hypermedia, in addition to this capability, a spatial area is exploited, in the way that the positioning of some nodes on an empty area may imply some kind of relation among them depending on their distance. Secondly, nodes may imply linking based on visual cues. Nodes being nearer and/or using the same border line or color have stronger relations than others. Finally, spatial hypermedia allows also the direct or indirect grouping of information.

Mobile learning through the use of wireless mobile technology allows anyone to access information and learning materials from anywhere and at any time. As a result, learners have control of when they want to learn and from which location they want to learn (Ally 2009). The technology that is needed for mobile learning to include both software and hardware that enabled the learning devices to be portable is described in Traxler (2009). Mobile learning offers features that allows learning "anytime and anywhere, just in time, just for me", and incorporates multimedia content and messaging (Shih and Mills 2007). One major challenge of web based learning systems is that the learner does not only have to acquire knowledge, but must navigate through the system to reach the desired content. Users should also exploit orientational and navigational aids provided by mobile devices, as well as visual communication elements, pictographic and typographic marks. Mobile environments support may also support the presentation of educational multimedia content uniformly in various displays and sizes (Acosta et al. 2003).

In this paper, the potential of using smartphones and spatial interfaces for educational applications is explored. Firstly, through a Resources List, the rationale for the selection of the characteristics of a potential spatial educational environment for mobile devices is presented. In Sect. 3, the list of characteristics is defined and then in Sect. 4, some desktop environments with spatial interfaces are presented that share some of these characteristics that could be applied in mobile educational environments. Section 4 describes how every environment supports these characteristics resulting in a comparison and determination of how these features should be supported in educational environments for mobile devices in Sect. 6. Finally, Sect. 7 concludes the paper and presents some future work.

2 List of Resources

This section presents some technology aspects that are necessary for the development of spatial multimedia educational environments for mobile devices. Content organization and structure are the core of every educational system, whereas navigational and visual aids orients students towards accessing the desired content. The exploitation of mobile devices' spatial areas is also a value added feature toward this direction. As mobile users are accustomed to use social applications in various devices, educational applications for these devices can integrate the advantages of their use.

More specifically, in a study examining pedagogical knowledge representation through concept mapping for educational reasons (Koc 2012), results indicated that concept mapping helped them prepare for class lessons and examinations, understand complex issues and reflect on their understandings. Concept maps relate to knowledge representation and its permanent, structured storage. The importance of structured content and feasible methodologies for its use for designing and developing courses are shown in Acosta et al. (2003). The paper shows how structured method and XML technology can be applied for a successful design of hypermedia documents components in the educational context.

Apart from content, one major challenge of web based learning systems is that the learner does not only have to acquire knowledge, but must navigate through the system to reach the desired content. Towards this direction, support by orientational and navigational aids is necessary (Brunstein et al. 2004), as well as search subsystems. Use of visual communication elements such as pictographic and typographic marks can be also important in the communication of intent and meaning (Fehr 2010). Responsive design is another factor that design methodologies for mobile devices should take into account (Holtzblatt et al. 2014).

The use of games in educational environments is an important means for disseminating knowledge. An authoring framework for the cultural heritage domain that aims to provide structured support, from content design to final implementation is presented in Bellotti et al. (2013). The model defines games that are set in realistic virtual worlds enriched with embedded educational tasks. Tangible Viewpoints (Mazalek et al. 2002) provide a physical means for interacting with multimedia educational stories.

Other hypermedia environments that could be used in education such as Palette (Sire et al. 2008) and Hypersea (Styliaras and Christodoulou 2012) point the need for part of the environments to embed the functionality of other applications and to be embedded in other applications in order to provide their own functionality. Eg. embed Hypersea in a blog. Map-based tools are also employed in educational environments, as in Cavanaugh and Cavanaugh (2008) where online courses used interactive geographic maps as a form of dialogue to reduce students' sense of transactional distance during the course, build their skills with Web 2.0 media and increase their motivation.

The need for exploiting space is pointed in Katifori et al. (2008) showing that users are still dissatisfied with their information organization and the challenge is to provide tools that support rather than replace the users' flexible and creative use of the current desktop. Similarly, Enabled Space (Christos Sintoris et al. 2007) is an architecture that permits efficient and effective co-existence of objects in the physical space and digital multimedia content. In other words, it is a physical space that has the ability to interact with mobile or ubiquitous devices and to serve its visitors in completing their tasks. Concerning other mobile device features, since the early years of using mobile devices in education (Scherp and Boll 2004), zoom, touch and drag operations were available and still are in modern devices along with support for multimedia playback of multiple types. Other workshops (e.g. Tse et al. 2010) discuss key HCI issues that next generation education systems face ranging from whole class interactive whiteboards, small group interactive multitouch tables, and individual personal response systems in the classroom. Authors in Garzotto and Gonella (2011) signify the importance of tangible mediums. The paper describes an open-ended environment that supports the creation and customization of tangible learning experiences for disabled children. The use of online educational space, made possible by free and readily available web 2.0 and open source applications, has been analyzed by three teacher-researchers of a Singapore's elementary school (Lye et al. 2012). The online space was meant to complement the physical learning space allowing learning activities not possible in the typical classroom setting. Results revealed positive learning experience.

Concerning social media and educational applications, authors in Shaw and Krug (2013) state that applications to engage youth must both allow for and encourage participation, communication, and collaboration. Social media sites, popular with youth and supporting sociocultural understandings of learning and identity development, could offer an excellent model for the design of such spaces. Millard et al. (2013) states that the successful design, development and delivery of high quality online education requires a coordinated agile effort from cross-functional teams of subject matter experts, instructional design consultants, and IT support professionals. The whole procedure includes issues such as the evaluation and selection of appropriate enterprise social media tools for team collaboration, best practices for organizational implementation and use, as well as challenges and future considerations of the enterprise social media space.

3 Classification Criteria

Based on the previous discussion, in this section a list of characteristics is presented that spatial educational environments should support, especially in mobile devices. Criteria are grouped as content, functionality and add-ons, interaction and technical issues.

Content

1. The environment's **content** may be created from scratch in the environment or imported into it from other sources. Content should be multimedia rich, of different kinds such as text, sound, video, image and animation. The procedure for importing content should be transparent to various user categories, as they are not expected to have much technical expertise.
2. Content **relations** are also a crucial factor for the success of a mobile educational environment, as they can exploit space and touch operations in order to define and represent both explicit relations through links and implicit ones through position. On the other hand, expressing relationships among content

portions is necessary for education material. A typical scenario for representing content in a spatial environment is to divide it in autonomous modules, characterize it by discrete and more descriptive properties and represent every module by a node, a page or other self-contained structure.

3. **Referential integrity** enhances content consistency as various users, content experts and instructors, insert, edit and delete content from the environment. It is necessary for content modules to remain related when needed through valid rules.
4. **Typed nodes** permit the environment's content to have an internal structure in the form of attributes and properties, depending on the environment's target. Examples include maps, music, encyclopedic lemmas etc. Content nodes may appear as autonomous on an educational space but should belong to a broader content structure. This is a different perspective compared to the ephemeral content modules of Flash applications or PowerPoint presentations.
5. **Edit** operations should allow altering the environment's content modules and relations depending on user access levels. Lower level users should be allowed less destructive operations on content. Furthermore, it would be useful to provide two operation modes: an editing mode and a read-only, playback mode, where content cannot be altered.
6. **Annotations** on an educational environment could permit users to leave ephemeral comments on content modules, single or thread-based, without interfering or altering the main content.

Functionality—Add-ons

7. **Visual cues** add semantics to educational content representations. Mobile users are accustomed to them in the form of notifications in their devices or in web 2.0 applications in general. Use of various colors, border sizes and fonts may differentiate the meaning of content modules at a glance, without requiring further explanations or instructions. Content modules may be differentiated according to their kind and type, access rights and origin.
8. **Game** integration permits users to test the knowledge acquired through their browsing of educational content. Games may be developed upon the multimedia features of mobile devices and, therefore, they can include touch and drag operations, content playback and sound interaction.
9. **Map-based** operations are usually necessary in educational content, especially in courses such as History and Geography. On the other hand, mobile devices support adequately map-based operations, both browsing maps and finding objects based on the users' current location. Therefore, mobile educational environments can integrate such functionality and gain from it.
10. **Multimedia playback** is a common operation in mobile devices. It is needed in educational environments to describe a scientific experiment, a historic event etc. Mobile devices differentiate on the diversity of supported multimedia content types.

Interaction

11. **User registration** includes the procedure that a new user joins the environment and is assigned a role. There should be some access levels regarding user roles such as, student, instructor, content expert and administrator. Content experts may prepare new content whereas instructors may combine content portions in order to prepare new activities and presentations. Instructors and content experts may be the same person. An overall administrator may coordinate content and environment parameters. Another aspect in user registration could be the exploitation of their accounts in social applications.
12. **Zoom, Touch and Drag** operation enables users to focus on specific aspects of educational content or activities. Due to small sizes of mobile devices screens, users should be able to focus on, select and move content that every time interests them, such as a specific content module, a module's relation with other content, playback of multimedia files or an educational game.
13. **Physical objects interaction** permits the enrichment of users' interaction with extra mediums than just a mobile device's touchscreen and maybe some sensors. Objects such as pawns and cards along with body actions may be integrated in an environment's interactivity and offer a more realistic experience to users. Another parameter related to physical interaction is the haptic feedback offered by devices that also enhance the user experience, such as vibration of mobile devices.
14. **Responsive design** should be followed during the design of an educational environment as it allows the environment to adapt smoothly on displays of different sizes and ratios. Users should be able enjoy the same functions from the environment regardless of the sizes of devices.
15. **Search and Navigation history** could support identifying content based on criteria and going back and forth among the content module creation process. The latter could express the logic behind the creation of a certain content structure e.g. explain step by step how to gather material for a historic event.

Technical Issues

16. The environment's **platform** depends on the implementation of the user interface. Environments oriented for mobile devices are usually web-based and integrate web 2.0 content and/or applications. HTML 5 should be the technology used for the implementation, enriched with its extensions such as scalable graphics (SVG). Flash or other custom applications tend to disappear.
17. **Persistency** allows content structure, placement and relations to be saved for later editing and shared to other users through social networks and other mediums.
18. **Embedding** could operate in two ways: Firstly, embed the educational environment in a web page or web 2.0 application. On the other hand, content from other applications and web pages can be embedded in the educational environment.

4 Exemplary Environments

For comparison purposes, 12 environments of different functionality and characteristics have been chosen. Some of them are educational, the rest could be used in education. Operation platforms also vary, as also the interaction mediums and the spatial properties. Selected environments could cover as a whole the characteristics presented in the previous section. In this way, these are complementary environments that could result in the design of a full-featured spatial educational environment for mobile devices. For every environment, a short description is presented along with a screenshot displaying its operation.

1. **Cantos and Contos** (de Oliveira et al. 2008) is a differentiated virtual learning environment for the development of children's skills in individual and collaborative modes. The first step for navigating the environment is to access the project homepage where a virtual environments is shown. Then a discipline or activity is selected that leads to a virtual room, e.g. Sciences Room, Language Room and Literature Room. In the room, pupils can access suitable multimedia content and comment it using audio with other pupils represented by avatars. They can also evaluate the material visited. Distance among user avatars signifies intent for communication (see Fig. 1).
2. **Canyon** (Alexandra Ion et al. 2013) is a map based environment that provides at real time location awareness of multiple moving objects on Large Displays. Users can focus on certain subsets of the map and still may track objects outside the map through the use of a special folding-like shadowed interface. For enhancing location awareness, the view in focus is not altered, a user is always aware of how an off-view object is related to the current view, the distance of an off-view object to the current view should be indicated and fast comparison of off-view objects related to each other is enabled (see Fig. 2).

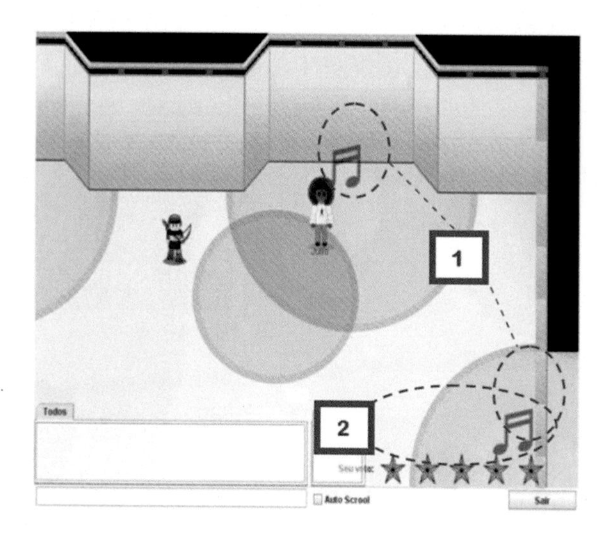

Fig. 1 Cantos and contos (de Oliveira et al. 2008)

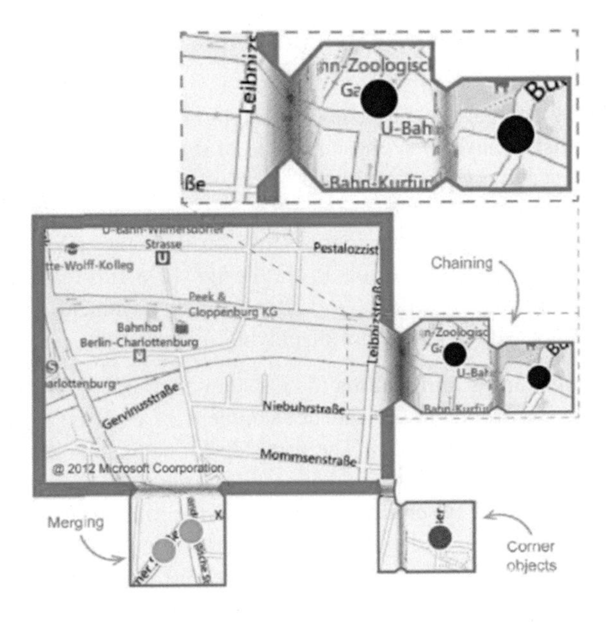

Fig. 2 Canyon (Alexandra Ion et al. 2013)

3. **Clui** (Pham et al. 2012) is a spatial platform where widgets to rich objects can be placed, called webits. Webits can provide handles to people, names, contact information, addresses, and so on. Dragging a webit to a web page results in the webit being uploaded in its entirety, without loss of information. Clui enables web applications or plugins to modify the default drop behavior of a webit. A webit combines many attributes that describe the represented object. Users may override what is pasted by inspecting the webit's metadata panel and dragging the desired piece of metadata (see Fig. 3).
4. **Cmaptools** (http://cmap.ihmc.us) is an offline Java environment where users can create, edit and share logical concept maps. When creating a new map, an empty space appears where concept nodes and arrows among nodes and other arrows can be defined. Users can drag nodes and reposition them and can zoom to the area of the concept map they are more interested in (see Fig. 4).

Fig. 3 Clui (Pham et al. 2012)

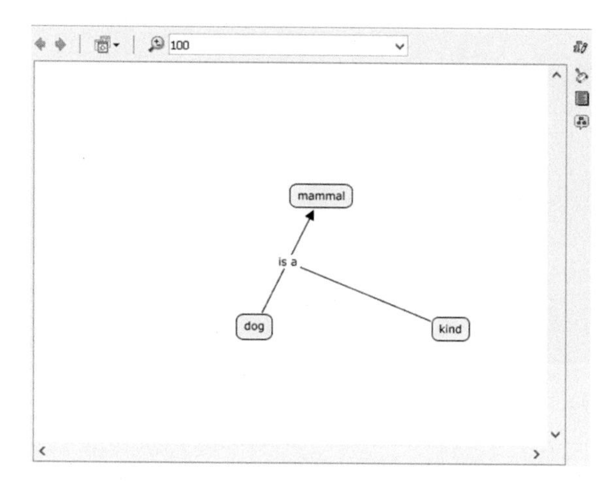

Fig. 4 Cmaptools (http://cmap.ihmc.us)

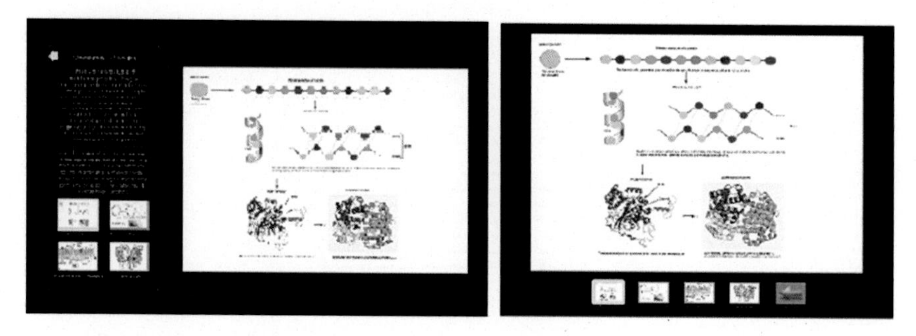

Fig. 5 CThru (Jiang et al. 2009)

5. **CThru** (Jiang et al. 2009) is a video-based educational environment that executes on multiple displays and exploits space in order to show video content and related information at the same time. It is a self-guided environment that allows the user to view and control the playback of a video clip on a central screen. At any time, the user can pause the video and multimedia information related to the point the video has been paused appears on the central and other secondary screens. The user controls the browsing of related information on the central screen and the information unfolds on a secondary screen in bigger dimensions. Related information explains aspects of the central video (see Fig. 5).
6. **Dipity** (www.dipity.com) is an online tool for creating dynamic timelines. The tool exploits space in order to define time intervals and attach nodes from various online sources. When clicking on a node, the user navigates to the respective source. The user may also drag the timeline horizontally in order to explore older and newer nodes and respective events. Timelines may be embedded in other sites and commented by other users (see Fig. 6).

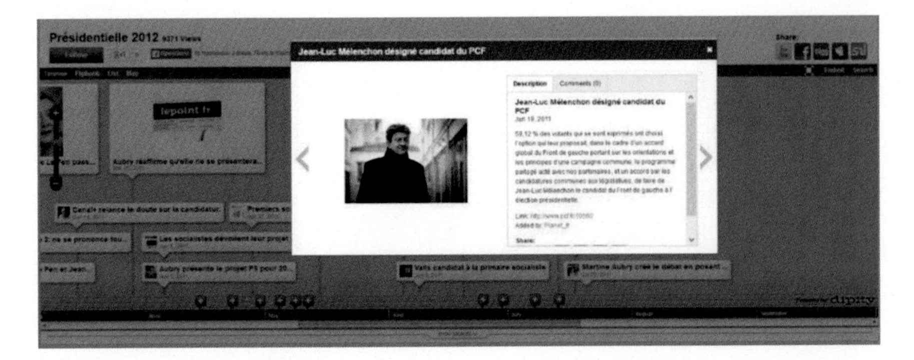

Fig. 6 Dipity (www.dipity.com)

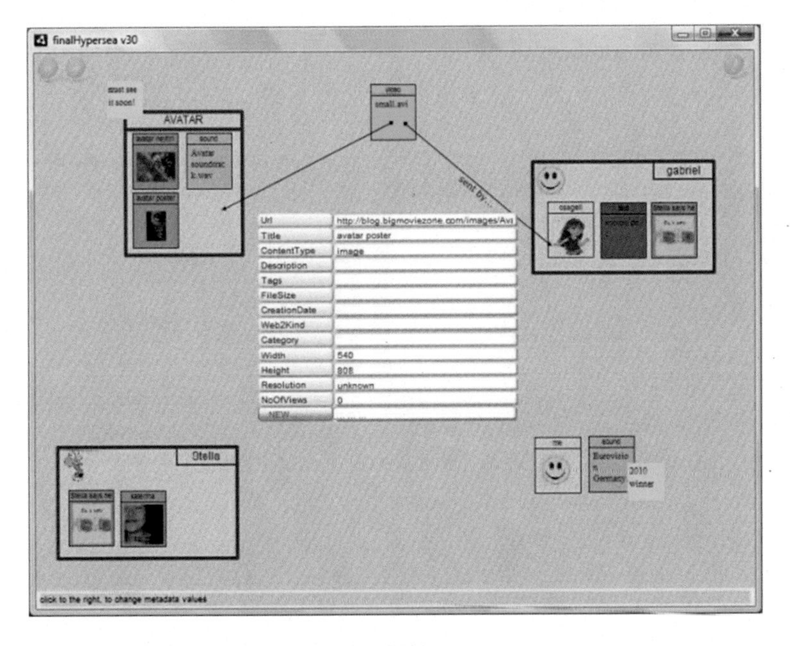

Fig. 7 Hypersea (Styliaras and Christodoulou 2009)

7. **Hypersea** (Styliaras and Christodoulou 2009) is an environment for collecting, organizing and presenting web 2.0 content. The environment allows a single user or many users to organize their information sources in one large space, called Archipelago. During editing mode, the user can drag a multimedia file or a web page and drop it into the environment in order to create a new node. All nodes have the same size, but their color, placement and border denote their origin, importance and relationships. During navigation mode, users can interact with nodes, move them and view their content and attributes. They can view, edit and insert new annotations (see Fig. 7).

Fig. 8 Padlet (http://padlet.com)

8. **Padlet** (http://padlet.com) is an online tool where nodes from other web sites may be dragged and form nodes on an initially empty space. For every node, a user may attach a title and description while controlling other formatting options (see Fig. 8).
9. **Palette** (Sire et al. 2008) is an online social learning space that integrates a lot of web 2.0 tools and other applications in order to provide users a combined functionality of these tools in a unified environment where users can explore, relate and share educational information. More specifically, Palette integrates a semantic Wiki, a structured document editor, an ontology based content classification editor, an activity oriented shared space and a knowledge sharing and argumentative reasoning workspace. Palette supports transparent user access, content that can be searched across services, reusable and sharable content and embedding some of these functionalities in other services.
10. A platform for implementing **Social Learn Analytics** on UK's Open University online learning space SocialLearn is proposed in Ferguson and Buckingham Shum (2012). The platform consists of a dashboard on which a user can overview at a glance five metrics regarding social aspects of learning modules. These metrics include social learning network analytics, social learning discourse analytics, social learning content analytics, social learning disposition analytics and social learning context analytics. Dynamic patterns, diagrams and graphs are employed to visualize the information available to learners, mentors or teachers and group leaders or administrators in order to identify behaviors and patterns within a learning environment that signify effective process (see Fig. 9).
11. **Tagxedo** (www.tagxedo.com) is an online environment where spatial word clouds can be created based on some initial sources that a user enters. Sources can be web sites, twitter accounts, RSS feeds or Del.icio.us IDs. After entering the initial sources, the environment crawls and discovers the most popular

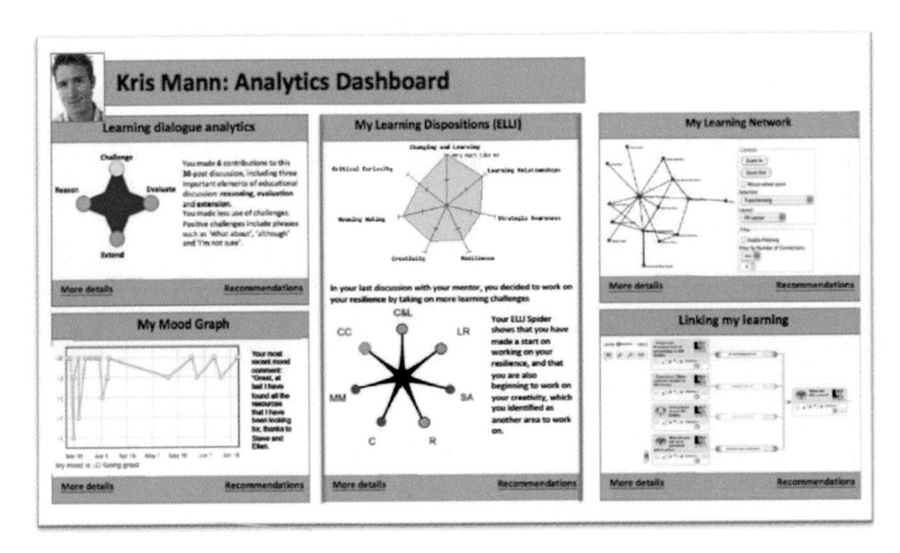

Fig. 9 SLA (Ferguson and Buckingham Shum, 2012)

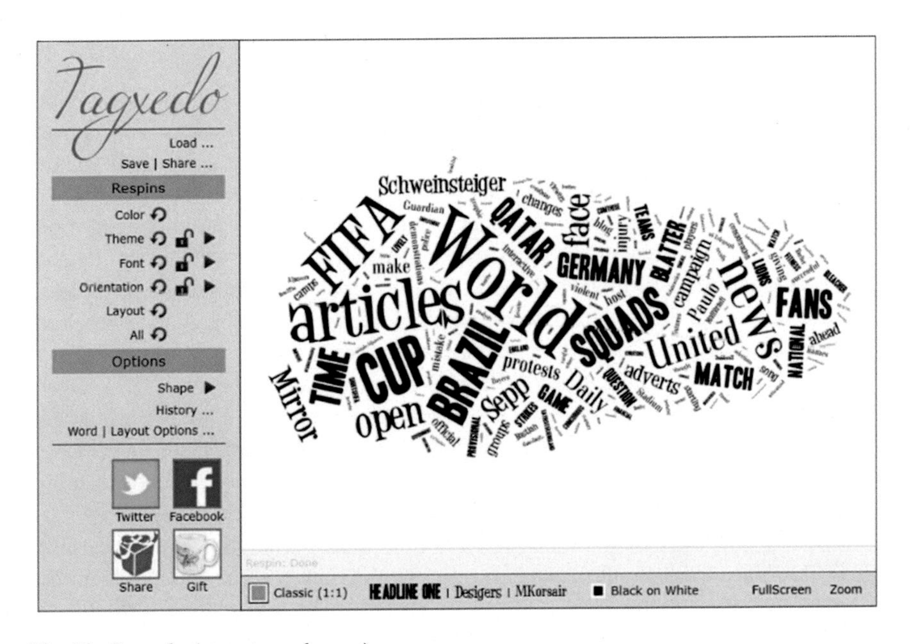

Fig. 10 Tagxedo (www.tagxedo.com)

words and their relations. It forms a cloud of these words, in which, as happens in all word clouds, most popular words have bigger size. Tagxedo's innovation is that these clouds are custom designed and navigated. The user can focus on a certain area of the cloud and discover words that are more closely related (see Fig. 10).

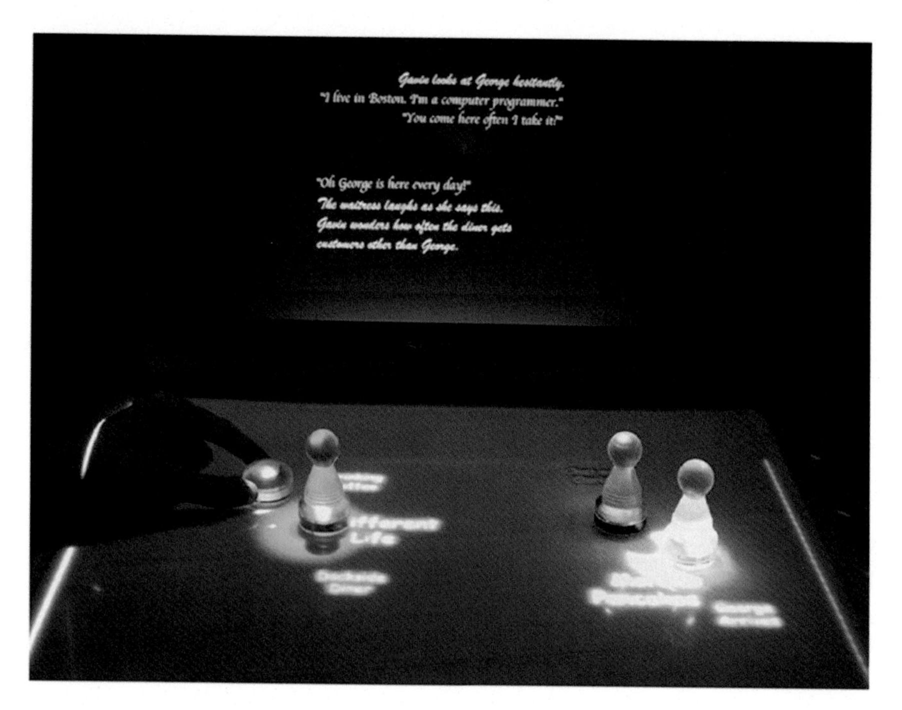

Fig. 11 Tangible Viewpoints (Mazalek et al. 2002)

12. **Tangible Viewpoints** (Mazalek et al. 2002) is a multimedia environment for educational games that employs physical pawns to control interaction. Parallel stories of an educational story unfold and are projected around a player's pawns. Pawns are multicolored and control the game's evolution by moving them. The nearer some pawns are, the more related are their respective stories. As pawns move on the interactive area, story segments with larger weights are bigger in size and brighter, while those with smaller weights are smaller in size and faded (see Fig. 11).

5 Comparison

In this section, every environment is evaluated against the characteristics that a spatial educational environment for mobile devices should support. Every environment is rated indicatively from 0 to 5, depending on how well the environment satisfies the needed functionality of the criterion. Then a short discussion follows that summarizes the environments' support for the certain criterion. The only meaning of rates is to display the diversity and the strengths of the environments, regarding their support for the above criteria. This is why rates are not summed in the end.

Criterion 1 Content source

ENVT	Characteristics	Rate
CANTOS	Text, images, offline audio streams that get louder when a user/avatar gets close to them	4
CANYON	Objects/points displayed on maps	2
CLUI	Custom structured data imported from web sites and web 2.0 applications	5
CMAP	Cannot import content, may only define new nodes and relations from scratch	0
CTHRU	Video as main content, images, texts and 3D modules for explaining main video, no importing feature	4
DIPITY	Web sites that are represented as nodes on a timeline	2
HYPERSEA	Web sites, text, image, video, sounds that are represented as structured nodes on a surface	5
PADLET	Web sites and other files that are represented as nodes on a surface	3
PALETTE	Embedded tools support creating content, relations and knowledge modules from scratch	0
SLA	Import and analyze online learning sources	2
TAGXEDO	Import text and perform word analysis from news sites, RSS feeds and web 2.0 accounts	3
TANGIBLE	No interface for custom content import	0

Most environments support multimedia content import from various online and offline sources. Some of them just display rich multimedia content without offering end users an interface to import new content.

Criterion 2 Content relations

ENVT	Characteristics	Rate
CANTOS	Content modules are independent, there are only implicit spatial relations	2
CANYON	N/A	0
CLUI	Webits, the content nodes of Clui, are independent from each other	0
CMAP	Content nodes are strongly interrelated. Users may define annotated links among relations and nodes.	5
CTHRU	There are relations between certain video frames and multimedia content.	3
DIPITY	No relations among content nodes except from their implicit temporal relation	2
HYPERSEA	Content nodes are related via explicit and implicit links	4
PADLET	There can be no relations among content nodes	0
PALETTE	There are only implicit among content from various content sources	2
SLA	Different analytics try to find implicit relations among learning modules	2
TAGXEDO	There are implicit, spatial relations among words regarding their origin	3
TANGIBLE	In the narrative engine, there are relations among pieces of stories	4

There are three different angles in this criterion. Some environments do not support any relations among content, some support only implicit relations based mainly on space and there are environments where relations are explicit and based on the underlying content structure.

Criterion 3 Referential integrity

ENVT	Characteristics	Rate
CANTOS	No clear support in the environment's description	0
CANYON	N/A	0
CLUI	No clear support in the environment's description	0
CMAP	Users cannot retain a relation after deleting a node that the relation is part of	5
CTHRU	No clear support in the environment's description	0
DIPITY	N/A	0
HYPERSEA	After deleting a node, all its relations are deleted as well	4
PADLET	N/A	0
PALETTE	No clear support in the environment's description	0
SLA	N/A	0
TAGXEDO	N/A	0
TANGIBLE	N/A	0

Most of the environments do not support referential integrity, either because it is not needed for their functionality, or because they have not been designed to support it.

Criterion 4 Typed nodes

ENVT	Characteristics	Rate
CANTOS	No clear support in the environment's description	0
CANYON	N/A	0
CLUI	Content modules are typed based on their origin and they have a clear internal structure	5
CMAP	Implicit support for creating typed nodes. Users should structure nodes appropriately.	3
CTHRU	No clear support in the environment's description	0
DIPITY	Nodes have a type regarding the content's origin and a structure holding title, description and comments	4
HYPERSEA	Nodes have a typed structure holding the content's origin, technical details and custom fields	4
PADLET	Nodes have a simple typed structure holding the content's origin, a title and description	4
PALETTE	No clear support in the environment's description	0
SLA	N/A	0
TAGXEDO	Words that appear in the cloud retain their origin	2
TANGIBLE	No clear support in the environment's description	0

Many environments support typed nodes, in which types denote the content's origin and form, e.g. music, web site, person etc.

Criterion 5 Edit

ENVT	Characteristics	Rate
CANTOS	Educators can create their own virtual environment and place multimedia content appropriately	3
CANYON	N/A	0
CLUI	Users can create webits and reuse them in other applications	5
CMAP	Users create concept maps from scratch, having full control on them	4
CTHRU	No clear support in the environment's description	0
DIPITY	Users can create their own timelines, insert and annotate nodes, leave comments	5
HYPERSEA	Users can create their own content nodes, edit, relate and group them and leave annotations	5
PADLET	Users can create their own content nodes, edit and annotate them	4
PALETTE	Users can use editing options of the tools that compose the environment	2
SLA	Analytics provide mainly output to users based on the learning modules of users	1
TAGXEDO	Users control the origin of words in a cloud, may skip some words or change their appearance	4
TANGIBLE	Not clear how story pieces and respective content is entered in the database	0

Most environments provide support for editing content, both during its initial import in the environment and after importing, while navigating.

Criterion 6 Annotations

ENVT	Characteristics	Rate
CANTOS	The environmrnt depends on user leaving comments and annotations through text and sound	5
CANYON	N/A	0
CLUI	There is support for adding comments throughout the reuse of webits	3
CMAP	No clear support in the environment's description	0
CTHRU	No clear support in the environment's description	0
DIPITY	Users can leave comments on a every timeline node and share timelines through other applications	5
HYPERSEA	Users can leave comments on a every content node	4
PADLET	Walls can be shared and commented through other social applications	3
PALETTE	Users can use annotation options of the tools that compose the environment	3
SLA	N/A	0
TAGXEDO	Word clouds can be shared and commented through other social applications	3
TANGIBLE	No clear support in the environment's description	0

Most environments provide support for annotations and comments either through their interface or through other social applications.

Criterion 7 Visual Cues

ENVT	Characteristics	Rate
CANTOS	Rich visual cues through virtual rooms, avatars, multimedia icons	5
CANYON	Folding options to represent remote objects, objects placement, distance indication visually	5
CLUI	Nodes appearance represent their origin	4
CMAP	Basic visual cues to represent nodes and relations	3
CTHRU	Related content and paused video are clearly represented	4
DIPITY	Many cues for nodes, zoom options, comments	5
HYPERSEA	Differently colored nodes according to their origin and type (e.g. content node, annotation)	3
PADLET	Differently colored nodes according to their origin	3
PALETTE	W3C Widget API use	3
SLA	Visual cues used in various analytics express learning modules' relations and perspectives	4
TAGXEDO	Visual semantics employed for word importance and relations	4
TANGIBLE	Various light intensity around pawns denotes closeness in story pieces	4

Depending on their functionality, all environments employ some kind of visual cues, in the form of shape, icons, colors, border and line use or shadow.

Criterion 8 Educational activities

ENVT	Characteristics	Rate
CANTOS	This an educational environment containing navigation to and interaction with educational content	5
CANYON	N/A	0
CLUI	N/A	0
CMAP	No direct support, could be used for educational purposes	2
CTHRU	Educational environment providing access to video and explanatory material	5
DIPITY	No direct support, could be used for educational purposes	2
HYPERSEA	No direct support, could be used for educational purposes	2
PADLET	No direct support, could be used for educational purposes	2
PALETTE	No direct support, could be used for educational purposes	2
SLA	Social analytics that identify behaviors and patterns within learning modules	5
TAGXEDO	No direct support, could be used for educational purposes	2
TANGIBLE	Educational activities in form of stories	4

Some environments have an educational orientation per se, other are not pure educational environments but could be used and have been evaluated for use in educational settings.

Criterion 9 Map operations

ENVT	Characteristics	Rate
CANTOS	Implicit support for navigating in custom virtual rooms	2
CANYON	Trace of on view and off view objects on maps	5
CLUI	No clear support in the environment's description	0
CMAP	N/A	0
CTHRU	No clear support in the environment's description	0
DIPITY	No clear support in the environment's web site	0
HYPERSEA	No clear support but a map could be placed behind nodes	2
PADLET	No clear support in the environment's web site	0
PALETTE	No clear support in the environment's description	0
SLA	N/A	0
TAGXEDO	N/A	0
TANGIBLE	Pawns could be placed on maps	3

Few environments support map-based navigation and objects placement. Few other environments could support maps, while there are others that map navigation is not an option.

Criterion 10 Multimedia playback

ENVT	Characteristics	Rate
CANTOS	Support for all education material and audio comments	5
CANYON	N/A	0
CLUI	Support for content represented by webits	4
CMAP	No support in the current version	0
CTHRU	Support for playback of the main video as well related content that explains main video's aspects	5
DIPITY	Users can playback images and related content to timeline's objects	4
HYPERSEA	Users can playback all multimedia content represented by nodes	5
PADLET	Users can playback all multimedia content represented by nodes	5
PALETTE	Playback of some content via the tools that compose the environment	2
SLA	N/A	0
TAGXEDO	No support for multimedia playback	0
TANGIBLE	All content related to the pieces of the stories may be accessed	3

Most environments support multimedia playback of all types. Some of them employ multimedia playback only when needed depending on their functionality.

Criterion 11 User registration

ENVT	Characteristics	Rate
CANTOS	Instructors, pupils and guests are differentiated	5
CANYON	No differentiation in functionality based on different levels of users	0
CLUI	Every user has their own webits that can be shared and exploited	3
CMAP	Every user has their own concept maps that can be shared and exploited	3
CTHRU	No differentiation in functionality based on different levels of users	0
DIPITY	Every user has their own timelines that can be shared and exploited	3
HYPERSEA	Differentiation for users that can create nodes and relations and for those that access them	3
PADLET	Every user has their own walls that can be shared and exploited	3
PALETTE	Exploitation of users' account in different social applications	2
SLA	Social analytics are calculated for certain users	3
TAGXEDO	Every user has their own word clouds that can be shared and exploited	3
TANGIBLE	It is implied that some instructors prepare the stories	2

More work should be done for accessing educational environments by different categories of users, who should have appropriate rights for editing, navigating, commenting content etc.

Criterion 12 Zoom, Touch and Drag

ENVT	Characteristics	Rate
CANTOS	No such operations in the description of the environment	0
CANYON	Users may zoom to certain map areas and drag the map with a pen to focus on new areas	5
CLUI	Drag and drop is supported for defining and exploiting webits	3
CMAP	Users can drag nodes and relations	2
CTHRU	Multitouch operations that lead to dragging and zooming is supported on multiple screens	5
DIPITY	A web site where timeline events may be dragged. Users may also zoom to certain timeline areas.	4
HYPERSEA	All operations are performed by dragging and dropping and require minimal other operations	3
PADLET	Users may drag nodes on the wall	2
PALETTE	No specific support	0
SLA	No specific support	0
TAGXEDO	No specific support	0
TANGIBLE	Touching is performed by physical pawns	3

There is a variety of operations regarding zooming, touching and dragging. Some environments depend on dragging for visually performing content access and editing operations, whereas zoom operations focus on certain content aspects.

Criterion 13 Physical objects interaction

ENVT	Characteristics	Rate
CANTOS	No specific support	0
CANYON	A virtual pen is used for focusing on certain map areas and, generally, for interacting with a map	5
CLUI	No specific support	0
CMAP	No specific support	0
CTHRU	Multitouch on multiple screens that could be considered as objects interaction	2
DIPITY	No specific support	0
HYPERSEA	No specific support	0
PADLET	No specific support	0
PALETTE	No specific support	0
SLA	No specific support	0
TAGXEDO	No specific support	0
TANGIBLE	Interacting is performed by physical pawns	5

Only few environments support interaction via physical objects. In the mobile world, such operations should be enhanced as mobile devices provide rich interaction means, including gestures.

Criterion 14 Responsive design

ENVT	Characteristics	Rate
CANTOS	No specific support	0
CANYON	Maps and foldings are adjusted to the screen's size	3
CLUI	No specific support	0
CMAP	No specific support	0
CTHRU	Multimedia content unfolds correctly on various screens	3
DIPITY	Web site design take into support the screen's size	3
HYPERSEA	No specific support	0
PADLET	Web site looks correct when changing screen size	3
PALETTE	No specific support	0
SLA	Design of dashboard seems to take into account the screen size	2
TAGXEDO	No specific support, just a full screen mode	2
TANGIBLE	N/A	0

There is poor, implicit support for responsive design. Environments for mobile devices should resolve this issue as there are various screen sizes on these devices.

Criterion 15 Search and Navigation history

ENVT	Characteristics	Rate
CANTOS	No specific support	0
CANYON	No specific support	0
CLUI	Webits may be saved and navigated later	2
CMAP	Concept maps may be saved and accessed on demand	3
CTHRU	No specific support	0
DIPITY	Timelines may be saved	3
HYPERSEA	Users can go back and forth the islands creation process	3
PADLET	No specific support	0
PALETTE	No specific support	0
SLA	No specific support	0
TAGXEDO	There is a history option for the various word clouds	3
TANGIBLE	Stories can be saved and played back	3

Same as the previous criterion, there is poor support for searching the environment's content and navigating back and forth.

Criterion 16 Platform

ENVT	Characteristics	Rate
CANTOS	Custom implementation	2
CANYON	Custom implementation	2
CLUI	Web technologies	4
CMAP	Java platform	3
CTHRU	Custom implementation	2
DIPITY	Web technologies	4
HYPERSEA	Flash/Adobe Air	3
PADLET	Web technologies	4
PALETTE	Web 2.0 technologies	4
SLA	Custom implementation	2
TAGXEDO	Web technologies	4
TANGIBLE	Custom implementation	2

Rates in this criterion show how open is the environment's implementation. Half of the environments are based on Web technologies, which means that they are more suitable for running on mobile devices.

Criterion 17 Persistency

ENVT	Characteristics	Rate
CANTOS	No specific support	0
CANYON	N/A	0
CLUI	Webits are stored and may be reused	5
CMAP	Concept maps are stored and may be reused	5
CTHRU	No specific support	0
DIPITY	Timelines are stored and may be reused	5
HYPERSEA	Content nodes and their placement may be stored and reused	5
PADLET	Walls can be saves as images, PDF, Excel etc.	5
PALETTE	Depends on tools that constitute the environment	3
SLA	N/A	0
TAGXEDO	Word clouds can be saved as pictures and HTML	5
TANGIBLE	Stories, as carved out d by pupils, can be saved	5

Most of the environments allow saving user's content. A lot of them may share this content to Web 2.0 applications.

Criterion 18 Embedding

ENVT	Characteristics	Rate
CANTOS	No specific support	0
CANYON	No specific support	0
CLUI	Webits are meant to import and export content from and to other web sites	5
CMAP	Underlying content of concept maps is structured and may be reused	2
CTHRU	No specific support	0
DIPITY	Timelines may be shared to other Web 2.0 applications and embedded in other sites	4
HYPERSEA	Underlying XML ontology is structured and may be reused	2
PADLET	Walls may be embedded in other Web 2.0 applications	3
PALETTE	Palette is designed so as to embed and be embedded in other social applications	4
SLA	N/A	0
TAGXEDO	Word clouds may be embedded in other Web 2.0 applications	3
TANGIBLE	N/A	0

Few environments provide straightforward support for embedding their content and functionality in other applications or to embed other applications. Most environments store content in such way that could be exploited in other applications.

6 Results

The presented environments excel in numerous but diverse aspects concerning the characteristics of a potential spatial educational environment for mobile devices. Such an environment should be based on strongly structured content that different categories of users should be allowed to edit, comment and view with rich interaction mediums and attractive design on different platforms. Towards this direction, in this section, some strong features of the previously presented environments are noted.

Concerning text, Cantos and Contos specifies diverse content such as text, images and offline audio streams. Clui provides complete content import from other web sites and web 2.0 applications. A video clip is the central information means in CThru, whereas images, texts and 3D material further elaborate on the video. Hypersea supports importing content such as links to web pages and offline multimedia material such as text, image, video and sounds. Apart from content diversity, relationships are also important. Content nodes in concept maps are interrelated by definition. These relations are extended by annotations. In Hypersea, content nodes are related via explicit and implicit links. In Tangible Viewpoints, the narrative engine supports relations among pieces of stories. Referential integrity constitutes another factor in content relationships. Concept Maps by default support integrity as users cannot retain a relation after deleting a node that the relation is part of. The same thing happens in Hypersea when deleting a node by dragging it out of the environment: The respective relational links disappear from both the space and the underlying XML-based ontology. Regarding types in content nodes, Clui has the best support as webits of information have a clear internal structure according to their origin. Nodes in Dipity are displayed and structured according to their origin. A title, some description and comments may also be added. Nodes in Hypersea have a typed structure holding the content's origin, technical details and custom fields.

Content editing in educational environments addresses the needs of instructors. Users in Clui can create webits from scratch. In the same way, users create logical maps in Concept Maps, timelines in Dipity, spaces of content nodes in Hypersea and Padlet and word clouds in Tagxedo. In all cases, users create content nodes and representations and may reuse them in other applications. Commenting and annotating is also an important process in the educational procedure. In Padlet, Hypersea and Dipity, users may leave comments on content nodes even if they are not they are not the owners of the original content. Apart from text, Cantos and Contos allow also the recording of audio comments.

Adding semantics to content representation through visual cues helps students identifying the right information. In Cantos and Cantos, there are diverse visual cues through virtual rooms, avatars and multimedia icons. In Canyon, the invention of folding interface is useful for representing remote, off view objects in maps. In Clui and Dipity, content nodes are carefully designed according to their origin, whereas in CThru, related content for specific video frames are clearly represented. Visual semantics, such as size and color, are employed in Tagxedo to express the importance of every word in the cloud. In Tangible Viewpoints, light intensity around pawns denotes closeness in story pieces.

Mobile devices offer and depend on operations such as touch and drag for navigating and interacting with content. Canyon employs zoom operations in order to focus on certain areas of the map. A lot of dragging and zooming is used in CThru for controlling the main video clip and supported multimedia content. Dragging is the main interface medium in Dipity for accessing different parts of the timeline. Regarding interaction with physical objects, users in Canyon employ a virtual pen in order to focus on certain areas of the map. In Tangible Viewpoints, players move physical pawns in order to focus on certain parts of the stories.

Among the environments examined, some of them have pure educational orientation, such as Cantos and Contos, where educational multimedia content is accessed and commented through virtual rooms. CTrhu presents a theme through a central video clip that is annotated by multimedia content projected in other screens. Tangible Viewpoints support storytelling through physical navigation. On the other hand, only Canyon supports map-based navigation and real-time identification of objects on a map. However, there a lot of map-based navigation applications for mobile devices. A lot of environments support multimedia playback. Cantos and Contos support for all education material and audio comments. Clui, Dipity, Padlet and Hypersea can playback content represented by hypermedia nodes, such as timeline nodes. CThru's main interaction scheme is the playback of a video clip, whereas related multimedia content unfold in other screens.

In other issues, as mentioned before, more work needs to be done in defining user categories with different privileges. Among the environments examined, only Cantos and Contos supports different rights for instructors, pupils and guests. An educational environment's platform needs to be based on widely used technologies. Towards this direction, Clui, Dipity, Padlet, Palette and Tagxedo are implemented with Web technologies, which enables them to execute on mobile devices of different operating systems. Concerning importing and exporting content, most environments allow persistent storage of content created by users. Clui's webits permit content exchange with other web sites. Finally, in the environments examined, there is poor, implicit support for responsive design and searching the environment's content and navigating back and forth.

7 Conclusions/Future Work

This paper has set the issue of deploying an educational environment for mobile devices. As yet, there is no complete environment that exploits the features of mobile devices and supports educational needs. In other words, although a lot of technologies are present such as search engines, responsive design in HTML 5, sensors for identifying user actions, multimedia playback, more work has to be done in the educational domain in order for these technologies to cooperate seamlessly.

Towards this direction, through the review of related work in both educational and multimedia areas, the needed features have been identified and a number of desktop environments have been chosen in which scattered features are supported. By presenting and comparing these environments, appropriate features are denoted that could be ported on a mobile environment. This review may lead to the design and implementation of a comprehensive educational environment executed on mobile devices covering the needs of both educators and students.

References

Acosta, P., Monguet, J.M., Rodriguez, R.: Educational hypermedia applications: design based on content models. In: Lassner, D., McNaught, C. (eds.) Proceedings of World Conference on Educational Multimedia, Hypermedia and Telecommunications 2003, pp. 292–295. AACE, Chesapeake, VA (2003)

Alexandra Ion, Y.-L., Chang, B., Haller, M., Hancock, M., Scott, S.D.: Canyon: providing location awareness of multiple moving objects in a detail view on large displays. In: Proceedings of the SIGCHI Conference on Human Factors in Computing Systems (CHI'13), pp. 3149–3158. ACM, New York, NY, USA (2013)

Ally, M.: Mobile Learning: Transforming the Delivery of Education and Training. Athabasca University Press, Edmonton (2009)

Bellotti, F., Berta, E., De Gloria, A., D'ursi, A., Fiore, V.: A serious game model for cultural heritage. J. Comput. Cult. Herit. **5**(4), 27 (2013)

Brunstein, A., Naumann, A., Krems, J.F.: Processing educational hypertext: support by orientational and navigational aids. In: Cantoni, L.., McLoughlin, C. (eds.) Proceedings of World Conference on Educational Multimedia, Hypermedia and Telecommunications 2004, pp. 3828–3835). AACE, Chesapeake, VA (2004)

Cavanaugh, T.W., Cavanaugh, C.: Interactive maps for community in online learning. Comput. Schools **25**(3), 235–242 (2008)

de Oliveira, F.S., Tavares, T.A., Chagas, A.A., Burlamaqui, A., Brennand, E.G., de Souza Filho, G.L.: Experiences from the use of a shared multimedia space for e-learning in Brazil primary schools. In: Proceedings of the 3rd International Conference on Digital Interactive Media in Entertainment and Arts (DIMEA '08), pp. 91–98. ACM, New York, NY, USA (2008)

Fehr, M.C.: Culturally responsive teaching awareness through online fiction. Multi. Educ. Technol. J. **4**(2), 113–125 (2010)

Ferguson, R., Buckingham Shum, S.: Social learning analytics: five approaches. In: Buckingham Shum, S., Gasevic, D., Ferguson, R. (eds.) Proceedings of the 2nd International Conference on Learning Analytics and Knowledge (LAK'12), pp. 23–33. ACM, New York, NY, USA (2012)

Garzotto, F., Gonella, R.: An open-ended tangible environment for disabled children's learning. In: Proceedings of the 10th International Conference on Interaction Design and Children (IDC'11), pp. 52–61. ACM, New York, NY, USA (2011)

Holtzblatt, K., Koskinen, I., Kumar, J., Rondeau, D., Zimmerman, J.: Design methods for the future that is now: have disruptive technologies disrupted our design methodologies? In: Extended Abstracts on Human Factors in Computing Systems (CHI EA'14), pp. 1063–1068. ACM, New York, NY, USA (2014)

Jiang, H., Viel, A., Bajaj, M., Lue, R.A., Shen, C.: CThru: exploration in a video-centered information space for educational purposes. In: Proceedings of the SIGCHI Conference on Human Factors in Computing Systems (CHI '09), pp. 1247–1250. ACM, New York, NY, USA (2009)

Katifori, A., Lepouras, G., Dix, A., Kamaruddin, A.: Evaluating the significance of the desktop area in everyday computer use. In: Proceedings of the First International Conference on Advances in Computer-Human Interaction (ACHI'08), pp. 31–38. IEEE Computer Society, Washington, DC, USA (2008)

Koc, M.: Pedagogical knowledge representation through concept mapping as a study and collaboration tool in teacher education. Austr. J. Educ. Technol. **28**(4), 656–670 (2012)

Lye, S.Y., Abas, S., Tay, L.Y., Saban, F.: Exploring the use of online space in an elementary school. Educ. Media Int. **49**(3), 155–170 (2012)

Mazalek, A., Davenport, G., Ishii, H.: Tangible viewpoints: a physical approach to multimedia stories. In: Proceedings of the Tenth ACM International Conference on Multimedia (MULTIMEDIA'02), pp. 153–160. ACM, New York, NY, USA (2002)

Millard, M., van Leusen, P., Whiting, J.: Enterprise social media tools for effective team communication and collaboration. In: Bastiaens, T., Marks, G. (eds.) Proceedings of World Conference on E-Learning in Corporate, Government, Healthcare, and Higher Education 2013, pp. 2457–2463. AACE, Chesapeake, VA (2013)

Pham, H., Mazzola Paluska, J., Miller, R., Ward, S.: Clui: a platform for handles to rich objects. In: Proceedings of the 25th Annual ACM Symposium on User Interface Software and Technology (UIST'12), pp. 177–188. ACM, New York, NY, USA (2012)

Scherp, A., Boll, S.: Generic support for personalized mobile multimedia tourist applications. In: Proceedings of the 12th Annual ACM International Conference on Multimedia (MULTIMEDIA'04), p. 178. ACM, New York, NY, USA (2004)

Shaw, A., Krug, D.: Youth, heritage, and digital learning ecologies: Creating engaging virtual museum spaces. In: Herrington, J., et al. (eds.) Proceedings of World Conference on Educational Multimedia, Hypermedia and Telecommunications 2013, pp. 673–679. AACE, Chesapeake, VA (2013)

Shih, Y.E., Mills, D.: Setting the new standard with mobile computing in online learning. Int. Rev. Res. Open Distance Learn. **8**(2) (2007)

Shipman, F.M., Marshall, M.M.: Spatial hypertext: an alternative to navigational and semantic links. ACM Comput. Surv. **31**(4), Article 14 (1999)

Sintoris, C., Raptis, D., Stoica, A., Avouris, N.: Delivering multimedia content in enabled cultural spaces. In: Proceedings of the 3rd International Conference on Mobile Multimedia Communications (MobiMedia'07), Article 7, 6 pages. ICST (Institute for Computer Sciences, Social-Informatics and Telecommunications Engineering), ICST, Brussels, Belgium (2007)

Sire, S., Vanoirbeek, C., Karacapilidis, N., Karousos, N., Tzagarakis, M., Latour, T.: What Makes a Software Socializable?. In: Luca, J., Weippl, E. (eds.) Proceedings of World Conference on Educational Multimedia, Hypermedia and Telecommunications 2008, pp. 5077–5082. AACE, Chesapeake, VA (2008)

Styliaras, G., Christodoulou, S.: HyperSea: towards a spatial hypertext environment for web 2.0 content. In: Menczer, F. (ed.) Proceedings of the 20th ACM Conference on Hypertext and Hypermedia, pp. 35–44. ACM, New York (2009)

Styliaras, G.D., Christodoulou, S.P.: Organizing personal web 2.0 content with Hypersea. In: do Nascimento, R.P.C. (ed.) Proceedings of the 6th Euro American Conference on Telematics and Information Systems (EATIS'12), pp. 223–230. ACM, New York, NY, USA (2012)

Traxler, J.: The evolution of mobile learning. In: Guy, R. (ed.) The Evolution of Mobile Teaching and Learning, pp. 1–14. Informing Science Press, Santa Rosa (2009)

Tse, E., Schöning, J., Rogers, Y., Shen, C., Morrison, G.: Next generation of HCI and education: workshop on UI technologies and educational pedagogy. In: Extended Abstracts on Human Factors in Computing Systems (CHI EA'10), pp. 4509–4512. ACM, New York, NY, USA (2010)

Author Biography

Georgios D. Styliaras was born in Agrinio, Greece in 1974. He obtained his engineering diploma in computer engineering and informatics from the University of Patras, Greece in 1996. He received his Ph.D. degree in computer engineering and informatics from the University of Patras, Greece in 2001. His major fields of study cover multimedia applications, multimedia systems, hypermedia and educational applications. He worked for the Hellenic Ministry of Culture as a project manager for multimedia applications projects. From 2006 until 2009, he was a lecturer at the Department of Plastic Arts and Art Sciences (Univeristy of Ioannina, Greece). Since 2009, he is an assistant professor at the Department of Cultural Heritage Management and New Technologies, University of Patras, Greece.

Science Teachers' Metaphors of Digital Technologies and Social Media in Pedagogy in Finland and in Greece

Marianna Vivitsou, Kirsi Tirri and Heikki Kynäslahti

Abstract In this study we draw from the interviews of four science teachers, one from Finland and three from Greece, and two science education experts from Finland in order to discuss and analyze pedagogical decisions and choices when the learning space is enriched with social networking environments, and digital and mobile technologies. Our research interest departs from concerns that technological pervasiveness generates and an observed suspension of belief in science. This suspension seems to be rooted in the risks that post-industrial societies are facing nowadays. By examining the study participants' experiences, we aim to trace the intersection of science and technology with pedagogy and, in this way, to gain an insight into the possible futures of science education. The qualitative analysis of the data indicates instances of both deductive and inductive logic that show up through views of science as way of thinking and as method. The analysis of participants' speech also reveals recurrent underlying conceptions of science and related issues.

Keywords Science educators · Pedagogical thinking · Science · Digital technology · Metaphors

1 Introduction

As technological pervasiveness becomes more and more apparent in life and the school nowadays, this study departs to investigate the intersection of digital technology with science education. To this end, we will discuss and analyze the pedagogical decisions and choices of a Biology teacher from Finland and three Computer Science and Technology teachers from Greece. These teachers enrich the learning space by integrating digital and mobile technologies for pedagogical purposes. In addition, we will also analyze the interviews of two Science education

M. Vivitsou (✉) · K. Tirri · H. Kynäslahti
Department of Teacher Education, University of Helsinki, Helsinki, Finland
e-mail: marianna.vivitsou@helsinki.fi

G.A. Tsihrintzis et al. (eds.), *Intelligent Computing Systems*,
Studies in Computational Intelligence 627, DOI 10.1007/978-3-662-49179-9_7

experts from Finland discussing the pedagogical integration of digital technologies for science education research. By examining the study participants' experiences, we aim to trace the intersection of science and technology with pedagogy and, in this way, to gain an insight into the possible futures of science education through an analysis of teachers' metaphors.

Our interest in Finnish and Greek science teachers and experts' metaphors as expressions of pedagogical thinking results from the consideration of a number of parameters. One such parameter is our firm belief that, ultimately, science education should aim to empower learners for democratic participation and citizenship. Another is the high-low relationship in terms of performance in Mathematics and Science between Finland and Greece. According to the results for the 2009 Program for International Student Assessment (PISA, OECD 2010), Finland ranks high (i.e., 6th and 2nd in Mathematics and Science respectively) while Greece scores below average in both cases (39th and 40th respectively). In terms of guidelines for teaching science in schools we trace similarities in the principles underlying the curricula in both countries. These build upon fundamental values such as human worth, justice, honesty, freedom, peace and democracy (Code of Ethics for Finnish Teachers 1998, Greek National Curriculum, 2003). In addition to these, our research interest is based on considerations of trends resulting from research findings (e.g., Schreiner and Sjøberg 2004; Sjøberg and Schreiner 2005) indicating that, when it comes to science education, it is society that feeds values and attitudes toward science and technology into the classroom, and not the other way around.

Certainly, we do not take a position that beliefs of science are culture-free or context-independent. What we argue, however, is that these trends are tied with a suspension of belief in science and are rooted in the risks that post-industrial societies are facing nowadays. Some examples of these are ecological hazards, global warming, dangers from nuclear, chemical and genetic technology, are made-made and are the by-products of scientific and technological advancement.

To raise student consciousness of environmental hazards curricula in Finland and Greece have introduced concepts such as environmental sustainment in the jargon of science and science-related subjects (e.g., Biology and Geography). Additionally, cross-curricular teaching hours are allocated in the school timetable (Lavonen and Krzywacki 2011). In a discussion of findings of a study that examines student values and attitudes from 75 schools in Finland, Uitto et al. (2011) argue that participation in school practices and learning sustainability can promote environmental responsibility more effectively than traditional instruction. We consider that the trend to extend indoor curriculum to outdoor scientific classes that we observed in our search for up-to-date themes in science education conferences correlates positively with Uitto et al.'s (2011) finding. As environmental sustainability relates to science both directly and indirectly, changes in teaching science seem but inevitable in the future.

Developments in digital technologies are also products of scientific advancement. Another type of disbelief then relates to the digital technologies themselves. Despite the fact that digital, mobile and networking devices and environments connect individuals, they often become a world of their own that dis-connects (Dreyfus 1992, 2002) and, eventually, comes to, for instance, undermine young people's educational achievement and time management for learning and personal growth. Along with the fact that digital technologies change the ecology of the traditional learning environment these challenges seem even more pressing nowadays when it becomes more and more evident that people learn with the configurations of multiple technologies in concert. This comes as a development of what is termed by Tsihrintzis et al. (2012) pervasive computing. Pervasive computing or the third wave of computing technology allows users to interact with a variety of networked digital devices. In our study however, the emphasis is on the human element rather than the purely technological. We believe then that, in order to gain a better insight into the situation, it is important to understand how science teachers think, make decisions and act. In other words, it is important to understand teachers' pedagogical thinking (Kynäslahti et al. 2006).

By considering teachers' thinking in relation to their professional life decisions we position science teachers as 'beings-in-the-world' and, at the same time, explore the subtleties of their particular situations. This is an aspect that has been traditionally disregarded in, for example, artificial intelligence design (Dreyfus 1992). The research need to augment our knowledge of teachers' thinking is consistent with the need expressed by Chrysafiadi and Virvou (2013) for a design of e-learning models and systems that is informed of user knowledge state and characteristics.

It remains, therefore, to look into the study participants' speech and determine whether our assumptions are confirmed, or not, and to what extent. In this way, we will seek answers to our main research question, 'In what ways do science educators use digital technologies and social media to promote science-related concepts and literacies?'

We will do so through the analysis and discussion of pedagogical decisions and choices as appear in the participants' speech and in relation to our review and discussion of the relevant literature.

2 Theoretical Background

In this study, we hold the view that the prevailing definition of science education changes into one that blends both deductive and inductive modes of thinking in the teaching of science and technology, with technology. By this, we mean that the transposition involves an insight into (natural) phenomena and related concepts that complements or even precedes an approach that relies upon laws, formulas and calculations. This way of blending a more 'naturalistic' with the core 'scientific' approach positions organically the study of science-related concepts into the living experience and enriches student learning with digital technologies, whether this occurs in the classroom, the school laboratory or outdoors. A twofold issue seems to arise. One concerns the logic of approaching science. The other relates to the type of science-technology relationship that becomes visible out of the pedagogical integration.

2.1 Approaching Science

One way to approach science links with what Polkinghorne (1983) calls the 'received view' that was framed as a result of the commitment of natural sciences to the investigation for a theory explaining events in a clear and precise manner. At that time reality was connected to an understanding of knowledge as certainty and the principles that underlie a deductive system of inquiry. Guiding perspectives (Polkinghorne 1983, 90) of this system are:

- Knowledge, as opposed to opinion, is contained in statements that are descriptions of direct observations or that are deductively linked to descriptions of direct observation.
- The goal of science is a network of knowledge statements linked together by the necessity of deductive logic generated from axiom statements and grounded ultimately in observation statements.
- The statements free from metaphysical overtones and personal bias are those grounded in observation and belonging to the axiomatic system. All sciences are to limit their assertions to these kinds of statements, including the sciences of human phenomena.

This so-called ideal model implies a system that connects observational statements and general statements of law or theory with statements of greater and greater specificity (Polkinghorne 1983, 71). These are derived from axioms until a statement of fact is implied; one can test whether the fact appears to observation. The model has the form of a deductive argument where the conclusion is guaranteed by the premises. This certainty is the aim and the reason for the law-deductive type of explanation that is applicable both to events that have happened and to those that have not happened yet. When the event has already occurred, the argument is an explanation; otherwise, the argument is a prediction.

One of the difficulties with the deductive system is that this logic dispenses with the time concept and treats the actual processes of thought as occurring in an eternal present in order to avoid contradictions between successives. The fact, being an instance of a general law, is non-temporal. This non-temporal scientific logic sought to classify thought in such a way that a uniform understanding of reality (Olsen 2010, 45) becomes possible. As Olsen (2010) argues, scientific postulations that view the physical outside time express a kind of deterministic rationality where reality appears as frozen. This is in opposition with the reality of transitory properties, or reality in 'becoming' (Olsen 2010, 49). As developments in science have shown, nowadays we need to admit that there is always some 'noise' in the quietness of a so-called uniform universe. Even if dissipative forces make a pendulum reach equilibrium and stop, it will not be at absolute rest but perform Brownian movements about the position of equilibrium. The energy of movement is so little that it seems to disappear. And yet, although not apprehensible, movement is still there (Olsen 2010, 55).

Overthrowing principles of axiomatic certainty, inductive logic describes supportive arguments that reason to a conclusion about all the members of a class from an examination of only a few members of the class. Induction, therefore, cannot guarantee that the conclusion statement is true, as the inductive argument lends only to probable support. The difference lies in that only the rules of the system within which the deduction is made are needed. Contrary to the deductive system, inductive logic takes account of factual information in the process of confirming a hypothesis. Factual information, however, can radically change the degree of confirmation. It seems that while certainty is the standard for acceptable knowledge on the one hand, no certain grounds for an inference that proves truthfulness can be provided on the other (Polkinghorne 1983, p. 100). In opposition to the commonly held view, therefore, the difference between deductive and inductive logic does not lie in whether a hypothesis is a kick-off point, or not. Actually, it seems that every scientific inquiry departs from an initial hypothesis. The difference lies in the degree of certainty that confirms, or not, initial assumptions.

One such certainty seems to underlie the way we apprehend our relation to technology.

2.2 The Relationship Between Science and Digital Technology

The prevalent instrumental account of technology as means to an end is rooted in the anthropological explanation of technology as an act of human artifice. The instrumental-anthropological account points toward one ultimate goal: humans should try to master technology and use it in the most profitable manner (Heidegger 1977). In his work, Heidegger attempts to capture the essence of technology as a way in which truth happens (Riis 2010, 125–6). According to Riis (2010), this is where the interest in Heidegger's philosophy of technology lies, that is in seeking answers to the question: 'How does the world appear when disclosed through modern technology?'

The essence of modern technology prepares nature to stand at command and be able to deliver what is ordered from it (Heidegger 1977, 320). Following this, the world is captured in terms of a resource that must be describable in quantitative terms that make resources easier to count and control (Riis 2010, 126). The technologically disclosed world is potentially controllable object and so everything in this framework appears as something, as object rather than subject, Riis argues. What humans fail to notice is that in this disclosure of the world they are themselves treated as resources that produce and secure even further resources. It is not enough, therefore, to take technology at face value. What technologies actually encourage us to do is to reason in terms of ends, means and objects. According to Heidegger, this is calculative reasoning, or thinking.

Considering pressing needs (e.g., environmental risks) for changes in the way we think and act upon the world, we agree with the position that calculative thinking is not enough. What is significant nowadays is to consider how education can best respond to the responsibilities and requirements the current era generates. The latter are tied with notions such as responsibility for other humans, other species and the environment, and respect for otherness. As such, they are also intertwined with the aims and goals of science education, both directly and indirectly. Our insights from research in pedagogical mediated publics (i.e., where teaching and learning experiences are enhanced with social networking and digital technologies; Vivitsou and Viitanen 2015) indicate that a space opens up for learning that builds upon human relations in an ongoing, interpretive dialogue. In this technologically enhanced space young people connect with peers and construct knowledge also by developing an understanding of not only surrounding environments but of the objects and the activities found in these.

Following the inductive logic as discussed in Polkinghorne (1983, 108–9), through the analysis of the participants' talk we aim to discover the values that are held to be true among them, as members of the science educators' community. In accordance with the principles of induction, we do not aim to generalize across a population. What we aim, however, is to generalize across the phenomenon, which, in this study, is science educators' pedagogical thinking.

To gain insight into how the participants view and integrate digital technologies for learning we aim to discuss the metaphors that will emerge out of the analysis of their speech. According to Ricoeur (1976), a metaphor makes sense only through interpretation, or through a metaphorical twist, an extension of meaning. This allows our sense making to take place where a literal interpretation would be nonsensical. By digging into the study participants' metaphorical meaning, therefore, we aim to understand their experiences of the present moment as well as their expectations of the future (Ricoeur 1986, p. 214). Certainly human speech is mostly metaphorical, at times presenting more and at other times less innovative twists of meaning, or metaphors (Vivitsou et al. 2014). Pedagogical thinking is one such metaphor. Calculative thinking is another and inductive reasoning is a third one. In addition to their metaphorical meaning, these three have been used to denote directions in scientific thinking (e.g., pedagogical thinking has been a guiding principle concerning the discussion on developments in education in Finland; e.g., Kynäslahti et al. 2006) and in philosophy (e.g., calculative thinking has been tackled in the work of philosophers such as Heidegger 1977 and Arendt 1958). In order, therefore, to understand the directions science education is taking, it is important to look into science educators' speech, and trace and interpret prevalent metaphors.

Considering all these, we will seek answers to our evolving research question, 'What main metaphors emerge in science educators' speech discussing the use of digital technologies and social media to promote science-related concepts and literacies?'

3 The Study

3.1 Aims & Methods

To achieve the aims of this qualitative study we will discuss and analyze the content of data resulting from semi-structured interviews. In these, the Finnish (N = 1) and the Greek science teachers (N = 3) and the science education experts (N = 2) from Finland discuss their experiences of integrating digital and mobile technologies, and social networking environments in the teaching of science and science-related subjects. Through content analysis we aim to gain a deep insight into the study participants' views and experiences and to offer valid research outcomes.

3.2 The Context and the Participants

As we mentioned above, in this study data sets result from semi-structured interviews that cover 3 phases of collection. The first is held in May 2011 and involves two one-hour long discussions. One with a female Biology and Geography teacher from an upper secondary school in Helsinki, Finland and another with a male Computer Science teacher in Northwest Greece. Both respond to questions about the ways they integrate web-based and digital environments and tools into the pedagogical practices for learning. The interview reveals the need for the Greek teacher to circumvent limitations imposed by the lower secondary Computer science curriculum. In this way, the teacher deals with issues relating to obsolete content of learning and insufficient subject teaching time.

By running after-school, project-based activities and integrating seventh with eighth and ninth graders, this teacher and his colleagues work on a voluntary basis and, eventually, construct a 'parallel', flexible curriculum. Participation depends on student choices among themes that draw from human and natural sciences, is technology enhanced and uses digital environments for real time and asynchronous communication with students and teachers from different locations. Considering this, we decided to hold a second round of interviews in the following year (October 2012). In addition to the Computer Science teacher, the group of study participants is enlarged with his volunteer colleagues, a female Technology and a male Computer Science teacher. They were interviewed separately for approximately 20 min each and questions were more focused on whether technologies can fail the overall teaching plan; how this can be amended; and what this whole pedagogical scheme means for students and the learning process, as well as for the teachers' professional development.

The final round of data collection takes place in November 2013 and involves two female science education experts at the University of Helsinki, one postdoctoral and one doctoral researcher. The interview comes, in reality, as an informal discussion (Denzel and Lincoln 2005) between three colleagues, i.e., the Finnish

Table 1 Participants, year of interview, technologies and learning content

Study participants (N = 6)	Year of interview	Technologies and content of learning
SCIE Fi T1 (Science, Finland, Teacher 1)	2011	This female teacher uses social networking environments to teach Biology and Geography in an Upper Secondary School, Helsinki
SCIE Gr T1	2011, 2012	These male teachers work with two thematic student groups. One examines illusions and dangers hidden in social networking. Another looks into environmental hazards and possible solutions for sustainability (Lower secondary, Northwestern Greece)
SCIE Gr T2	2012	
SCIE Gr T3	2012	Like her colleagues, this female teacher uses networked spaces and digital and mobile technologies to teach Technology and to look into Human Relationships with a group of students. The students read books, discuss ideas and, in order to express their emotions and tell stories with posters and pantomime in project-based after school activities (Lower secondary, Northwestern Greece)
SCIE Fi Xp1	2013	These female Science education experts use technologies to study teaching and learning Physics and Chemistry with digital and mobile technologies in 2 Finnish primary schools
SCIE Fi Xp2		

researchers and the lead author where the former elaborate on the experience of design and research into learning Physics and Chemistry with digital and mobile technologies in two primary schools in Finland. Water, air and motion are the natural phenomena under investigation. Interview questions mainly ask about the ways science teachers teach science nowadays, how technology and networked spaces enhance student understanding of scientific concepts and in what ways the latter engage in digitally enhanced activities.

Overall, Table 1 below presents the study participants and the technologies they use to teach or study teaching and learning science and science-related subjects.

4 Findings

Out of the process of analyzing and reviewing categories two focused codes have emerged. In one science comes up *as way of thinking*. In another science is *method* that appears as digital technology used for the teaching of science or science-related concepts and literacies. As Fig. 1 below shows, there is a third code that is recurrent in the participants' talk. This brings forward *underlying conceptions* or meanings the participants attribute to science and related issues. These do not directly associate with science as way of thinking or as method. They express, however, values that seem to influence them.

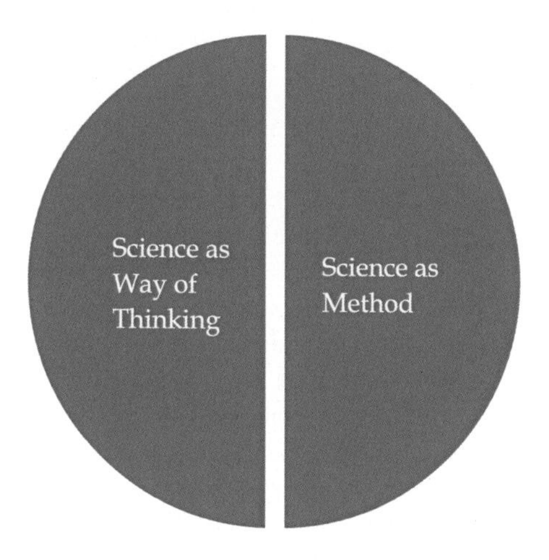

Fig. 1 The axial codes of the analysis

The axial codes (i.e., science as way of thinking and as method) are further divided into categories and subthemes. One such category is science as thinking in an abstract and analytical way. As Fig. 2 below shows, the notion of 'movement from the general to the specific' comes to delineate the notion of science as thinking in an analytical and abstract way. Although the former indicates an instance of deductive logic, argumentation further clarifies that such 'movement' requires looking at the phenomenon as whole (i.e., in order to get its 'qualitative sense') before experimentation in order to understand aspects of the phenomenon. 'Digital spaces for growth and development' is another main category that results from the axial code 'science as method'. In the sections that follow we will discuss findings in more detail.

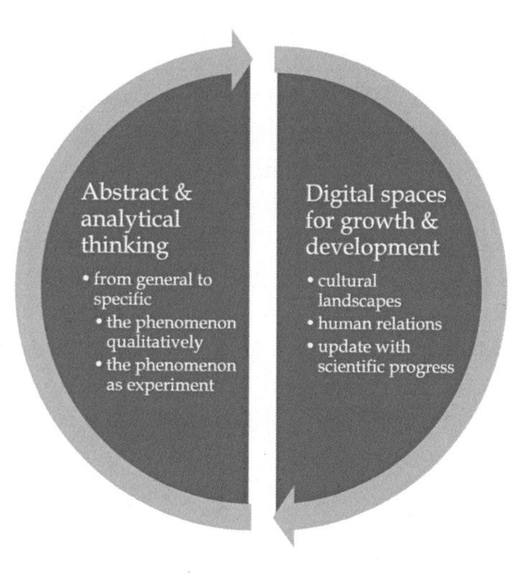

Fig. 2 Categories and subthemes

4.1 Science as Way of Thinking

In this study, science as way of thinking seems to be a theme in the talk of, almost exclusively, female participants. As categories emerge through the analysis of interviews, the Finnish teacher's statement that knowledge building is a movement from the whole picture to details expresses a pedagogical belief on the one hand. On the other, the scientific stance for synthetic and analytical thinking also comes up and relates with the need for teaching natural phenomena as a whole, '*not just objects*', as Xp1 explains. To further delineate the argument, Xp1 presents her view of learning science as starting from the qualitative sense of the phenomenon and ending up with the quantitative aspect. This is a perspective that Xp2 seems to support as well. In addition, knowledge construction in science requires experimentation, measurement, gathering data and deriving rules instead of using calculations and mathematical formulas.

Science is goal-oriented, takes hard work and creates tensions, as both Xp1 and Xp2 strongly argue. The Greek T3 seems to be thinking along similar lines. This relation shows up in T3's argument that it is hard for students to translate '*scientific knowledge into practical thinking about space and time*'.

At this point a number of opposing pairs appear. One lies in what Xp2 terms as '*divide in teaching science*' where the need for learning with experiments is juxtaposed to established practices of teaching and learning science by the book. In Xp1's view, the latter can explain students' pre-conceptions of what learning science means. According to Xp1 and Xp2's observations, high-achievers consider learning science with technologies as '*waste of time*'. As Xp2 elaborates, '*Learning the rules from the book and listening to the teacher was good for them. They found this* [i.e., learning with videos] *was more complicated*'.

Another oscillation (i.e., movement between extreme points like the one that occurs in a pendulum) starts with the statement that science is everywhere in daily life. As Gr T3 explains, the theory of relativity lies behind every single activation of a cell phone's global positioning system (GPS). And yet, despite this notion of science as 'everywhere', T3 negates the previous statement with the argument that it is too hard for students to grasp what '*vulgarization of science really means*'. It seems, therefore, that, although all around, science is not for all. The notion of science as hard work introduced by Xp2 and supported by Xp1 corroborate the elitist view posed by T3. As Xp2 argues, young people become less interested in science at about the age of 12 when study becomes more and more cognitively demanding. On the other hand, young female students seem unprepared to successfully deal with fear for labs and experiments. We trace a similar view about women in science in the literature. As Nussbaum (2000) argues, women 'lack opportunities for […] the cultivation of their imaginative and cognitive faculties' and 'fewer opportunities than men to live free from fear'. However, as the arguments link decrease in interest with age and gender, in addition to the requirement for hard work, science seems to become rather exclusive than inclusive.

The notion of 'hard work', extrapolates onto the field of humanities. As Xp1 explains, although we need '*imagination and creativity when doing some kind of lab works... we also need hard work while doing those. And it takes time to do this homework, i.e., calculating results'*. According to Xp2's view, '*It takes some effort to learn, repetition and practice in different sort of ways than in humanities'*. While another oscillation shows up between natural and human sciences, imagination and creativity seem not to be part of the hard work equation.

Ultimately, it seems that, according to our study participants' views, abstract thinking required in science equates with thinking analytical and calculative and is separate from thinking imaginatively and creatively. The latter seems to be left for the humanities where things are easier, since they, allegedly, have been divorced from science. This notion possibly explains the Greek T3's shift from the language of science to the language of emotions. As T3 claims, it is an attainable goal to aim for student gaining insights into literature rather than into the abstract history of Technology.

In the definition of her identity as scientist the Finnish teacher points out that, in addition to an adherence to facts, she also cares not to feed own values to students. This remark actually brings forward the paradigmatic opposition between the value free nature of sciences and the value laden-ess of humanities. It also brings forward that, eventually, the study participants seem to be neither value free nor theory free.

More importantly, these theories do not come out value free.

4.2 Science as Method

It is surprising how absent the notion of science per se is from Greek male teachers' speech. Both are Computer Science teachers in lower secondary education where the one-hour weekly time allocated has consequences not only on how the teachers view the content of learning but how they perceive their own position in the school as well. Teachers who specialize in Computers and Technology, like T1, T2 and T3 do, are frequently faced with work and travel to more than two schools in distant areas. In a sense they are 'road teachers'. It only comes natural that T2 categorizes the subject as second-rated in comparison to others more favored in the curriculum. One way to extend allocated time is by organizing after-school programs with student participation on voluntary basis. The programs involve inter-institutional, home and international partnerships and relations that are enhanced with web-based and digital technologies.

All study participants view social environments and digital technologies as spaces for student growth and development. The Greek teachers, mainly the male ones (T1 and 2), discuss these as spaces where opportunities open up for students to build and strengthen relationships with peers from other schools and countries. This approach attributes an interpretive property to technology. In this sense, student connecting with peers is a meaning making process, by becoming familiar with peers' everyday lives, what their views on specific subjects are etc., in web-enhanced discussions or through digital stories. In this way, students get to know peers and their cultural landscapes.

One after school theme concerns the environmental problems and risks the area is faced with. Another aims to enhance student understanding of illusions hidden in social networking sites. As both plans link with the teachers' background field, it is evident that opting for science-related topics is no accident at all. They combine this with a 'human relationship' oriented approach. As T2 argues, they (i.e., teachers) want to share views and opinions with students, in order to better understand how they think and feel about things. T1 argues that changes in, particularly, marginalized student attitudes toward schooling are visible. Developing a sense of responsibility, working in groups, learning and using technologies as pathways for communication seems to be the key of success in terms of student involvement. Although T3 does not overtly take this stance, her participation in the scheme points toward the same direction.

The Finnish experts also take a stance in favor of the view of 'digital technologies for learning that draws from cultural landscapes toward improved human relationships'. This becomes more obvious when Xp1 stresses out how important it is '*that learning science goes to their home and discuss with parents and siblings and friends'*. In this way students can learn not only through cables, says Xp1. Although this dimension is conflicting with the Greek view of building relations through cables the bottomline is more convergent than divergent: human relations do matter when learning about or with science.

Similarly, the Finnish teacher expresses a view of networked technologies for growth and development, and as ways to enrich the content of learning and keep up with rapid scientific progress and science-related discussions. Moreover, social sites, such as blogs, can accommodate provocative ideas. However, her background as teacher in Upper secondary seems to pose time limitations and imposes a priority for subject-centered material coverage. Building identities in Upper secondary seems to be entrusted to popular sites that create space for students to socialize and grow there (Boyd and Marwick 2011, Marwick and Boyd 2011).

During the discussion, the Finnish teacher's scientific background in Biology becomes visible while she stresses out how important it is to keep with 'facts'. In this sense, the Internet infrastructure blurs the limits between what is essential, or fact, and what is not, which allows the teacher's skepticism to show up. She is not alone there. The Finnish experts bring forward the challenges of pedagogical integration of digital technologies, a point where especially the Greek male teachers are in agreement. Although launched from different kick-off points (e.g., reliability of information, bandwidth, software compatibility and update etc.), invisible systems of technology, like Pandora's box (Vivitsou and Viitanen 2015), seem to package both hopes and fears into the experiences of the study participants.

Professional development, in terms of teaching science and using technologies to teach science, seems to be a dark hole, a nowhere-leading path, for almost all. For the female Greek T3, it is a good way to plan a state-of-the-art lesson but, otherwise, it is mostly for students to learn and grow in the process. The male teachers are not very opimistic either. Other than becoming familiar with people from other cultures and perspectives, they do not see much personal learning or development taking place. The Finnish teacher does discuss practices with colleagues in the natural sciences but teaching science per se is like recipe to her. One of the Finnish experts (Xp2) expresses the somewhat different view that teaching with technologies can enhance insights into learning by video recording and observing the process.

5 Conclusions

Overall, it seems that science as way of thinking is one main metaphor in the study participants' speech. Further than this, the science educators' thinking reflects instances of deductive logic. In this sense the study participants view science connected with regularities like those of universal laws inferred axiomatically (Olsen 2010; Polkinghorne 1983). The other main metaphor, however, of science as method presents digital environments as spaces where young people build knowledge and identities. In such learning spaces, getting connected with peers enhances development and growth through an appreciation of the world. This metaphor, therefore, frees the view of technology from the almost exclusively adopted, current means-to-ends, instrumental approach (Heidegger 1977; Riis 2010). As teachers use digital media to create opportunities for learning science and science-related concepts by building relationships and communicating with others, they also attribute an interpretive property to technology (Ihde 1990). This changes the scenery and turns away from the tool-oriented view.

Certainly, the approach to digital media as interpretive, relational spaces does not happen without implications for both the design and the content of learning. This kind of learning does not aim for the ideal certainty. On the contrary, it builds certainty gradually as young people gain insights through the multiple interpretations of peers' and own cultural landscapes. As this view reflects the logic of argumentation that supports a conclusion-making process, it indicates an instance of inductive reasoning. Anyway, as Polkinghorne (1983, p. 128) puts it (referring to Peirce's words), although 'there is no hope for eliminating all error' in order to reach the ideal, an agreement between what constitutes belief and reality can be reached through experience, in a process of negotiation with others of what the real is over time. The metaphors in their speech reveal that science educators are in such process of negotiation between what used to be 'known' and what is 'new'; or, as Kansanen et al. (2000, 65–66) would put it, a negotiation toward a new empirical reality.

In addition to instances of inductive and deductive logic, teachers' talk presents underlying conceptions of science. These are made visible where, among others, the decreasing interest in science is related to age and gender, where a pessimistic view of development as scientist–educator shows up and where imaginative and creative thinking is seen as separated from hard work and the natural sciences.

We believe that these conceptions, as expressed by the participants, disclose wider socially rooted values. As such, these phenomena generate new research challenges and call for further investigation.

References

Arendt, H.: The human condition. University of Chicago Press, Chicago (1958)
Boyd, d., Marwick, A.: Social privacy in networked publics: Teens' attitudes, practices, and strategies. Paper presented at the Oxford Internet Institute Decade in Internet Time Symposium, 22 Sept 2011
Chrysafiadi, K., Virvou, M.: Student modeling approaches: A literature review for the last decade. Expert Syst. Appl. **40**, 4715–4729 (2013)
Denzel, N., Lincoln, Y.: Handbook of Qualitative Research. Second Edition. California: Sage Publications, Thousand Oaks (2005)
Dreyfus, H.L.: What computers still can't do: a critique of artificial reason. MIT Press, Cambridge (1992)
Dreyfus, H.L.: Anonymity versus Commitment: the dangers of education on the internet. Educ. Philos. Theor. **34**(4), 369–378 (2002)
Heidegger, M.: The question concerning technology and other essays. Harper Torchbooks, New York (1977)
Ihde, D.: Technology and the lifeworld: from garden to earth. Indiana University Press, Bloomington and Indianopolis (1990)
Kansanen, P., Tirri, K., Meri, M., Krokfors, L., Husu, J., Jyrhämä, R.: Pedagogical thinking. Theoretical landscapes, practical challenges. Peter Lang, New York (2000)
Kynäslahti, H., Kansanen, P., Jyrhämä, R., Krokfors, L., Maaranen, K., Toom, A.: The multimode programme as a variation of research-based teacher education. Teach. Teach. Educ. **22**, 246–256 (2006)
Lavonen, J., Krzywacki, H.: Finnish secondary school teacher education in mathematics, physics and chemistry: assumptions behind the programme. The professional and pedagogical training of secondary school teachers: A critical and comparative approach, 23 p, 1st edn. University of Athens, Athens (2011)
Marwick, A., Boyd, d.: The drama! Teen conflict in networked publics. Paper presented at the Oxford Internet Institute Decade in Internet Time Symposium, 22 Sept 2011
Nussbaum, M.: Women's capabilities and social justice. J. Hum. Dev. **1**(2), 219–247 (2000)
Olsen, J.K.B.: Becoming through technology. In: Olsen, J.K.B., Selinger, E., Riis, S. (eds.) New Waves in Philosophy of Technology, pp. 40–61 (2010)
Polkinghorne, D.: Methodology for the human sciences: systems of inquiry. State University of New York, Albany (1983)
Riis, S.: The question concerning thinking. In: Olsen, J.K.B., Selinger, E., Riis, S. (eds.) New Waves in Philosophy of Technology, pp. 123–45 (2010)
Ricoeur, P.: Interpretation theory: discourse and the surplus of meaning. Texas Christian University Press, Fort Worth (1976)
Ricoeur, P.: From Text to Action: Essays in Hermeneutics II, trans. Kathleen Blamey and John B. Thompson, Northwestern University Press, Evanston (1986)
Schreiner, C., Sjøberg, S.: Sowing the Seeds of ROSE. Acta Didactica, 4, ILS, Oslo (2004)
Sjøberg, S., Schreiner, C.: Student perceptions of Science and Technology. Connect: UNESCO International Science, Technology & Environmental Education Newsletter **30**(1/2), 3–8 (2005)
Tsihrintzis, G., Virvou, M., Hatzilygeroudis, I.: Guest editorial. Int. J. Artif. Intell. Tools **21**(2), 1–4 (2012)
Uitto, A., Juuti, K., Lavonen, J., Byman, R., Meisalo, V.: Secondary school students' interests, attitudes and values concerning school science related environmental issues in Finland. Environ. Educ. Res. **17**(2), 167–186 (2011)

Vivitsou, M., Viitanen, K.: The pedagogies of the future: Through young people's eyes in storytelling experiences with the digital in Finland and Greece. In: Zlitni, S., Lienard, F. (Eds.) Electronic Communication: Political, Social and Educational uses. Bern: Peter Lang Europäischer Verlag der Wissenschaften, pp. 110–123 (2015)

Vivitsou, M., Tirri, K., Kynäslahti, H.: Social Media in Pedagogical context: a study on a Finnish and a Greek Teacher's Metaphors. Int. J. Online Pedagogy and Course Des. (2014)

Data Driven Monitoring of Energy Systems: Gaussian Process Kernel Machine for Fault Identification with Application to Boiling Water Reactors

Miltiadis Alamaniotis, Stylianos Chatzidakis and Lefteri H. Tsoukalas

Abstract Energy production units are large complex installations comprised of several smaller units, subsystems, and mechanical components, whose monitoring and control to secure safe operation are high demanding tasks. In particular, human operators are required to monitor a high volume of incoming data and must make critical decisions in very short time. Although they are explicitly trained in such situations, there are cases that may not be able to identify a gradually developing crucial faulty state. To that end, automated systems can be used for monitoring operational quantities and detecting potential faults in time. The field of machine learning offers a variety of tools that may be used as the ground for developing automated monitoring and control systems for energy systems. In the current chapter, we present an approach that adopts a single Gaussian process learning machine in monitoring high complex energy systems. The Gaussian process is a data-driven model assigned to monitor a set of operational parameters. The values of the operational parameters at a specific instance comprise the system's operational vector at that time instance. The operational vector consists the input to the individual Gaussian process machine whose task is to classify the operation of the system either as normal (or steady state) or match it to a faulty state. The presented approach is benchmarked on a set of experimentally data taken from the Fix-II test facility that is a representation of a Boiling Water Reactor. Obtained results exhibit the potential of Gaussian processes in monitoring highly complex systems such as nuclear reactors, by identifying with high accuracy the faults in system operation.

M. Alamaniotis (✉) · S. Chatzidakis · L.H. Tsoukalas
School of Nuclear Engineering, Purdue University, West Lafayette, IN 47907, USA
e-mail: malamani@ecn.purdue.edu

S. Chatzidakis
e-mail: schatzid@ecn.purdue.edu

L.H. Tsoukalas
e-mail: tsoukala@ecn.purdue.edu

G.A. Tsihrintzis et al. (eds.), *Intelligent Computing Systems*,
Studies in Computational Intelligence 627, DOI 10.1007/978-3-662-49179-9_8

Keywords Gaussian processes · Kernel learning machines · Data-driven monitoring · BWR

1 Introduction

Operating experience from energy installations (either conventional or nuclear) indicates that monitoring the health of their individual systems and components is rather an intensive task. System health among all depends on various non-well known physical mechanisms and may be affected by dynamically varying factors. Thus, there is a need for new automated technologies that perform system monitoring, detect abnormalities and subsequently identify operational faults (Alamaniotis et al. 2014b).

Monitoring and fault-prognostic tools are both essential parts of implementing condition-based maintenance strategies in complex energy systems (Jarrell et al. 2004). In particular, they allow the a priori scheduling of maintenance and/or replacement of components by considering both its current and anticipated future condition (Alamaniotis et al. 2014a). Condition-based maintenance could optimize operational costs in highly complex systems and critical infrastructures such as nuclear power plants by avoiding long term operational pauses and thus preventing severe accidents (Jardine et al. 2006; Jin et al. 2011; Alamaniotis et al. 2010).

To that end, there are three basic categories of monitoring and fault identification methods based on the type of modeled information used in prediction-making: Data-based, Stress-based, and Effects-based methods (Hines and Usinyn 2008); the data based methods use historic experimental or simulated data, stress-based use environmental stresses, while effects-based methods use measurable degradation parameters. Regarding data-based methods, the use of sufficient number of representative historical datasets is essential for training the models employed for system monitoring and fault identification. Historical datasets consist of measurements obtained from prior operation cycles of the same or similar systems; datasets are collected either from real system operation or controlled experiments or advanced simulators. However, there are cases in which collection of historical data is difficult or even impossible. For instance, there is no long term recorded operation for nuclear power plants components; in such a case models will only depend on knowledge derived from simulations (Agarwal et al. 2013).

Data driven methods developed for monitoring and fault identification in energy systems make use of artificial intelligence or statistical tools. To that end, a pure data driven method named *symbolic dynamic* is presented in Ray (2004), and its application to nuclear power plants in Jin et al. (2011). In addition, a combination of symbolic dynamic with wavelets is introduced in Rajagopalan and Ray (2006), while the integration of symbolic dynamic with ultrasonic signals for anomaly detection in Gupta et al. (2008). A hybrid neural network-wavelet approach is introduced in Wang and Vachtsevanos (2001), while a backpropagation method is proposed in Huang et al. (2007). Other reported data driven methods include the use of neurofuzzy approaches (Chinnam and Baruah 2003; Ikonomopoulos et al. 1993), particle filtering (Orchard and Vachtsevanos 2009), principal component analysis (Choi et al. 2005), empirical mode decomposition (Wu and Huang 2009), and fuzzy logic (Xu et al. 2007). Furthermore, a more complex method based on fuzzy logic and support vector regression is presented in (Alamaniotis and Agarwal 2014), an evolved ensemble of support vector regressors in Alamaniotis et al. (2012a), and probabilistic kernels in Alamaniotis et al. (2012b). Despite the plethora of proposed methods, there is room for new and more sophisticated methods that are applicable in all types of energy systems.

In this chapter, the application of a data driven approach based on machine learning for monitoring and identifying faults in complex energy systems is introduced (Uhrig and Tsoukalas 1999). In particular, the proposed approach employs Gaussian process (GP) kernel machines (Rasmussen and Williams 2006) for processing measured signals and subsequently indicating the operational system state, which may be characterized either as normal (steady state) or non-normal (transient). In case of non-normal state, the GP based approach also diagnoses the type of fault (Isermann 2006). The proposed approach is applied on set of data obtained from a highly complex system, and more specifically a boiling water reactor (BWR) (Tsoulfanidis 2013). Our approach exploits the ability of GP in learning and classifying sparse signals under high uncertainty.

The chapter is organized as follows. Section 2 gives a short introduction to Gaussian processes classification and Sect. 3 provides a description of the proposed approach. Section 4 discusses the application of the proposed approach in BWR monitoring and fault identification, while Sect. 5 summarizes and concludes the chapter.

2 Gaussian Process Kernel Machines

In this section Gaussian processes in the context of machine learning are presented. Particularly, the kernel based notion of Gaussian processes is introduced and its framework derived from simple linear regression is given.

Kernel modeled Gaussian Processes belong to parametric models expressed by the Bayes formula (Papoulis and Pillai 2002) and require the adoption of a prior probability distribution, which expresses prior information (Bishop 2006). In the machine learning realm kernel methods are a class of parametric models that can be formulated using the dual representation:

$$k(x_1, x_2) = f(x_1)^T f(x_2) \tag{1}$$

where f(x) can be any valid mathematical function, and $k(x, x')$ represents a kernel. So, every method that encompasses a simple kernel or a valid composition of kernels is called kernel machine (Rasmussen and Williams 2006; Bishop 2006). In addition, kernel machines formulated in the context of Bayesian statistics are known as probabilistic kernel machines. Probabilistic kernel machines can be used either for classification, where they assign an unknown vector to a class among two or more targets (i.e. categories), or for regression where they make predictions over continuous values of a parameter.

For the purposes of the current work the regression form of kernel based Gaussian processes is introduced. Furthermore, Gaussian processes as a Bayesian method provide a predictive distribution instead of single point estimation. In order to derive the predictive distribution we start from the simple linear regression:

$$y = \sum_{i=1}^{M} w_i \phi_i(\mathbf{x}) \tag{2}$$

where w_i are the regression coefficients, $\phi_i()$ are the basis functions and M is the number of regression coefficients. In the next step, we assign a normal distribution over the regression coefficients w_i having a mean value 0 and share the same variance denoted as σ_w^2. If we consolidate all coefficients in a vector **w** then the normal distribution is expressed by:

$$P(\mathbf{w}) = N(0, \sigma_w^2 \mathbf{I}) \tag{3}$$

with **I** representing the *M*x*M* identity matrix. The choice of distribution in (3) is convenient since in general we have no a priori information over the regression coefficients. By using (3) and the Bayes formula, Gaussian Process takes the following form:

$$GP \sim N(m(\mathbf{x}), \mathbf{C}(\mathbf{x}_1, \mathbf{x}_2)) = N(0, \mathbf{C}(\mathbf{x}_1, \mathbf{x}_2)) \tag{4}$$

where C(x, x) is a covariance function which can be taken equal to the Gram matrix **K**:

$$\mathbf{C}(\mathbf{x}, \mathbf{x}') = \mathbf{K} \tag{5}$$

where **K**'s entries are given by:

$$K_{ij} = k(x_i, x_j) = \sigma_w^2 \varphi^T(x_i) \varphi(x_j) \tag{6}$$

with $k()$ being a predetermined kernel function. Therefore the "trick" in Gaussian processes is to identify a suitable kernel function which expresses the modeller's belief over the distribution of the data. Furthermore, in the context of regression we assume that the targeted values **t** are comprised of two components, i.e. the real value y and the additive noise ε, whose variance is denoted as σ_n^2. The latter provides a normal prior distribution over the target vector **t**:

$$P(\mathbf{t}) = N(0, \mathbf{K} + \sigma_n^2 \mathbf{I}). \tag{7}$$

Assuming that there is an availability of N training points, i.e. known targets $\mathbf{t}_N$ for known inputs $\mathbf{x}_N$, then the predictive distribution over an unknown target t_{N+1} for the input x_{N+1} becomes:

$$P(t_{N+1}|\mathbf{t}_N) \propto \exp\left[-\frac{1}{2}[\mathbf{t}_N t_{N+1}]\mathbf{C}_{N+1}^{-1}\left[\begin{array}{c}\mathbf{t}_N^T\\ t_{N+1}\end{array}\right]\right]. \tag{8}$$

In the next step, the matrix $\mathbf{C}_{N+1}$ is decomposed into four blocks as shown below:

$$\mathbf{C}_{N+1} = \begin{bmatrix} [\mathbf{C}_N] & [\mathbf{k}] \\ [\mathbf{k}^T] & [k] \end{bmatrix} \tag{9}$$

where the matrix $\mathbf{C}_N$ is an NxN matrix, $\boldsymbol{k}$ is a vector of length N with entries computed by the kernel $k(\mathbf{x}_m,\mathbf{x}_{N+1})$ with $m = 1,\ldots,N$ and k is a scalar taken equal to kernel $k(\mathbf{x}_{N+1},\mathbf{x}_{N+1})$. Finally, by using (9) it can be found that the predictive distribution over $\mathbf{t}_{N+1}$ for an unknown input $\mathbf{x}_{N+1}$ is a normal distribution with mean and (co)variance respectively (Bishop 2006; Gibbs 1997):

$$m(\mathbf{x}_{N+1}) = \mathbf{k}^T \mathbf{C}_N^{-1} \mathbf{t}_N, \tag{10}$$

$$\sigma^2(\mathbf{x}_{N+1}) = k - \mathbf{k}^T \mathbf{C}_N^{-1} \mathbf{k}. \tag{11}$$

Therefore, it is apparent from Eqs. (10) and (11) that the predictive distribution strongly depends on the form of kernel function.

3 Methodology

The backbone of the proposed methodology is the use of a kernel modeled multivariate Gaussian process for processing information coming from BWR sensors. The block diagram of the methodology is depicted in Fig. 1, where we observe that there are two paths: the training and classification.

From a time point of view, the training process precedes classification. The goal of training is evaluating the GP model parameters so that the GP model possesses the capability of identifying faults. To that end, the training process mandates use of datasets that are representative of system faults. Training datasets, which consist of known inputs for known outputs, may be acquired either by simulation or experimentally.

At the end of the training process, the GP model should be able to associate unknown inputs to one of the systems states, i.e., fault or normal. In the current work, training datasets consist of: i) vectors that contain a number of measurements

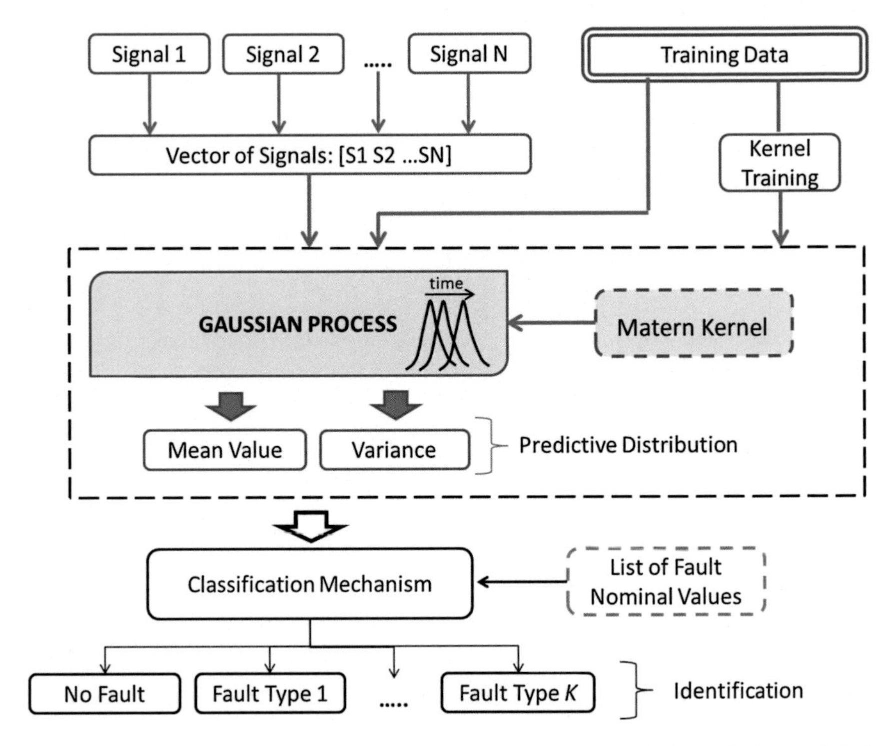

Fig. 1 Block diagram of the kernel modeled Gaussian process based monitoring and fault identification methodology

associated with system faults and normal operation, and ii) scalars that are the label (or nominal value), which is selected to stand for that respective fault or normal operation.

With regard to classification process, initially, the sensor measurements are acquired and put together into a single vector that is called the *vector of signals*. This vector is forwarded to the trained multivariate Gaussian process model. The Gaussian process is an independent kernel machine equipped with a kernel function. The use of kernel function allows implicit modeling of covariance among the system variables. Examples of kernel functions that may be employed are the *Gaussian*, *Matérn*, and *Neural Network* kernels, whose analytical formulas may be found in Rasmussen and Williams (2006). The output of the GP model is a nominal value that is forwarded to the classification mechanism. The latter gets the nominal value from the GP model and classifies it either as a normal operation or as one of the faulty ones. Classification is performed by simply computing the absolute distance of the GP output with the nominal values of each of the fault and normal operations (Ikonomopoulos et al. 2013). The steps of classification mechanism are depicted in Fig. 2.

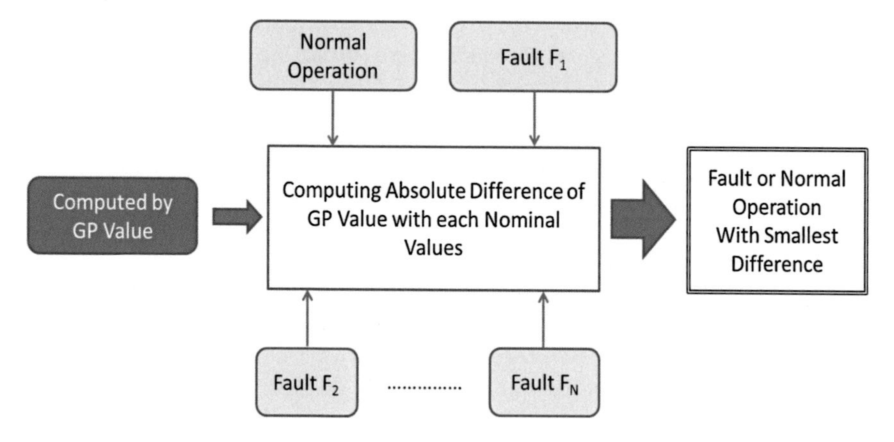

Fig. 2 Classification mechanism for indicating normal operation of faults

The computed absolute differences are compared and the smaller one is retained while the rest ones are rejected. Thus, it is this fault or normal operation that gives the smallest absolute difference that is promoted as the final output of the classification mechanism. The output of the classification mechanism coincides with the overall output of the presented GP based approach.

4 Application to Monitoring Complex Energy Systems: The Boiling Water Reactor (BWR) Case

4.1 Problem Statement

The GP based monitoring methodology is tested on experimentally data taken from the Fix-II test facility. That specific facility represents the operation of a Swedish BWR with external pumps (Nilsson 1982). It is a 1:777 scaled volume, and its pressure vessel has 36 rods and a spray condenser at the top to allow steady state operation (Alamaniotis et al. 2015).

In the current work, the GP model is set to monitor the signals and identify three types of faults:

1. an intermediate size split break in one of the four recirculation pumps (Label 1),
2. a large size split break in one of the four recirculation pumps (Label 2), and
3. a guillotine break in one of the four main recirculation lines (Label 3).

If none of the above three faults are identified then the operation is classified as normal (Label 4). The identification process is based on processing measurements obtained from a set of three sensors. In particular the three sensors measure:

(i) *the differential pressure at reactor bundle bottom* (Nilsson 1982),
(ii) *the temperature at reactor downcomer pipe (inside the wall)* (Nilsson 1982), and
(iii) *the pressure at the top of the steam dome* (Nilsson 1982).

The measured signals by sensor (i) during the experimentation time for faults (1)–(3) are depicted in Fig. 3. It should be noted that initiation of faults takes place at time 0 s; negative times are employed as a convention to express the normal operation of the reactor system. Details regarding the structure of the experimental facility and the placement of instrumentation can be found in Nilsson (1982), while a brief summary of the experimental event sequence is also given in Alamaniotis et al. (2015).

The experimental datasets at our disposal are divided into two subsets: the training subset, and the test subset. The training subset is utilized for model training while the test subset is used for testing the GP classification process. In the current work the training datasets included 60 measurements for each variable, thus providing a matrix of 60×3 dimensions. Therefore the GP model is trained by using three vectors of 60 entries long, where each entry is a vector of 3 points long. Similarly, the test data were comprised of a vector of thirty entry-long, and thus it is a matrix of dimension 40×3. In particular, the measured signals (among them those depicted in Fig. 3) are sampled uniformly and the training datasets are created. In order to create the test dataset 10 samples from each fault are randomly selected, and ten samples regarding normal operation are also randomly picked. Therefore, we create a test dataset of 40 samples that consist of 10 samples from fault (i), 10 samples from fault (ii), 10 samples from fault (iii), and 10 samples from normal (steady state) operation.

The Gaussian process model is equipped with the *Matérn kernel* whose analytical formula is as given below (Rasmussen and Williams 2006):

$$k(\mathbf{x}_1, \mathbf{x}_2) = \left(\frac{2^{1-\theta_1}}{\Gamma(\theta_1)}\right)\left[\frac{\sqrt{2\theta_1}|\mathbf{x}_1 - \mathbf{x}_2|}{\theta_2}\right]^{\theta_1} K_{\theta_1}\left(\frac{\sqrt{2\theta_1}|\mathbf{x}_1 - \mathbf{x}_2|}{\theta_2}\right) \tag{12}$$

that is modeled as a function of two parameters θ_1, θ_2 that should be non-negative, the modified Bessel function $K_{\theta 1}()$, and the gamma function $\Gamma()$. It should be noted that θ_1 is taken equal to 3/2 (Rasmussen and Williams 2006).

4.2 Testing Results

The numerical results of the classification are provided in Table 1. We observe that in the majority of the tested datapoints the GP model identifies correctly the fault in the BWR system. The testing for both the *Large break* and the *Intermediate break* is 100 % correct. However, it is observed a false identification of 10 % for the *Guillotine break*; the Guillotine break was misidentified as a Large break.

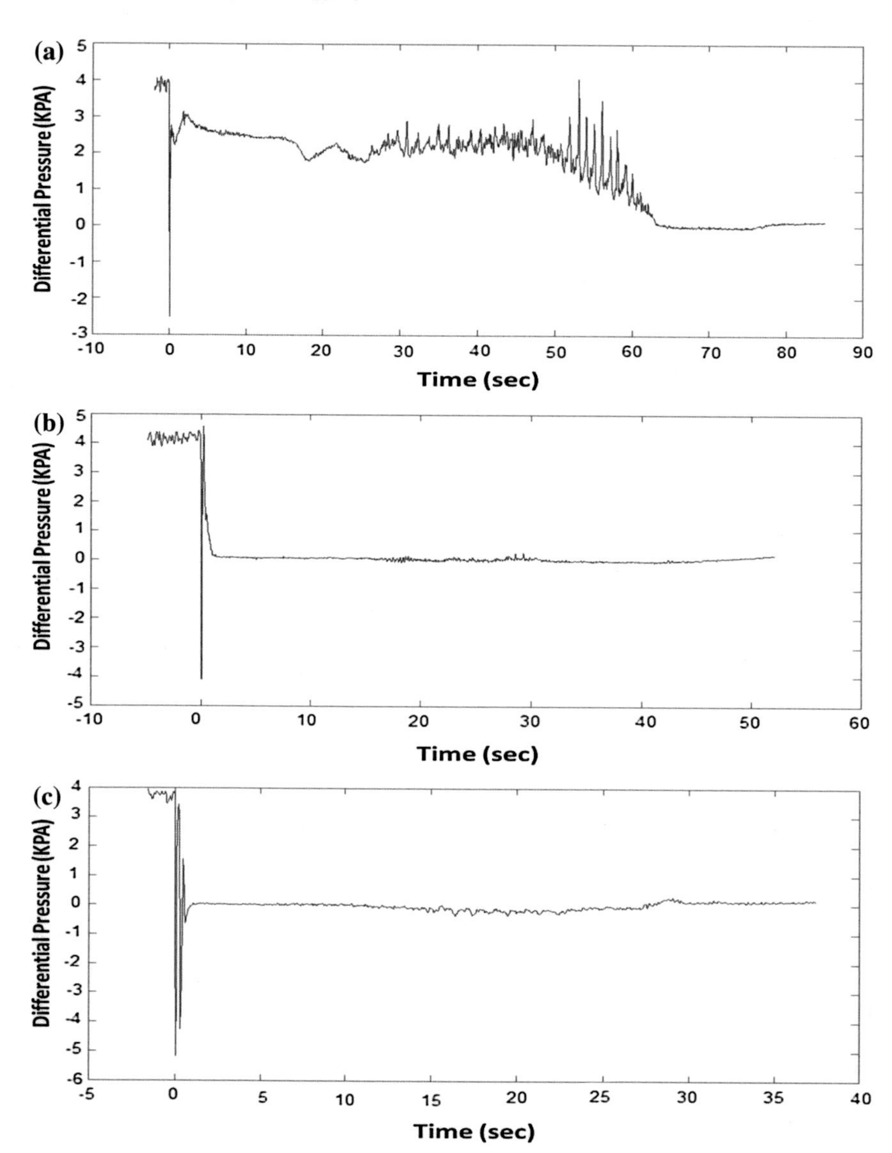

Fig. 3 Experimental measurements for the differential pressure at the reactor bundle bottom for **a** intermediate size break, **b** large size split break, and **c** guillotine break. Faults initiate at 0s, while negative time denotes normal operation

The reason for this misclassification is that the signal of the differential pressure at the reactor bundle bottom regarding the intermediate break is close to that of the large break and the system failed to correctly identify the fault for that specific datapoint. In addition, the normal operation datapoints are also correctly identified with 100 % rate. Overall, the Matérn kernel provides a very high identification

Table 1 Numerical results from BWR fault identification in 40 test datapoints

Data points	Real fault	# Correct identification	Correct identification (%)	False identification (%)
1–10	Normal operation	10/10	100	0
11–20	Intermediate break	10/10	100	0
21–30	Large break	10/10	100	0
31–40	Guillotine break	9/10	90	10 (identified as large break)

performance. Of course it should be noted that one of the reasons for such high correct classification rate is the appropriate selection of sensor signals. The selected signals have very low correlation rate. So we conclude that appropriate selection of signals boosts the GP based approach performance, no matter the small size training datasets.

5 Conclusions

A kernel based method for monitoring and fault identification in Boiling Water Reactors has been presented. A probabilistic kernel machine, particularly a Gaussian process equipped with a Matérn kernel, has been initially used for classifying incoming input signals into known faults or normal operations. In particular, a vector of signals is formed and forwarded to the GP model that gives a nominal value as an output. In the next step a classification mechanism uses the GP output to associate it with a specific fault or normal operation.

Testing the methodology on sensor signals from a boiling water reactor, and more particularly on experimentally data taken from the Fix-II test facility, exhibited a high accurate identification rate in the majority of the tested cases. Future work will focus on two directions: testing the GP model in larger datasets from BWR, and testing performance of other than Matérn kernel functions.

References

Agarwal, V., Lybeck, N.J., Pham, B.T., Rusaw, R., Bickford, R.: Online monitoring of plant assets in the nuclear industry. In: Annual Conference of the PHM Society, New Orleans, pp. 1–7, Oct 2013

Alamaniotis, M., Agarwal, V.: Fuzzy integration of support vector regressor models for anticipatory control of complex energy systems. Int. J. Monit. Surveill. Technol. Res. **2**(2), 26–40 (2014)

Alamaniotis, M., Ikonomopoulos, A., Tsoukalas, L.H.: Gaussian processes for failure prediction of slow degradation components in nuclear power plants. In: Proceedings of the European Safety and Reliability Conference 2010 (ESREL 2010), Rhodes, Greece, pp. 2096–2102, Sept 2010

Alamaniotis, M., Ikonomopoulos, A., Tsoukalas, L.H.: Optimal Assembly of support vector regressors with application to system monitoring. Int. J. Artif. Intell. Tools **27**(6), 1250034 (2012a)

Alamaniotis, M., Ikonomopoulos, A., Tsoukalas, L.H.: Probabilistic kernel approach to online monitoring of nuclear power plants. Nucl. Technol. **177**(1), 132–144 (2012)

Alamaniotis, M., Agarwal, V., Jevremovic, T.: Anticipatory monitoring and control of complex energy systems using a fuzzy based fusion of support vector regressors. In: Proceedings of the 5th International Conference on Information, Intelligence, Systems and Applications, Chania, Greece, pp.1–5, July 2014

Alamaniotis, M., Grelle, A., Tsoukalas, L.H.: Regression to fuzziness method for estimation of remaining useful life in power plant components. Mech. Syst. Signal Process. **48**(1), 188–198 (2014)

Alamaniotis, M., Jin. X., Ray, A.: On-line condition monitoring of boiling water reactors using symbolic dynamic analysis. In: 9th International Topical Meeting on Nuclear Plant Instrumentation, Control, and Human Machine Interface Technologies (NPIC&HMIT 2015). American Nuclear Society, Charlotte, NC, USA (2015)

Bishop, C.: Pattern Recognition and Machine Learning. Springer, New York (2006)

Chinnam, R.B., Baruah, P.: A neuro-fuzzy approach for estimating mean residual life in condition-based maintenance systems. Int. J. Mater. Prod. Technol. **20**(1–3), 166–179 (2003)

Choi, S.W., Lee, C., Lee, J.M., Park, J.H., Lee, I.B.: Fault detection and identification of nonlinear processes based on kernel PCA. Chemometr. Intell. Lab. Syst. **75**(1), 55–67 (2005)

Gibbs, M.N.: Bayesian Gaussian processes for regression and classification. Ph.D. thesis, Department of Physics, University of Cambridge (1997)

Gupta, S., Singh, D.S., Ray, A.: Statistical pattern analysis of ultrasonic signals for fatigue damage detection in mechanical structures. NDT and E Int. **41**(7), 491–500 (2008)

Hines, J.W., Usinyn, A.: Current computational trends in equipment prognostics. Int. J. Comput. Intell. Syst. **1**(1), 94–102 (2008)

Huang, R., Xi, L., Li, X., Liu, R., Qiu, H., Lee, J.: Residual life predictions for ball bearings based on self-organizing map and back propagation neural network methods. Mech. Syst. Signal Process. **21**, 193–207 (2007)

Ikonomopoulos, A., Tsoukalas, L.H., Uhrig, R.E.: Integration of neural networks with fuzzy reasoning for measuring operational parameters in a nuclear reactor. Nucl. Technol. **104**, 1–12 (1993)

Ikonomopoulos, A., Alamaniotis, M., Chatzidakis, S., Tsoukalas, L.H.: Gaussian processes for state identification in pressurized water reactors. Nucl. Technol. **182**(1), 1–12 (2013)

Isermann, R.: Fault-diagnosis Systems: An Introduction from Fault Detection to Fault Tolerance. Springer Science & Business Media, Berlin (2006)

Jardine, A.K., Lin, D., Banjevic, D.: A review on machinery diagnostics and prognostics implementing condition-based maintenance. Mech. Syst. Signal process. **20**(7), 1483–1510 (2006)

Jarrell, D.B., Sisk, D.R., Bond, L.J.: Prognostics and condition-based maintenance: a new approach to precursive metrics. Nucl. Technol. **145**, 275–286 (2004)

Jin, X., Guo, Y., Sarkar, S., Ray, A., Edwards, R.M.: Anomaly detection in nuclear power plants via symbolic dynamic filtering. IEEE Trans. Nucl. Sci. **58**(1), 277–288 (2011)

Nilsson, L.: FIX-II LOCA Blowdown and Pump Trip Heat Transfer Experiments, Experimental Results from Pump Trip Experiments of Test Groups Nos. 1, 2, 8 (STUDSVIK/NR-83/324) (1982)

Orchard, M.E., Vachtsevanos, G.J.: A particle-filtering approach for on-line fault diagnosis and failure prognosis. Trans. Inst. Measur. Control **31**(4), 221–246 (2009)

Papoulis, A., Pillai, S.U.: Probability, Random Variables, and Stochastic Processes. Mc-Graw Hill, New York (2002)

Rajagopalan, V., Ray, A.: Symbolic time series analysis via wavelet-based partitioning. Sig. Process. **86**(11), 3309–3320 (2006)

Rasmussen, C.E., Williams, C.K.I.: Gaussian processes for machine learning. MIT Press, Boston (2006)

Ray, A.: Symbolic dynamic analysis of complex systems for anomaly detection. Sig. Process. **84** (7), 1115–1130 (2004)

Tsoulfanidis, N.: The Nuclear Fuel Cycle. American Nuclear Sosiety, New York (2013)

Uhrig, R.E., Tsoukalas, L.H.: Soft computing technologies in nuclear engineering applications. Prog. Nucl. Energy **34**(1), 13–75 (1999)

Wang, P., Vachtsevanos, G.: Fault prognostics using dynamic wavelet neural networks. Artif. Intell. Eng. Des. Manuf. **15**, 349–365 (2001)

Wu, Z., Huang, N.E.: Ensemble empirical mode decomposition: a noise-assisted data analysis method. Adv. Adapt. Data Anal. **1**(01), 1–41 (2009)

Xu, L., Chow, M.Y., Taylor, L.S.: Power distribution fault cause identification with imbalanced data using the data mining-based fuzzy classification e-algorithm. IEEE Trans. Power Syst. **22**(1), 164–171 (2007)

A Framework to Assess the Behavior and Performance of a City Towards Energy Optimization

Stella Androulaki, Haris Doukas, Evangelos Spiliotis, Ilias Papastamatiou and John Psarras

Abstract A Smart City Energy Assessment Framework (SCEAF) is introduced to evaluate the performance and behavior of a city towards energy optimization, taking into consideration multiple characteristics. The SCEAF aims to provide to city authorities a systematic and independent evaluation means of the actions taken towards energy efficiency in parallel with the transition to become a "Smart City". The framework consists of indicators that are structured on three major assessment axes (1) Political Field of Action, (2) Energy & Environmental Profile, (3) Related Infrastructures-Energy & ICT. The framework can be designed generally for the whole activities spectrum of a city, but it can also be customized per sector, providing more focused information.

Keywords Local authorities · Smart cities · Energy optimization · Energy assessment framework · Multidisciplinary data sources

1 Introduction

Energy is an essential component of life in cities, since it supports the whole spectrum of economic activities and provides high living standards to residents. Cities in the EU are essential actors for the fruition of the EU's short-, interim and long-term energy and climate objectives and the transformation of the EU into a low-carbon economy on a 2050 horizon (Energy Cities 2014). It is estimated that 80 % of the European population will live in urban areas by 2020 (European Commission 2010a) while urban areas are responsible for 80 % of energy consumption and CO_2

S. Androulaki · H. Doukas (✉) · I. Papastamatiou · J. Psarras
Decision Support Systems Laboratory, School of Electrical & Computer Engineering,
National Technical University of Athens, Athens, Greece
e-mail: h_doukas@epu.ntua.gr

E. Spiliotis
Forecasting and Strategy Unit, School of Electrical & Computer Engineering,
National Technical University of Athens, Athens, Greece

G.A. Tsihrintzis et al. (eds.), *Intelligent Computing Systems*,
Studies in Computational Intelligence 627, DOI 10.1007/978-3-662-49179-9_9

emissions (Joint Research Centre 2013). Thus cities hold an important but untapped potential for the improvement of energy efficiency. As a result, there is an urgent need for developing strategies capable of reducing energy consumption, exploiting green technologies and reducing CO_2 emissions.

Although local authorities in the 28 EU Member States have a different degree of legal powers and responsibilities in climate and energy policy, they greatly determine the relations between energy consumers, regulators, advisors, energy producers and buyers, enabling them to actively contribute to energy efficiency at the city level.[1] Local communities demonstrate their willingness to implement sound local sustainable energy policies, especially through their participation in initiatives such as the Covenant of Mayors (ERENET 2011).

Cities tend to become "Smarter", with emphasis given on improving conveniences and welfare of residents. One of the major challenges, within this scope, is the increasing energy demand patterns of cities. To assure the transition to "Smart Cities" in parallel with the fulfillment of EU targets for energy efficiency and performance, cities should systematically monitor their activities, both per activity sector (i.e. residential sector, municipal buildings sector, industrial sector, transportation, waste management) and for the city as a whole.

In this scope, Cities are gradually directed to the use of tools and methodologies to monitor and support their actions towards energy optimization and sustainability. Such tools commonly concern long-term actions and strategies adopted by the City.

Nevertheless, a wide variety of multidisciplinary data available at the city level could be additionally exploited to provide support to city authorities to enhance sustainability and energy optimization at the city level, regarding short-term energy action plans. Such data concern weather forecasting data, social data, energy prices, energy profiles and energy production data.

Figure 1 depicts the holistic approach of a "Smart City" with emphasis on the need to promote a sustainable-energy—oriented "Smart City" concept.

The building sector, responsible for about 40 % of the EU's total final energy consumption and CO_2 emissions, provides many cost-efficient opportunities, while contributing to the welfare of EU citizens.[2] The EU has emphasized on the importance of energy performance of public buildings by promoting the directive 2010/31/EU that sets public buildings in priority in achieving nearly-zero energy consumption goals by December 2018 (European Commission 2010b).

Within this scope, the current book chapter introduces a Smart City Energy Assessment Framework (SCEAF) to evaluate the performance and behavior of a city, taking into consideration multiple characteristics. The SCEAF aims to provide to city authorities a systematic and independent evaluation means of the actions taken towards energy efficiency in parallel with the transition to become a "Smart

[1]Covenant Of Mayors commitment. Available at: www.eumayors.eu/IMG/pdf/covenantofmayors_text_en.pdf.

[2]GENERATION: Green Energy Auditing for a low carbon Economy. White Paper, Energy Efficiency in Public Buildings Recommendations for Policy Makers.

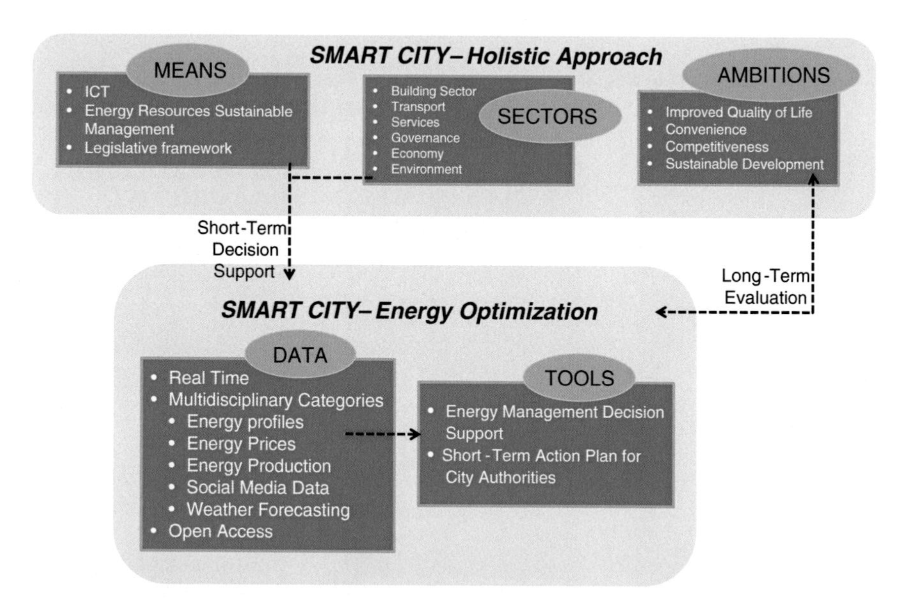

Fig. 1 "Smart-city" holistic and energy optimization approach

City". The added value of the prescribed SCEAF is that it is an assessment tool that clearly indicates underperforming sectors, providing to authorities a clear overview of the city performance per sector in order to be able to lead targeted energy action plans. The framework consists of indicators that are structured on three major assessment axes (1) Political Field of Action, (2) Energy & Environmental Profile, (3) Related Infrastructures—Energy & ICT.

This is done given the importance of acquiring a complete view of the City's behavior beyond pure energy performance measures, considering also its motivation in becoming "Smarter" with emphasis on energy efficiency. Each axis is further subdivided into specific pillars, and each pillar is described by one or more indicators. Figure 2 depicts the pillars per axis within the SCEAF structure. The additional axis 'General Characteristics' is incorporated to provide general information of the City, such as population and city area.

The set of indicators per pillar may differ between the *Whole City Level SCEAF* and the customized on the *Municipal Building Level SCEAF*. Through the SCEAF, the ex-ante and ex-post status of a Smart City, in relation to Energy Optimization issues can be assessed, in a coherent, transparent and integrated way, compared with the OPTIMUS city, which is the city that achieves the best performance in all proposed indicators.

The rest of the paper is structured as follows: Sect. 2 presents the relevant Policy Context while Sect. 3 provides a brief state-of-the-art survey, enumerating initiatives relevant to the ideas of optimizing the energy use in cities, and Sect. 4 analyzes the SCEAF methodology in terms of Axes, Pillars, and Indicators, as well as the aggregation method incorporated. Finally, Sect. 5 presents the Municipal

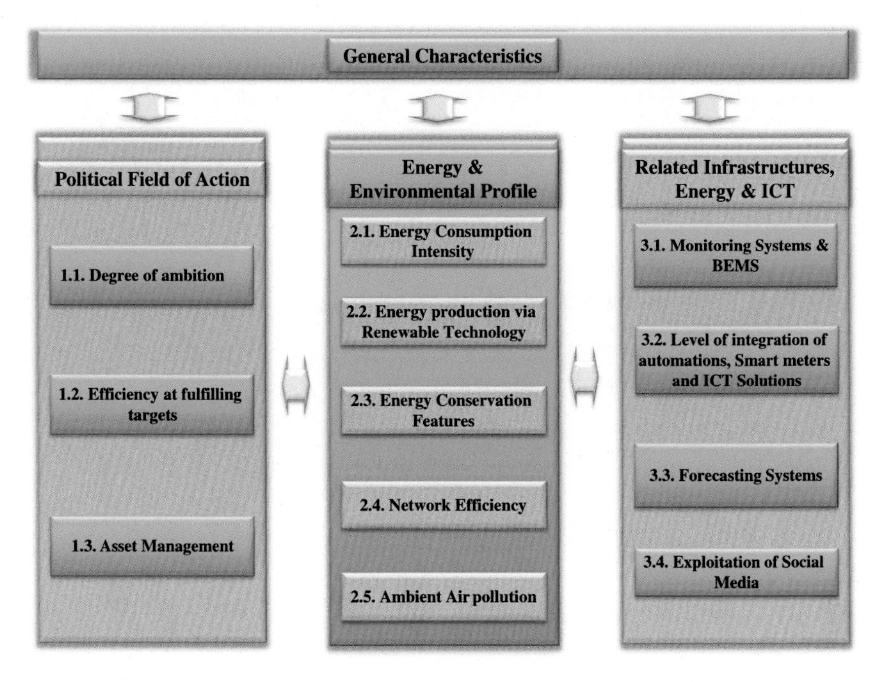

Fig. 2 SCEAF pillars and axes—whole city level and municipal building level

Building customized version of SCEAF, while Sect. 6 provides some conclusions with emphasis on the added value that the SCEAF gives to city authorities.

2 Policy Context

The EU tries to promote a shared European vision of sustainable urban development. European cities should be characterized by social progress and developed environmental consciousness, as well as economic growth based on a holistic integrated approach in which all aspects of sustainability are taken into account (European Commission 2011a). A number of European initiatives have been launched related to these ideas, as presented in the following paragraphs.

Cities, as consumers, demonstrate their willingness to implement sound local sustainable energy policies, especially through their participation in the Covenant of Mayors (CoM)[3] initiative. The CoM is a European initiative addressed to local and regional authorities, focused on the promotion of energy efficiency and renewable energy sources on their territories. By their commitment, Covenant signatories aim to meet the European Union 20 % CO_2 reduction objective by 2020.

[3]http://www.eumayors.eu/index_en.html.

The Energy-efficient Buildings Public Private Partnership in research (EeB PPP) (European Commission 2010c), launched under the European Economic Recovery Plan, represents the first step set by the industry to create more efficient districts and cities while improving the quality of life of European citizens. In addition, the Directive 2010/31/EU (EPBD recast) has indicated that the public sector has an important role in the field of energy performance of buildings, as new buildings occupied and owned by public authorities should be nearly zero-energy buildings, after 31 December 2018.

A new European Innovation Partnership (EIP) on Smart Cities and Communities (SCC) (European Commission 2012) was launched in July 2012 to promote the deployment of smart city solutions in Europe, focusing on the intersections of ICT, energy and transport. Cities themselves have also taken a pro-active role and launched the Green Digital Charter in 2009. The idea is that the cities signing up to the Charter commit to reduce the carbon footprint of their ICT and roll-out ICT solutions which lead to more energy efficiency in areas such as buildings, transport and energy.

The Smart Cities Stakeholder Platform[4] aims to support EU to achieve the target of 80 % reduction of GHG emissions by 2050 by promoting the development and market deployment of energy efficiency and low-carbon technology applications addressed to the urban environment.

The European Energy Research Alliance (EERA) Joint Programme on Smart Cities[5] was launched in September 2010. The main scope of the programme is to promote the development of tools and methods to enable a more sophisticated design, planning and operation of the energy system of a city in the near future.

Finally, two of the main pillars of Europe 2020 strategy are the Digital Agenda for Europe and the resource efficient Europe:

- Digital Agenda for Europe—DAE (European Commission 2010d): The Digital Agenda for Europe provides a policy framework aiming at delivering economic and social benefits from a digital single market based on internet and interoperable applications (Petrović et al. 2014). In this context, the EU promotes the use of ICTs in cities to enhance sustainability; buildings to become energy efficient; energy grids to transform into smart; climate change management. Cooperation between the public authorities and ICT sector is essential.
- Resource-efficient Europe (European Commission 2011b): The initiative for a resource-efficient Europe under the Europe 2020 strategy supports the shift towards a resource-efficient, low-carbon economy to achieve sustainable growth.

Based on the above, it is clear that EU places the issue of optimizing the energy-use in cities in the “heart” of its agenda. It is also treated as a necessary part of the research on Smart Cities. To this direction, the most suitable and available pilots for the first implementation of research are the municipal buildings.

---

[4]http://eu-smartcities.eu/.

[5]http://www.eera-set.eu/index.php?index=13.

3 Current Relevant Initiatives

Several initiatives have been identified in the international research scene, relevant to the ideas of optimizing the energy use in cities.

The technological approach of the SEMANCO initiative (Semantic Tools for Carbon Reduction in Urban Planning) is worth mentioning. The project is based on the integration of energy related open data, structured according to standards, semantically modelled and interoperable with a set of tools for visualizing, simulating and analysing the multiple relationships between the factors determining CO_2 production. The tools and methods in SEMANCO enable structuring energy related data, held in distributed sources and diverse formats, using data mining techniques (SEMANCO 2012).

The CitInES initiative (Design of a decision support tool for sustainable, reliable and cost-effective energy strategies in cities and industrial complexes) aims to design innovative energy system modelling and optimization algorithms to allow end-users to optimize their energy strategy using data sources on local energy generation, storage, transport, distribution and demand, including demand-side management and functionalities enabled by smart grid technologies. CitInES integrates information about local renewable energies, smart grid integration and demand-side management, as well as fuel price uncertainties (Page et al. 2013).

The ICT 4 E2B FORUM (European stakeholders forum crossing value and innovation chains to explore needs, challenges and opportunities in further research and integration of ICT systems for Energy Efficiency in Buildings) aims to the creation of a strategic research roadmap for ICT supported energy efficiency in construction, by bringing together all relevant stakeholders involved in ICT systems and solutions for energy efficiency in buildings, at identifying and reviewing the needs in terms of research and systems integration as well as at accelerating implementation and take-up. The roadmap produced (ICT 4 E2B FORUM 2012) consists of the Vision, the Strategic Research Agenda and it also provides suggestions and implementation activities.

In the framework of IREEN (ICT Roadmap for Energy Efficient Neighbourhoods) initiative the ways that ICT for energy efficiency and performance can be extended beyond individual homes and buildings to the wider context of neighbourhoods and communities are examined (IREEN 2013).

NiCE (2012) (Networking intelligent Cities for Energy Efficiency) promotes the implementation of commitments to the Green Digital Charter while in the framework of FINSENY (Future Internet for Smart Energy) initiative Future Internet technologies are exploited for the development of Smart Energy infrastructures, enabling new functionality while reducing costs (Fluhr and Williams 2011).

ENPROVE (Energy consumption prediction with building usage measurements for software-based decision support) provides an innovative service to model the energy consumption of structures supported by sensor-based data. The service makes use of novel ICT solutions to predict the performance of alternative energy-savings building scenarios in order to support relevant stakeholders in the procedure of identifying optimal investments for maximizing energy efficiency of an existing building (ENPROVE 2013).

The INTENSE[6] initiative (From Estonia till Croatia: Intelligent Energy Saving Measures for Municipal housing in Central and Eastern European countries) provides a holistic approach for planning energy optimized housing. The project comprises an analysis of legal preconditions, experience exchange on best practice examples, pilot planning activities at partner municipalities, and public awareness raising.

i-SCOPE (Interoperable Smart City services through an Open Platform for urban Ecosystems) delivers an open platform on top of which it develops, within different domains, a series of "smart city" services, based on interoperable 3D Urban Information Models (UIMs) (Patti et al. 2013).

In the framework of RESSOL-MEDBUILD[7] (RESearch Elevation on Integration of SOLar Technologies into MEDiterranean BUILDings) simulation models are developed for the optimization of building energy management and energy performance (RESSOL-MEDBUILD 2013).

The ENRIMA initiative (Energy Efficiency and Risk Management in Public Buildings)[8] specifies the development of a Decision Support System (DSS) engine for integrated management of energy-efficient sites, promoting adaptation of the DSS on buildings and/or spaces of public use (Cano et al. 2013).

TRACE (Tool for Rapid Assessment of City Energy) is a DSS tool designed and implemented by WorldBank within the Energy Sector Management Assistance Program (ESMAP) to assist municipal authorities in identifying and prioritizing Energy Efficiency actions. The methodology incorporated within the TRACE project initially evaluates the city performance focusing on energy consuming municipal sectors (passenger transport, municipal buildings, water and wastewater, public lighting, power and heat, and solid waste), and then it prioritizes energy efficiency improvement actions and interventions for the most energy intense sectors (Energy Sector Management Assistance Program (ESMAP) and WorldBank 2010).

Table 1 summarizes the main scope of each of the relevant identified initiatives. The table, based on the available literature (Page et al. 2013; ICT 4 E2B FORUM 2012; Fluhr and Williams 2011; ENPROVE 2013; Patti et al. 2013; RESSOL-MEDBUILD 2013; Cano et al. 2013; ESMAP (Energy Sector Management Assistance Program) and WorldBank 2010) indicates the main objectives that each initiative addresses. Most of the Initiatives presented deal with more technical issues, such as Smart Grids and ICT. The least covered scopes are those of DSS and Information Exchange, meaning weakness in participatory approaches and in combination of criteria to support actions.

---

[6]INTENSE ENERGY: Project leaflet. Available at: http://www.intense-energy.eu/uploads/tx_triedownloads/INTENSE_Leaflet_v3_final_web_11.pdf.

[7]RESSOL-MEDBUILD—RESearch Elevation on Integration of SOLar Technologies into MEDiterranean BUILDings project. Available at: http://www.ressol-medbuild.eu/.

[8]Energy Efficiency and Risk Management in Public Buildings (EnRiMa) project. Available at: http://www.enrima-project.eu/.

Table 1 Initiatives and Objectives

	DSS	ICT	Stakeholder participation	Smart grids	Information exchange	Policy support
SEMANCO		x		x		x
CitInES	x			x		x
ICT 4 E2B FORUM		x	x	x	x	x
IREEN		x	x	x	x	x
NiCE		x	x	x	x	
FINSENY		x		x		x
ENPROVE		x	x	x		
INTENSE			x	x	x	x
i-SCOPE		x		x		
RESSOL-MEDBUILD		x		x		
ENRIMA	x	x	x	x	x	
TRACE	x	x		x		x

Managing the energy use in municipal buildings requires a holistic approach to ensure optimum results in terms of energy consumption and CO_2 emissions reduction (Center for Climate and Energy Solution 2009).

Optimising the energy use in city areas with different types of demand to enable local balancing, demand response services, variable tariffs and easy change of supplier is integral to mitigating the threats posed by energy scarcity and climate change to the European cities' development in the coming decades. Taking into consideration that a city authority has often control over a large number of buildings, equipment and facilities, a systematic approach is necessary, in order to ensure a coherent and efficient energy policy covering the entire energy consumptions over which the local authority exercises control.

4 Description of the Framework

The main aim of the SCEAF (Smart City Energy Assessment Framework) is to direct "Smart Cities" to energy optimization by highlighting the strengths, the vulnerabilities and the opportunities arising given the existing energy strategy, environmental policy, municipal facilities and related infrastructures of each city. The determination of the above is done through an evaluation of the city in terms of energy efficiency, CO_2 emission intensity and fund savings. The advantage of using such a methodological tool is that the progress of each city can be revealed by analyzing and assessing its status on a systematic basis. For example, the proposed framework can be applied in parallel with the application of any decision support tool or energy management strategy, in an ex-ante and ex-post basis, as well as independently. In this scope, ex-ante SCEAF will provide a baseline picture of the situation of the municipality while ex-post SCEAF will depict the situation of the municipality after the improvements actions applied. Thereby, all the actions a city authority applies towards energy optimization can be evaluated directly for its effects on environmental and energy issues by tracing the evolution of the score achieved in the respective fields.

The SCEAF consists of three main axes composed by a number of corresponding pillars. Each pillar includes a number of appropriately defined, exhaustive and non-redundant indicators which describe satisfactorily its content and are used as criteria for its evaluation. The evaluation of each pillar provides a total assessment for each axis, as well as for the city as a whole.

The 'Political field of action' axis evaluates the level of ambition and activity of the city, its efficiency at fulfilling the targets set, and its performance regarding its asset management. It assesses the awareness of the city regarding environmental and energy problems as well as its environmental consciousness and its degree of adoption of the EU directives. It also includes the use of basic financial schemes and structures in the direction of saving funds related to energy consumption. The second axis, 'Energy & Environmental Profile' evaluates the energy consumption intensity of the city, the penetration degree of renewable energy sources (RES), the efficiency of the city energy networks and its energy conservation features, as well as the city's emission intensity. The evaluation of these pillars is based on city-level data, including energy consumption of municipal buildings, households, commercial buildings, public transport, municipal fleet, water treatment units, recycling centers and other related facilities. The last axis, 'Related Infrastructure & ICT', evaluates the integration of ICT solutions, automations and smart meters in the city aiming to energy optimization. It also includes the exploitation of monitoring systems, Building Energy Management Systems (BEMS), weather and energy consumption forecasting systems, and social media.

While most of the criteria incorporated within this framework are quantitative (Yager 1988) and directly measurable (e.g. energy consumption per capita), indeed not all of them can be evaluated using a clear measuring scale. For example, it is not easy to assign a clear quantitative assessment method to criteria, such as the Exploitation of Monitoring Systems and Social Media. For such criteria, the proposed framework incorporates also a qualitative assessment scale, with the exploitation of linguistic variables (Kundu 1997). Thus, the criteria incorporated within the SCEAF are categorized between *numeric* (*N*) and *linguistic* (*L*), according to their measuring scale. The application of linguistic decision analysis in the development of the theory and methods in decision analysis is very beneficial because it introduces a more flexible framework, allowing analysts and decision makers to represent the information in a more direct informative and adequate way when impossible to express it precisely (Herrera and Herrera-Viedma 2000; Herrera et al. 2005).

Information derived from the 'General Characteristics' is used to normalize the indicators, so that all indicators from each measuring scale category (*N, L*) are normalized and expressed on a common basis. In this respect, indicators in each axis and pillar can be easily aggregated, with the use of an additive value system to reach conclusions regarding the performance per axis and/or pillar.

All criteria are of strict increasing preference. Additionally, for the sake of simplicity it is considered that no strong evidence of significance between the grade of each pillar and the individual values of the corresponding indicators exists. Therefore, all criteria contribute equally in the additive value system, so that the higher their values, the better the grade of the pillar. Besides, the existence of *Linguistic* indicators which values are defined quite approximately makes it impractical to seek accuracy by precisely determining the weights of criteria.

Based on the above, the *Smart City Energy Performance* (*SCEP*) can be evaluated as the weighted sum of the performance of the city on each of the three axes [*Political Field of Action* (*PFA*), *Environmental & Energy Profiles* (*EEP*) and *Related Infrastructures & ICT* (*I&I*)].

The weights should add to one and are determined by the Decision Maker involved, according to his preferences. Similarly, the city's performance on each of the three axes (*PFAP, EEPP, I&IP)* is a function of the corresponding pillars.

According to the described evaluation framework an *OPTIMUS City* is an ideal city that achieves the highest scores at all the three axes of evaluation and therefore at the aggregation evaluation function.

The *OPTIMUS City* can be used as a benchmark for every city to monitor its performance, focusing either on specific performance axes or on the whole Smart-City energy performance. Moreover, the proposed framework can be exploited to compare cities with one another. Consider the two hypothetical Cities X (*PFAPx* = 0.8, *EEPPx* = 0.3, *I&IIPx* = 0.4) and Y (*PFAPy* = 0.4, *EEPPy* = 0.8, *I&IPy* = 0.3). In both cases the whole Smart-City Energy Performance is *SCEPx* = *SCEPy* = 0.5. Both cities appear to score half of the OPTIMUS City, but City X demonstrates high scores in the *PFAP* axis, accompanied by not very satisfactory results in the other two axes, while City Y performs adequately regarding the energy profiles axis but it lags on the other two. Figure 3 depicts the performance of Cities X and Y in comparison also with the *OPTIMUS City.*

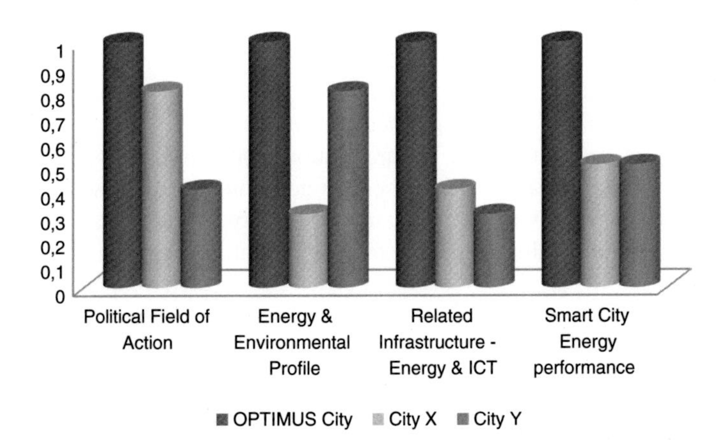

Fig. 3 Smart-city energy performance per axis, OPTIMUS city and cities *X* and *Y*

5 Municipal Building Level SCEAF

The building sector is responsible for about 40 % of the EU's total final energy consumption and CO_2 emissions and thus it provides many cost-efficient opportunities (see Footnote 2). The EU has emphasized on the importance of energy performance of public buildings by promoting the directive 2010/31/EU that sets public buildings in priority in achieving nearly-zero energy consumption goals by December 2018 (European Commission 2010b). In a similar direction, According to the EU Climate and Energy Package "20–20–20", energy savings will arise more directly if energy efficiency measures are applied on building level. Local authorities are the first to be activated by optimally managing loads of municipal buildings, acting that way as role models for citizens and helping transform cities into Smart energy ones.

Based on these facts, the customized on the *Municipal Building Level SCEAF (MB-SCEAF)*, designed to assist the evaluation of the municipal building sector, is outlined, with emphasis on the specific indicators per pillar and axis. An indication is provided on the measuring scale category per indicator as well as a description of each indicator, in terms of measuring scale.

As a result, important steps in energy efficiency of municipal buildings are made, as well as in energy monitoring in order to indicate smart solutions that will yield the desired improvements at the lowest possible cost.

MB-SCEAF consists from the axes of the *SCEAF*: *Political Field of Action, Environmental & Energy Profiles* and *Related Infrastructures & ICT*, but the pillars 'Network Efficiency' and 'Ambient Air Pollution' are not present. In addition, the pillars 'Monitoring Systems and BEMS' and 'Level of Integration of automations, Smart Meters and ICT solutions' have been merged into one pillar. A detailed structure of *MB-SCEAF* is presented in Table 2.

Indicators 1.1.1–1.1.3 evaluate the energy and environmental reduction targets of the municipal buildings set on a baseline year for 2020, which is the crucial year used by EU Climate and Energy Package "20–20–20" and the majority of word climate change organizations. The targets are expressed as a percentage of total energy consumption, emissions and penetration degree of RES respectively. The following three indicators (1.2.1–1.2.3) assess the efficiency of the city at fulfilling the above targets by calculating their coverage rate at the moment the framework is applied. The aim of pillar 1.2 is to ensure that no city will be highly graded thanks just to an environmental target issued, but also due to the efforts made in the direction of actually achieving it. Indicators 1.3.1 and 1.3.2 judge the management of funds devoted for energy needs. The first one normalizes the expenditures according to gross floor area of the municipal buildings, while the second assesses the switching ratio between energy providers based on price, consumption peaks and other data, a process which can become extremely profitable if organized

Table 2 Indicators of MB-SCEAF per pillar and axis

Key performance indicator		Type	Weather	Profile	Social media	Prices	Renewable production
(1) Political field of action		*Data input*					
1.1	CO_2 reduction target in municipal buildings till 2020: % of total emissions	N	✓				
1.2	Energy consumption reduction target in municipal buildings till 2020: % of total energy consumption (delivered energy)	N		✓			
1.3	Renewable energy sources in the final use target in municipal buildings till 2020: % of total energy consumption (delivered energy)	N		✓			✓
1.4	Medium term results for CO_2 reduction in municipal buildings: % of total goal (the goal refers to indicator 1.1)	N	✓				
1.5	Medium term results for energy consumption reduction in municipal buildings: % of total goal (the goal refers to indicator 1.2)	N		✓			
1.6	Medium term results for renewable energy sources in the final use in municipal buildings: % of total goal (the goal refers to indicator 1.3)	N		✓			✓
1.7	Funds devoted for renewable energy sources and energy efficiency: the funds to be given to energy efficiency and renewables' investments	N		✓			
(2) Environmental and energy profiles							
2.1	Energy consumption reduction in municipal buildings: % of energy saved compared to the benchmark year	N		✓			
2.2	Percentage reduction of fossil fuels: % reduction of fossil fuel energy consumption compared to the benchmark year	N		✓			
2.3	Average CO_2 emission factor: average emission factor of the building based on its energy mix	N		✓			
2.4	Renewable energy sources (RES) production intensity: % energy produced from RES per total energy consumption	N		✓			✓

(continued)

Table 2 (continued)

2.5	Ability of storing energy produced (thermal storage, electrical storage): % of total energy production	N		✓			
2.6	Cogenerating heat and power: % of total electricity consumption	N		✓			
2.7	Energy performance and envelope efficiency of the building: energy performance rate of the building based on its energy performance certificate (EPC)	L		✓			
(3) Related Infrastructures and ICT							
3.1	Monitoring systems, BEMS and BACS: evaluated based on the level of automation in the building according to prEN 15232:2006	L		✓			
3.2	Energy monitoring systems: % of energy consumption monitored with sub-meters compared to total energy consumption	N		✓			
3.3	Forecasting systems: existence of systems forecasting energy consumption, energy production and temperature	L			✓		
3.4	Level of switching energy providers (electricity and gas): flexibility of switching among energy providers based on price, consumption peaks, etc.	N		✓		✓	
3.5	Cost reduction for energy needs (gas, petroleum and electricity) in the municipal building: % reduction of energy cost compared to the benchmark energy bill records	N		✓		✓	

(continued)

Table 2 (continued)

3.6	Existence of social media (facebook, twitter): facebook page or twitter account for the municipal buildings	L			✓		
3.7	Building management action plans influenced by occupants'/inhabitants'/ citizens' preferences: average number of days per week, when social feedback was taken into consideration for the development of an action plan	L			✓		

correctly and an open energy market exists. Finally, indicator 1.3.3 examines the funds devoted by the municipality for RES and energy efficiency purposes by dividing them by the number of inhabitants for normalization reasons.

Within the second axis, energy consumption intensity is evaluated through three indicators: Indicator 2.1.1 measures the amount of energy saved per number of inhabitants, while indicators 2.1.2 and 2.1.3 detect the percentage of fossil fuels and electricity in the total energy mix of the buildings. For environmental reasons, high electricity and low fossil fuel share is desired and that is why the decrease of the first one is actually examined. Production intensity from RES is evaluated through indicator 2.2.1 having divided total energy production by the area covered by the photovoltaic panels. This indicates that in *MB-SCEAF* renewable energy sources are exclusively considered as photovoltaic because of the building based analysis. The energy conservation features of the building are measured based on its storing abilities (thermal through boilers and electrical through batteries and hydrogen), the amount of electricity produced through Cogeneration and the exploitation of weather conditions for optimizing energy performance. The last indicator is calibrated qualitatively, like the rest of the linguistic indicators, in seven-class climate (unacceptable, very bad, bad, neutral, good, very good and excellent).

The last axis consists of three linguistic criteria related to monitoring Systems and BEMS, weather and energy consumption forecasting systems and exploitation of social media. All these automations and ICT based solutions provide information about future energy needs of the municipal buildings, highlight potential peaks in energy consumption and help program energy production and schedule energy intensive procedures such as preheating, HVAC use, cooking etc. The inability to evaluate precisely the utility and efficiency of that kind of infrastructures and their uneven distribution in the various municipal buildings are the main reasons for choosing the qualitative assessment approach proposed.

6 Conclusions

Cities are characterized by a concentration of people, communities, activities, flows and impacts leading to severe sustainability challenges. Energy is an essential component of life in cities, as it supports the whole spectrum of their economic activities, and secures a certain level of quality of life to residents. Nowadays cities tend to become "Smarter", usually disregarding the issues of energy efficiency and sustainability.

Taking into consideration that a city authority has often control over a large number of buildings, equipment and facilities, a systematic approach is necessary, in order to ensure a coherent and efficient energy policy covering the entire energy consumptions over which the local authority exercises control. The current work provides City Authorities with an effective framework to assess the behavior and performance of a City, reflecting the clearly quantifiable energy related indicators, but also the related policy context performance and the integration of Smart infrastructure. Using appropriate indicators, the progress of a city in that direction can be revealed by analysing and evaluating its ex-ante and ex-post status across three axes: 'Political Field of Action', 'Energy and Environmental Profile' and 'Related Infrastructures and ICT'. The framework can be designed generally for the whole activities spectrum of a city, but it can also be customized per sector, providing more focused information. In this respect, the customized on the *Municipal Building Level SCEAF* (*MB-SCEAF*) *was also presented*, designed to assist the evaluation of the municipal building sector.

Based on the proposed framework, a computerized software can be developed and applied in a city and/or building level, in order to evaluate its usefulness in a "real life environment". An additional perspective for further research is to explore fusion methods and algorithms for merging multiple information, which is the case for many of the previously presented indicators.

Acknowledgment Part of the work presented is based on research contacted within the project "OPTIMising the energy USe in cities with smart decision support system (OPTIMUS)", which has received funding from the European Union Seventh Framework Programme (FP7/2007-2013) under grant agreement n° 608703. The content of the paper is the sole responsibility of its authors and does not necessarily reflect the views of the EC.

References

Cano, E.L., Javier, M.M., Ermolieva, T., Ermoliev, Y.: Energy efficiency and risk management in public buildings: strategic model for robust planning. Computational Management Science, forthcoming (2013)

Center for Climate and Energy Solution: Buildings overview, buildings and emissions, May 2009 (2009)

Energy Cities: Empowering local and regional authorities to deliver the EU climate and energy objectives (2014)

Energy Sector Management Assistance Program (ESMAP), WorldBank: Rapid assessment framework, an innovative decision support tool for evaluating energy efficiency opportunities in cities. Report No. 57685 (2010)

ENPROVE: Final Report (2013)

ERENET: Rural web energy learning network for action. In: The Network of Energy Sustainable Local Communities (2011)

ESMAP (Energy Sector Management Assistance Program), WorldBank: Rapid assessment framework, an innovative decision support tool for evaluating energy efficiency opportunities in cities. Report No. 57685 (2010)

European Commission: A digital agenda for Europe. COM(2010) 245 Final (2010d)

European Commission: A resource-efficient Europe—flagship initiative under the Europe 2020 Strategy. COM(2011) 21 (2011b)

European Commission: Cities of tomorrow—challenges, visions, ways forward. Brussels, Belgium (2011a)

European Commission: Directive on the energy performance of buildings (2010b)

European Commission: Energy-efficient buildings PPP: multi-annual roadmap and longer term strategy. Directorate-General for Research, Industrial Technologies, Prepared by the Ad-hoc Industrial Advisory Group, Brussels, Belgium (2010c). http://ec.europa.eu/research/industrial_technologies/pdf/ppp-energy-efficient-building-strategic-multiannual-roadmap-infoday_en.pdf

European Commission: Smart cities and communities—European innovation partnership. COM (2012) 4701 Final (2012)

European Commission: The European environment state and outlook, environment, health and quality of life (2010a)

Fluhr, J.W., Williams, F.: FINSENY: Future internet for smart energy. Unternehmen der Zukunft **2** (2011)

Herrera, F., Herrera-Viedma, E.: Linguistic decision analysis: steps for solving decision problems under linguistic information. Fuzzy Sets Syst. **115**, 67–82 (2000)

Herrera, F.L., Martınez, L., Sanchez, P.J.: Managing non-homogeneous information. Eur. J. Oper. Res. **166**, 115–132 (2005)

ICT 4 E2B FORUM: Final research roadmap (2012)

IREEN: Roadmap for European-scale innovation and take-up (2013)

Joint Research Centre: The covenant of mayors in figures—5-year assessment (2013). Available at: http://www.peer.eu/news-events/detail/print.html?tx_list_pi1[uid]=425

Kundu, S.: Min-transivity of fuzzy leftness relationship and its application to decision making. Fuzzy Sets Syst. **86**, 357 (1997)

Networking intelligent Cities for Energy Efficiency (NICE): Project Leaflet (2012). Available at: http://nws.eurocities.eu/MediaShell/media/NiCE%20Leaflet_11.2011.pdf

Page, J., Basciotti, D., Pol, O., Fidalgo, J.N., Couto, M., Aron, R., Chiche, A., Fournie, L.: A multi-energy modeling, simulation and optimization environment for urban energy infrastructure planning. In: Proceedings of BS2013, 13th Conference of International Building Performance Simulation Association, Chambery, France, 26–28 Aug 2013

Page, J., Basciotti, D., Pol, O., Fidalgo, J.N., Couto, M., Aron, R., Chiche, A., Fournie, L.: A multi-energy modeling, simulation and optimization environment for urban energy infrastructure planning. In: Proceedings of the 13th Conference of International Building Performance Simulation Association, Chambery, France, 26–28 Aug 2013

Patti, D., de Amicis, R., Prandi, F., D'Hondt, E., Rudolf, H., Elisei, P., Saghin, I.: iScope smart cities and citizens. In: Proceeding of the REAL-CORP 2013, Tagugnsband (2013)

Patti, D., de Amicis, R., Prandi, F., D'Hondt, E., Rudolf, H., Elisei, P., Saghin, I.: iScope smart cities and citizens. Proceedings of Conference on REAL-CORP 2013, Tagugnsband (2013)

Petrović, M., Bojković, N., Anić, I., Stamenković, M., Pejčić Tarle, S.: An ELECTRE-based decision aid tool for stepwise benchmarking: an application over EU digital agenda targets. Decis. Support Syst. **59**, 230–241 (2014)

RESSOL-MEDBUILD: Periodic report summary (2013). Available at: http://cordis.europa.eu/result/report/rcn/54817_en.html

SEMANCO: Project methodology report, 2012. Semantic Tools for Carbon Reduction in Urban Planning (2012)

Yager, R.R.: On ordered wighted averaging aggregation operators in ulticriteria decision making. IEEE Trans. Syst. Man Cybern. **18**, 183–190 (1988)

An Energy Management Platform for Smart Microgrids

Federico Delfino, Mansueto Rossi, Fabio Pampararo and Luca Barillari

Abstract This paper is relevant to the important issue of planning and management of a so-called smart microgrid, namely a group of interconnected loads and distributed energy resources (DER) with clearly defined electrical boundaries that acts as a single controllable entity with respect to the public grid. The Energy Management System (EMS) plays a crucial role in governing the power flows coming from the different generating sources inside such infrastructure and can resort to optimization procedure and algorithms in order to minimize the daily operating costs. The main features of the EMS platform suitably developed to control the University of Genoa test-bed facility called Smart Polygeneration Microgrid (SPM) are here illustrated and thoroughly discussed.

1 Introduction

The power delivery system, the concept of sustainable energy, and the use of innovative technologies for distributed generation are key issues of state-of-the-art research on smart grids and Microgrids (MGs). MG research fits very well with ongoing smart grid activities throughout the world and several challenges could be faced by means of the employment of pilot test facilities installed worldwide.

MG is an important technology to integrate distributed energy resources, including wind turbines, solar photovoltaic panels, and energy storage devices such as battery. A MG can connect and disconnect from the grid to enable it to operate in both grid-connected or island modes. MGs are receiving attention, due to the increasing need to integrate distributed generations and to ensure power quality and to provide secure energy to critical loads (Bracco et al. 2013, 2014).

F. Delfino (✉) · M. Rossi · F. Pampararo · L. Barillari
Department of Naval, Electrical, Electronic and Telecommunication Engineering, University of Genoa—Campus of Savona, Via A. Magliotto 2, 17100 Savona, Italy
e-mail: federico.delfino@unige.it

G.A. Tsihrintzis et al. (eds.), *Intelligent Computing Systems*,
Studies in Computational Intelligence 627, DOI 10.1007/978-3-662-49179-9_10

Energy Management System (EMS) of a MG is a comprehensive automated system primarily aimed at optimal resources scheduling: it is based on advanced IT technology and can optimize management of distributed power and energy storage devices within the microgrid.

In the specialized literature, many works are related to energy management system (EMS) for MGs. A centralized control system for a MG has been presented in (Tsikalakis and Hazizargyriou 2008). The controller is used to manage the operation of the MG during interconnected operation, i.e., the production of local generators and energy exchanges with the distribution network are optimized. Two market policies are assumed to offer options for the demand for controllable loads, and this demand-side bidding is incorporated into the centralized control system. In (Chakraborty et al. 2007), the energy management strategy of a renewable-based MG has been analyzed. A distributed power supply side model was proposed in (Marnay et al. 2008) where a comparison between the installation and running costs of distributed generation and supply from the main grid has been carried out.

In (Hajizadeh and Golkar 2007), an online power energy management for a hybrid fuel cell/battery distributed generation system is presented. The online architecture consists of three layers: the first one captures the possible operations modes, the second is based on a fuzzy controller for power splitting between batteries and fuel cells, and the last one regulates each subsystem. In (Teleke et al. 2010), a rule-based control strategy is designed for a battery energy storage system with photovoltaic arrays and a wind farm. The renewable sources can be dispatched hourly, based on forecasting of the solar and wind conditions. The rule-based controller determines the current reference for the converter that will charge/discharge the battery bank by using the state of charge (SOC) and the battery voltage. This system can deal with variability in the wind and solar generation.

The authors in (Westermann and John 2007) describe a combination of wide-area measurement and ripple control for Demand Side Management (DSM). The proposed control systems moderate the impact of increased renewable sources on adjacent transmission grids. In (Palma-Behnke et al. 2013), an energy management system based on a rolling horizon strategy for a renewable-based MG is proposed. The EMS provides online set points for each generation unit and signals for consumers based on a DSM mechanism. Moreover, the benefits of DSM are achieved by means of shifting demands to periods in more renewable resources are available.

In this paper, attention is focused on the operation of the University of Genoa Smart Polygeneration Microgrid (SPM). The SPM (Bracco et al. 2013, 2014) is the test-bed facility at the Savona Campus of the University of Genoa, aiming at producing and managing in an efficient way clean energy for the university loads and to operate as a test bed for research, testing and development of management strategies and devices. SPM basically consists of different sources of power generation, electrical storage systems and electric vehicle charging stations. All the components of the MG test-bed are equipped with interfaces, which are either compatible with the new smart grid protocol IEC 61850 or connected via appropriate gateways.

In this paper, also, the performance of Decentralized Energy Management System (DEMS) is analyzed. DEMS has been specifically developed by Siemens for monitoring, operation, control and management of virtual power plants and microgrids. DEMS has been installed in the SPM operation center (Control Room) and guarantees the SPM functionalities, monitoring and alarms management, through the connections with the RTUs (Remote Terminal Units) and local control panels of devices in the field.

The paper is organized as follows. In Sect. 2, the components of the SPM located in Savona Campus are described. The application and functionalities of DEMS are discussed in Sect. 3. In Sect. 4, the forecasted, monitored and scheduled results of DEMS are analyzed, while in Sect. 5 some conclusive remarks are drawn.

2 The Smart Polygeneration Microgrid Pilot Plant

The SPM Project begun in 2010 as a joint special project in the energy sector between the University of Genoa and the Italian Ministry of Education, University and Research (MIUR), which is the public body fully financing the initiative with 2,4 M€. During the years 2011 and 2012 the preliminary design, the final design and the working plan of the infrastructure have been developed, while in 2013 works started and were completed in February 2014 (Fig. 1). The SPM project is aimed at creating an R&D facility based on the use of both renewable and fossil sources to produce thermal and electrical energy in accordance to a distributed generation strategy.

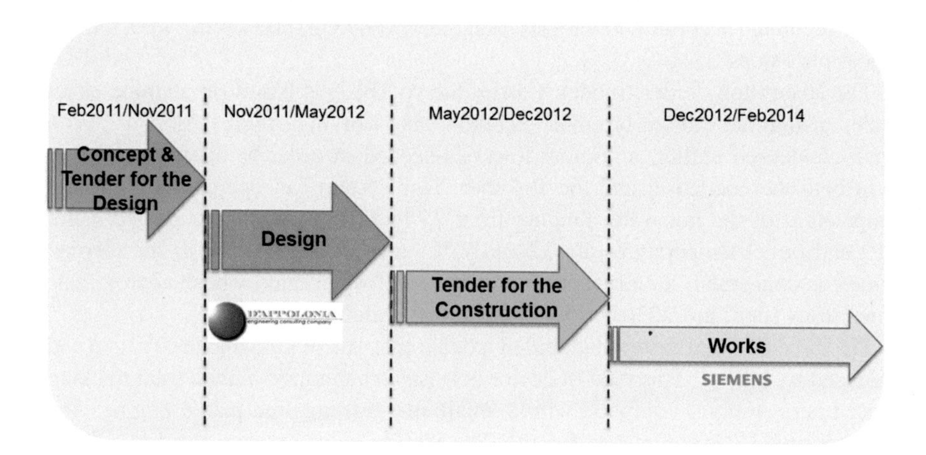

Fig. 1 SPM project evolution

SPM is a three phase low voltage (400 V line to line) distribution system composed of electrical and thermal generation, distribution and control subsystems (Bracco et al. 2014). An overview of the SPM, with the devices location in the Campus, and their main connections, is shown in Fig. 2.

The main equipment and generation units of the SPM are:

- 3 Combined Heat and Power (CHP) Capstone microturbines, namely a C65 dual-mode unit (65 kWe and 112 kWth), C65 Grid Conneted (65 kWe and 112 kWth) and a C30 unit (27 kWe, 54 kWth);
- 2 boilers (500 kWth each) fed with natural gas;
- thermal storage (3000 lt) and chiller (cooling power: 70 kWth): the chiller will allow to use the thermal power from the microturbine for contributing to the cooling of the Campus library in summer
- 3 Concentrating Solar Power (CSP) units (1 kWe, 3 kWth each) coupled with Stirling engines;
- Electrical storage (sodium–nickel SoNik batteries by FIAMM, rated capacity 141 kWh; installation of Li-ion and advanced lead acid batteries is also planned)
- a photovoltaic field (peak power 80 kWp);
- 2 recharging stations for electric vehicles.

The Capstone C65 and C30 microturbines are based on variable speed engines (up to 96,000 rpm), coupled with permanent magnet generator, connected to a rectifier/inverter system to convert their variable frequency, variable voltage output to the grid rated frequency and voltage. For each turbine, during start-up and shut down cycles, the converter system is reconfigured to act as a variable speed drive and the generator is used as a motor. A recuperator allows to extract part of the heat in the exhaust gas, thus making cogeneration possible. The electrical efficiency at rated power is 29 % for the C65 model and 25 % for the C30 model.

The two identical boilers are natural gas fired, of conventional type. Their output can be regulated according to 4 levels each (zero, 166, 333, 500 kWth), for a total of 8 possible "steps".

The absorbtion chiller (model: Carrier Sanyo 16LJ) is based on a single-effect cycle, using a lithium-bromide solution as working fluid. As this is a water-condensed chiller, a cooling tower is needed in order to dissipate the heat from both the condenser and the absorber. This system can operate with an inlet temperature of the hot water ranging from 75 to 110 °C, while the chilled water inlet and outlet temperatures are 12 and 7 °C, respectively. As far as the thermal storage is concerned, it consists of a stainless steel water tank, whose approximate dimensions (mm) are 2230 × 1610 × 2230 and weighting 3280 kg.

The three CSP are quite small, but of a rather innovative kind, named Trinum and produced by Innova. This kind of device consists in a thermodynamic solar tracking concentrator system, equipped with a small size Stirling free piston engine. The collector area is approximately 9.6 m^2. The Stirling engine is connected to a linear permanent magnet generator, working at network rated frequency, so no power electronic converter is needed. The heat extracted from the engine "cold side" can be

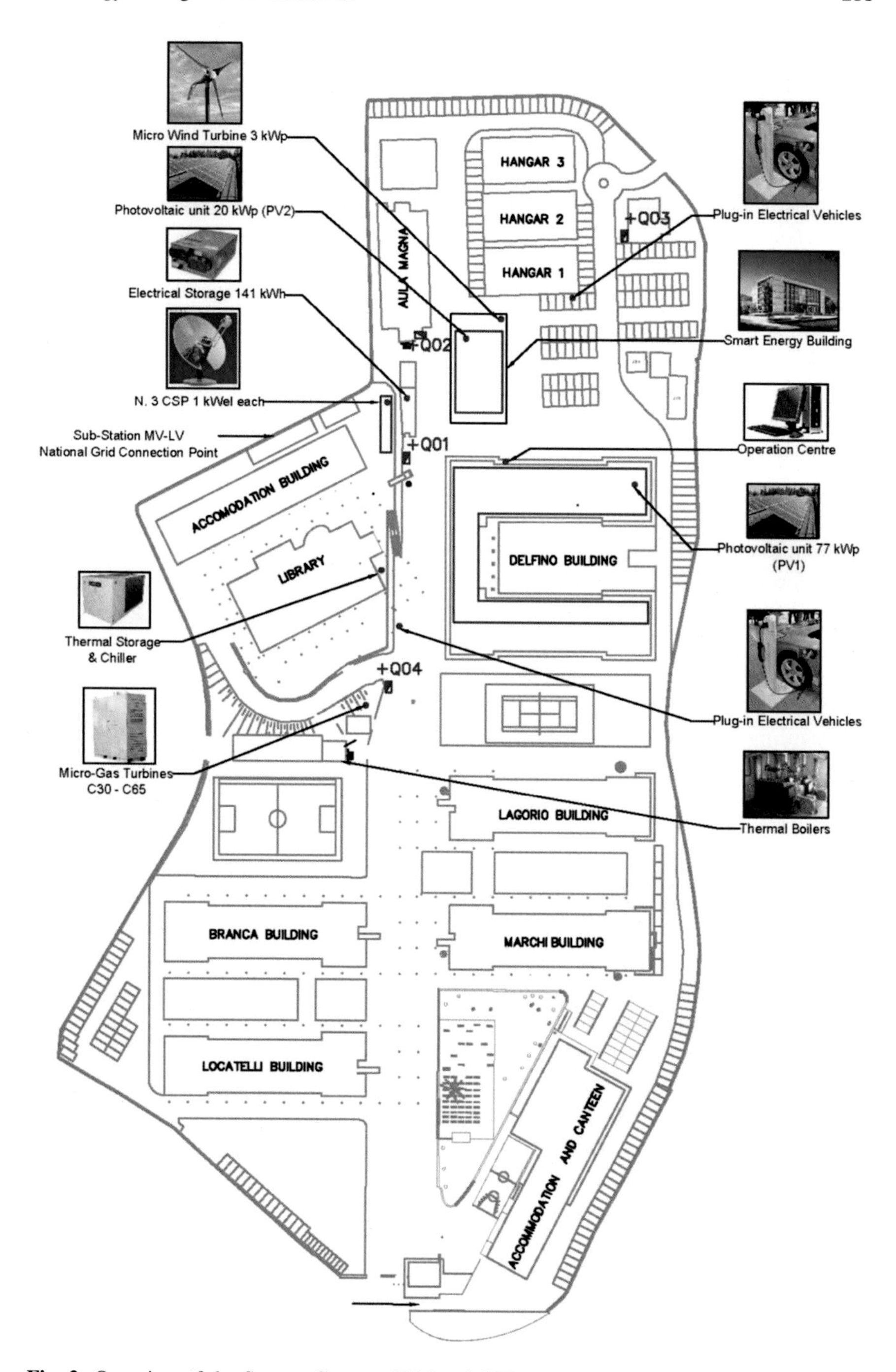

Fig. 2 Overview of the Savona Campus SPM and SEB

delivered to a heating network, so this CSP is actually a cogeneration system, with a rated electrical efficiency of 13.8 % and a thermal efficiency of 41.4 %.

The electrical storage system is based on SonNik batteries from FIAMM. The whole battery energy storage system (batteries, management system and power electronic converters) is hosted in a container. The commercial name SoNick identifies a line of high temperature, high voltage Sodium Nickel Chloride ($NaNiCl_2$) batteries, developed (in different models) both for automotive and energy storage applications. The batteries installed in the SPM are characterized by a DC operating voltage ranging from 420 to 648 V; each module has a nominal energy capacity of 23.5 kWh, a depth of discharge of about 80 % and round trip efficiency of about 85 %. Six modules are hosted in the container. The rated power of the inverter connecting this system to the grid is 63 kW. One of the main features of this storage system is that, being based on high temperature batteries, it can tolerate a wide range of ambient temperatures (−20 to 60 °C) with no or little need for conditioning. Furthermore, its typical charge/discharge times (in the order of hours), makes it ideal for application such as production shift and daily production optimization in a system with renewable sources and CHP generators, whose electrical production is constrained by the thermal demand. In the context of the RESILIENT project, two additional storage systems, one based on installation of Li-ion technology and one based on advanced lead acid batteries is also planned.

The photovoltaic field is composed by 320 polycrystalline panels, subdivided in 20 strings, and mounted on the roof of the Delfino building.

Regarding the inverters of these last two systems, they allow for the regulation of the reactive power injected into the grid (i.e., they are not limited to unit power factor, as it was the case with older installations for dispersed generation), thus making it possible to contribute to voltage control.

The recharge stations are controlled by a dedicated management system (connected to the overall management system of the SPM, described below). Two electric vehicles have been acquired in the context of the SPM project and will be used by the Campus personnel. The management system will allow exploring the possibility of using the recharging vehicles as controllable loads for demand response and even as supplementary storage resources.

As far as the network topology is concerned, a dedicated MV-LV transformer, linked to the MV busbar of the Campus MV/LV substation, connects the microgrid to the distribution network; the main Campus network (not shown in the diagram) is fed by other transformers connected to the same MV busbar. So, the two networks share the same point of common coupling with the external grid, but from an electrical point of view do not overlap.

One of the major issue in implementing the SPM project was to deal with the different protocols used by the various equipment of the smart microgrid to communicate. In addition, all the devices are typically arranged to be supervised by their own proprietary monitoring program and not by a third party SCADA. This problem has been fixed by exploited the capability of the Remote Terminal Units (RTUs) to act as gateways allowing the communication among a large number of protocols.

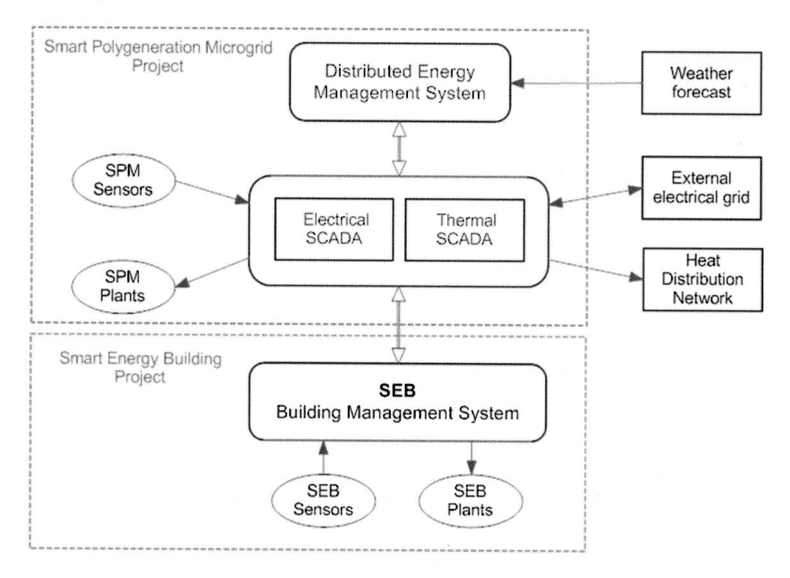

Fig. 3 The architecture of control system

Another peculiarity of the SPM is the close coupling between thermal and electrical systems, typically controlled and monitored by different kind of sensors and SCADAs. In this respect, SPM could be considered a pioneer project at international level.

Moreover, a new project named Sustainable Energy Building is defined in parallel with SPM project. SEB will be characterized by both renewable power plants and energy efficiency measures. More precisely, the building will be connected to the Smart SPM as an energy "prosumer". The SEB will be equipped with photovoltaic modules (PV2), thermal solar collectors and a horizontal axis wind turbine on the roof of the building, a geothermal heat pump, a controlled mechanical ventilation plant and low consumption lamps. The energy consumptions of the building and the energy production of its generation units will be monitored in real time, in order to evaluate the environmental and economic benefits consequent to have a sustainable building instead of a non sustainable one.

Figure 3 illustrates the interactions among the SEB Building Management System, the SPM EMS, and the different power production units.

3 The Energy Management Platform

The Energy Management platform governing the SPM is called Decentralized Energy Management System—DEMS and has been developed by Siemens. The purpose of DEMS is to operate a decentralized energy supply systems in an "optimized" way. "Optimized" does in this context mean that the operation shall be

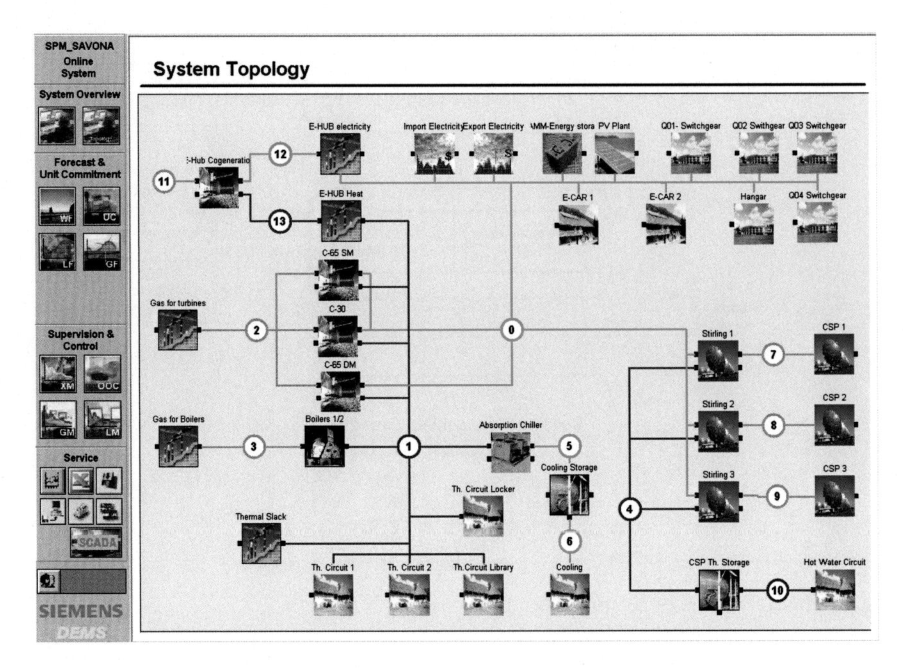

Fig. 4 DEMS software snap shot concerning with the architecture of the University of Genoa SPM

carried out according to optimized operation cost/profit and shall consider all technical, contractual and environmental constraints. A snap shot of DEMS software interface is shown in Fig. 4.

The functions of DEMS are categorized in two stages: (1) day-ahead operational planning and (2) real-time management. These two stages are described as follows:

- At the day-ahead operational planning (off-line) stage, DEMS uses costs and revenues functions, data concerning electric and thermal energy, gas consumptions, etc.…, takes into account operative constraints, and makes use of renewable units' production and load forecasting methods, to calculate an optimal schedule for dispatchable sources.
- In the real-time management (online) stage, the DEMS regulates the actual generation and consumption in the real-time where the prescribed profiles for power generation are met and, in case, takes corrective actions.

The DEMS architecture is based on a centralized intelligence that receives forecasts for renewable resources availability, demands, prices, and system state (for example the level of charge of the battery system), and, on the basis of an optimization model, performs a unit commitment for dispatchable sources. The optimization model considers performance indicators based on costs, emissions, and primary energy use, and constraints able to represent the overall system and the different technical requirements. Also, with a proper tuning of the mathematical model parameters, DEMS can take into account environmental aspects; in general, by using decentralized power generation at minimized operating costs, it helps to significantly reduce environmental pollution and the depletion of resources.

DEMS exploits all important information, such as weather forecasts, current electricity prices, and the energy demands. This data forms the basis for the optimized dispatch plan.

DEMS takes also into account the interconnections between electricity, thermal and cooling energy, gas, and other energy sources, as well as DSM concepts.

The Decentralized Energy Management System handles

- Contracts for power import and export (electricity, primary energy, reserves).
- Controllable, switchable, and non-controllable loads (electric, thermal, gas).
- Power plants—for example, micro-turbines and photovoltaic systems.
- Electric and thermal storage.

In the following, the main modules of DEMS will be outlined.

A. *Weather Forecast*

The DEMS Weather Forecast function is based both on inputs from weather service and historical data collected by the system by means of a local weather station. The availability of this latter data is exploited to refine the information provided by the weather service, usually intended as mean values on a quite wide area and thus usually affected by a certain degree of error if used to obtain local conditions on a, e.g., 15 min time scale. This is done by means of a moving average correction algorithm, which minimizes the difference of the deviation between external forecast and local measured weather data. The weather forecast is one of the inputs for the other planning functions.

B. *Load Forecast*

Load forecast is essentially based on weather, historical data, typical consumptions according to the day of the week, or holiday, etc...., and other variables, e.g. related to academic activity. The basic data needed for this task are the historical measurement of the demands of both electrical and thermal energy in the time resolution of the planning functions (15, 50 or 60 min).

The demand time behaviour is expressed as piecewise linear function of a number of selected "influencing variables" like day types (working/not working) and weather variables (such as temperature). The coefficients for this representation are estimated, based on the available measurements of demand and influencing variable behaviour in the past. Time records referring to a maximum of about 80 days in the past can be used. For each time interval of the day (e.g. 96 time intervals for a 15 min time resolution) a coefficient analysis is performed. The mathematical method used to calculate the coefficients is a Kalman Filter.

C. *Generation Forecast*

This is one of the most important DEMS function, as it provides the data used for the subsequent function (unit commitment): its purpose is to estimate the expected production of renewable energy sources, based on the forecasted weather conditions. The forecast algorithm is based on a piecewise linear transformation, expressing the power generated by the renewable plant as a function of two weather variables (e.g. wind speed and direction for wind power units, radiation and ambient temperature for photovoltaic systems, etc.). The coefficients of this transformation can be set according to the unit technical specifications and can be refined based on the actual measurements over time by means of an off-line analysis, exploiting a dedicated tool.

D. *Unit Commitment*

The DEMS Unit Commitment function calculates the optimal dispatch schedules for all flexible units. Flexible units includes:

- contracts (i.e. unidirectional or bidirectional electrical energy exchange with the grid and natural gas fed by the local gas utility, characterized by given prices that can be variable over time, e.g. taking into account peak/off peak hours. as in the case of electrical energy, and, eventually, constraints on maximum fluxes);
- dispatchable generation units (e.g. generators, characterized by given relation between output power—thermal for boilers, electrical for electrical generators—and fuel consumptions, and, for cogeneration units, characterized by given curves for electrical energy vs. fuel consumptions and thermal energy vs. electrical energy; furthermore, rated values, such as maximum value and maximum gradient for the generated power have to be specified);
- storages (both thermal and electrical, modeled taking into account their rated capacity and charge/discharge efficiencies);
- flexible demands (if any).

The energy from renewable sources, as well as non-flexible demands, are considered equal to their forecasted values. A source can be excluded from the optimization, for instance requiring its production to follows a given behaviour: among the component parameters, a flag is provided that identifies a source as not available, optimizable, must run, or fixed schedule; by choosing "optimizable" the source takes part to the optimization.

The objective function of the optimization is the profit, i.e. the difference between revenues and costs. Rated value of the devices, contract limits, and specified threshold on flows, if any, are included in the computation as constraints.

The scheduling algorithm considers the parameters of the model elements and their connection: several balancing nodes can be defined, both electrical and thermal, to which energy sources, demands and other nodes are connected. On each node the total energy flux (inward energy fluxes minus outward energy fluxes) must sum to zero.

The unit commitment uses Mixed Integer Linear Programming to calculate the results of the optimization problem.

The DEMS control applications are intended for the control and supervision of the units. Furthermore, a control scheme aimed at maintaining the electrical exchange with the utility network (in each time interval) as close as possible to the scheduled value is provided.

E. *Generation Management*

The DEMS Generation Management function performs control and supervision of all units. Also in this case, the actions that the system can perform on an equipment vary according to a flag defined for it:

- independent: the Generation Manager performs monitoring only on the unit;
- manual: the set point for the unit output (e.g. thermal power for boilers, electrical power for generators) is specified by an operator;
- schedule: the set point for the unit output is as scheduled by the unit commitment function;
- control: the unit can be used for regulation purposes.

According to the unit flag, the unit parameters (minimum/maximum power, power gradients, energy content), its actual state (such as in start-up phase or online) and the actual power output of the unit, the start/stop commands and power set points are calculated and sent to it.

The generation management function also monitors that the units follow the assigned set point and, in case of unexpected events, such as a unit fault, can trigger a new unit commitment calculation to force a rescheduling of the remaining units, taking into account the new conditions.

F. *Exchange Monitor*

The DEMS Exchange Monitor function performs the aforementioned task of maintaining the electrical exchange with the utility network (in each time interval, i.e. 15 or 30 or 60 min) as close as possible to the scheduled value.

Based on the actual energy consumptions and productions in the current time interval and the current trend for power interchange, the expected energy interchange at the end of the interval is estimated. The difference between this value and the scheduled one, divided by the remaining time to the end of the current time interval, represents the required correction to the total generated power, in order to fulfil the agreed energy interchange for this time interval. This correction is the input to the Online Optimization and Coordination function.

G. *Online Optimization and Coordination Function*

The DEMS Online Optimization and Coordination function dispatches the correction value to all controllable units. The distribution algorithm takes into account actual unit constraints and operates trying to reach the overall power correction value as fast as possible, minimizing at the same time the costs: this is achieved by subdividing the total power correction over the available units in inverse proportion to their incremental power costs at the current operating point. The power corrections are then passed as input to the Generation Management function, which send them to the units.

In addition, the functions of operator control and visualization, as well as customer specific additions, are also provided. Real-time data is exchanged with secondary automation systems in a number of ways, including protocols belonging to the SIMATIC automation world, OPC, and XML. Relevant data are saved manually or automatically while DEMS is being run.

4 The Supervisory, Control and Data Acquisition (SCADA) System

In Fig. 5, the SPM data acquisition, control and supervision system is sketched. This figure shows the main elements and devices of the ICT infrastructure:

- a control room, where the servers hosting the SCADA and DEMS are installed, together with clients for operators;
- a double fiber optic ring, connecting the server in the control room with the switch located in each switchboard;
- four TM 1703 ACP remote terminal units, one in each switchboard;
- I/O modules for the RTU, in the same switchboards.

In the following, this equipment will be briefly described.

Control room

The control room hosts two identical servers running the SCADA and DEMS, in a hot redundant configuration: one acts as master, actually in charge of the system supervision and control; the other one is in stand-by, but running and kept up to

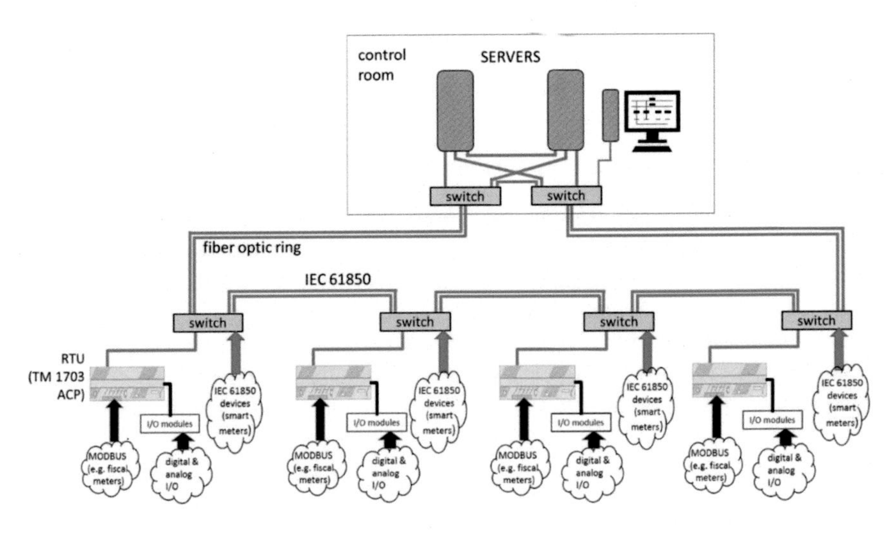

Fig. 5 SPM data acquisition, control and supervision system

date with the master in real time. In the event of a fault of the master server, the stand-by server takes over the control, without affecting the system operation.

Although the servers (installed in a rack) are equipped with a KVM (Keyboard Video Mouse) set, the operators usually access the system by one of four workstations by means of web interface made available by the WinCC environment. Users with different level of permissions for system supervision, parameters modifications, etc.…, are defined. In addition, full control over the system, with the possibility of modifying the DEMS configuration (e.g., adding/removing equipment, etc.…) is allowed from one of the workstations, via a remote desktop connection.

A NAS is also installed, in order to store data acquired from the field, and a dedicated UPS guarantees the supply for the control room.

Communication equipment

The communication system consists of two (redundant) switches installed in the control room, and one switch installed in each switchboard. These switches are connected via a double fiber optic ring (multimodal, 50/125 micron), thus obtaining a so-called dual counter-rotating ring configuration, with one fiber transmitting in one direction and the other in the opposite one. This kind of configuration is highly resilient, both thanks to the fact that there is a redundant physical ring, and because, in the event of a fault of a single node or link, the counter-rotating transmission ensures that any node can communicate with any other node in one direction or the other. The main characteristics of the switches are:

- compliant with Ethernet 100Base-FX protocol;
- compliant with IEC61850 protocol;
- optical interfaces 10/100 Mbps for the connection with the communication ring;
- standard RJ45 10/100 Mbps for the connection with the field or with the devices in the control room.

Power meters

Two sets of power meters are installed, with different purposes:

- smart meters, installed into each switchboard, for data acquisition and supervision;
- power meters for metering, installed to measure the energy produced by each generation unit and the energy absorbed by the auxiliary loads of each busbar: this is intended for the fulfillment of fiscal obligations.

5 Results and Discussion

In this section, the performance of DEMS is described by presenting monitored data and numerical results from a unit commitment example. One of the important input data for energy scheduling is solar radiation and power forecast data for the scheduling horizon. Figure 6 shows the forecasted and actual values of solar radiation in eight days of February. As shown, the prediction has been carried out with an acceptable accuracy.

The DEMS is able to monitor and measure the real-time data of distributed energy resources. For example, the output power of PV system during some days in February 2013 has been illustrated in Fig. 7. Moreover, the measured data is stored in DEMS database as historical data of SPM for future offline analysis.

Thermal power generations and demand are also monitored and stored. Figure 8 shows the thermal power of the two boilers.

Three CSP installed in the microgrid delivers both thermal and electrical energy. The thermal and electrical output powers of the CSPs have been illustrated in Figs. 9 and 10, respectively. It is worth to mention that the CSPs are protected against rainy days. So, in the rainy days they cannot deliver power.

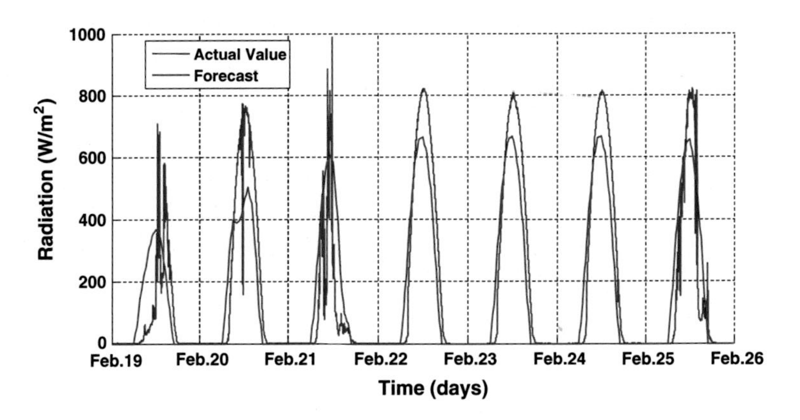

Fig. 6 Solar radiation (actual value vs. forecast) on 19–28 February 2014

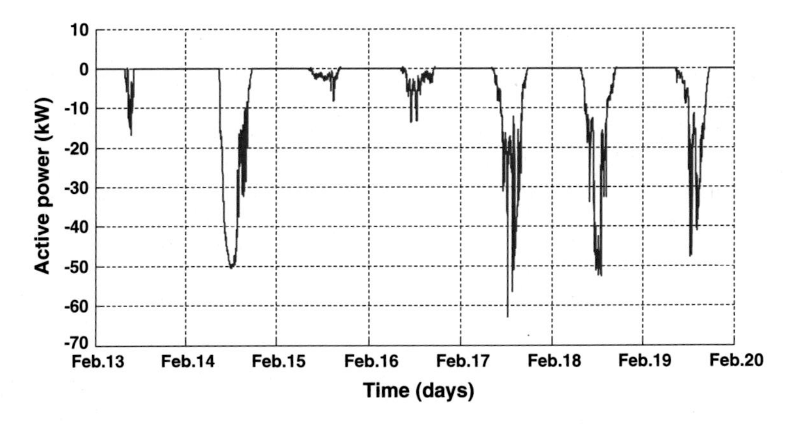

Fig. 7 PV electric power on 13–20 February 2014

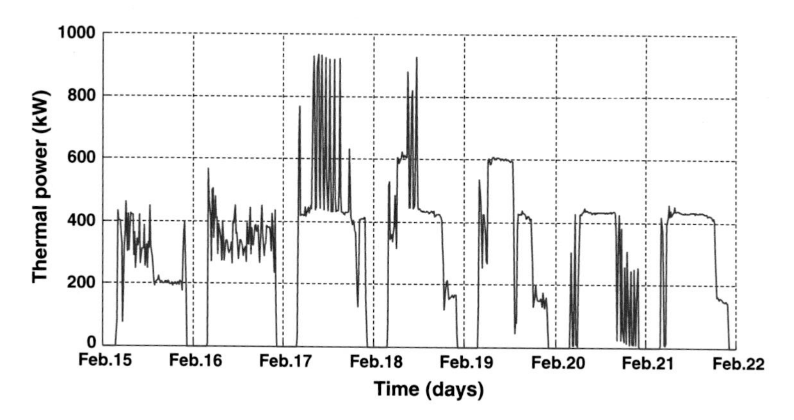

Fig. 8 Boilers thermal power on 15–22 February 2014

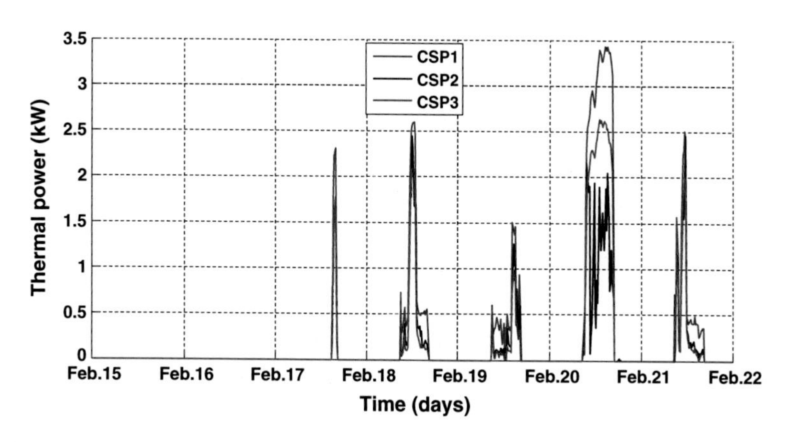

Fig. 9 CSPs thermal power on 15–22 February 2014

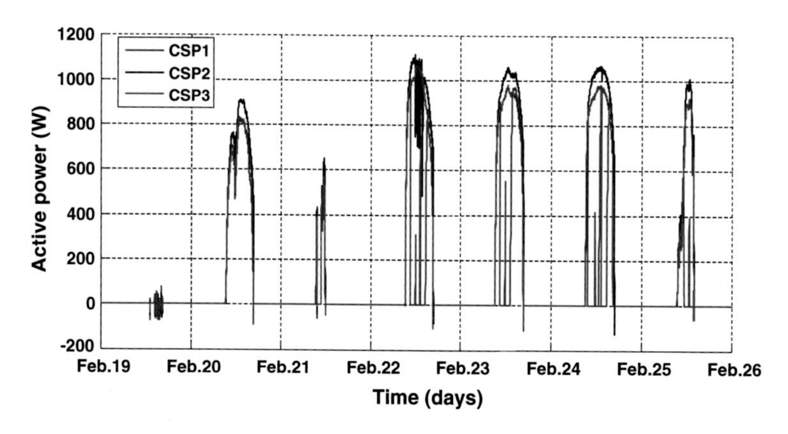

Fig. 10 CSPs electric power, from February 19 to February 25

As previously stated, the microgrid is also intended as a test bed for new devices: manufacturers involved in the development of new components and willing to test them in a real-word environment, can install these components in the microgrid and take advantage of the SPM control architecture to operate and monitor them, also remotely. As an example, during the first months of operation of the SPM, FIAMM performed a number of test on their new storage system: Fig. 11 shows the power exchanged by the storage system during one of those tests: the battery system has been discharged from 13:00 to 15:00 while it has been switched in charging mode during hours 18:00–21:00.

The active power of the SPM that is exchanged with the Campus network has been illustrated in Fig. 12. The negative values refer to absorbed power by the Campus loads or injected power to the external network.

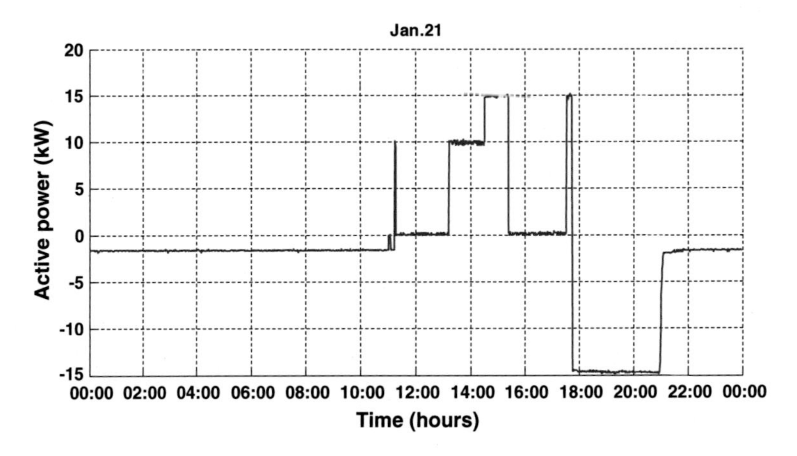

Fig. 11 Discharge at different power levels, followed by a recharge phase

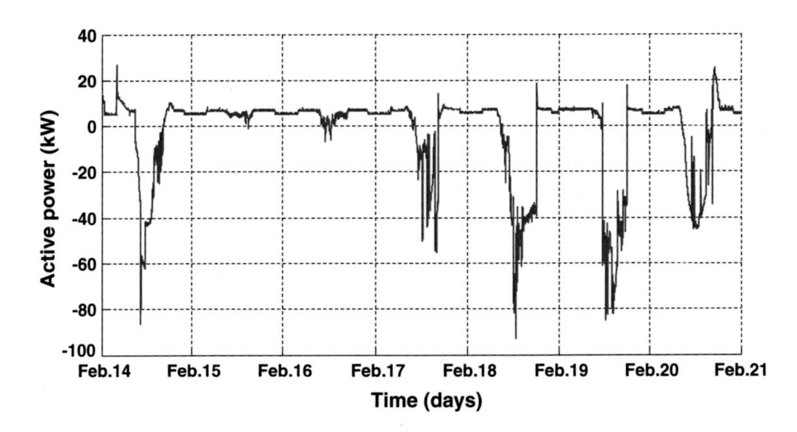

Fig. 12 Power exchange with the Campus grid, from February 14 to February 26

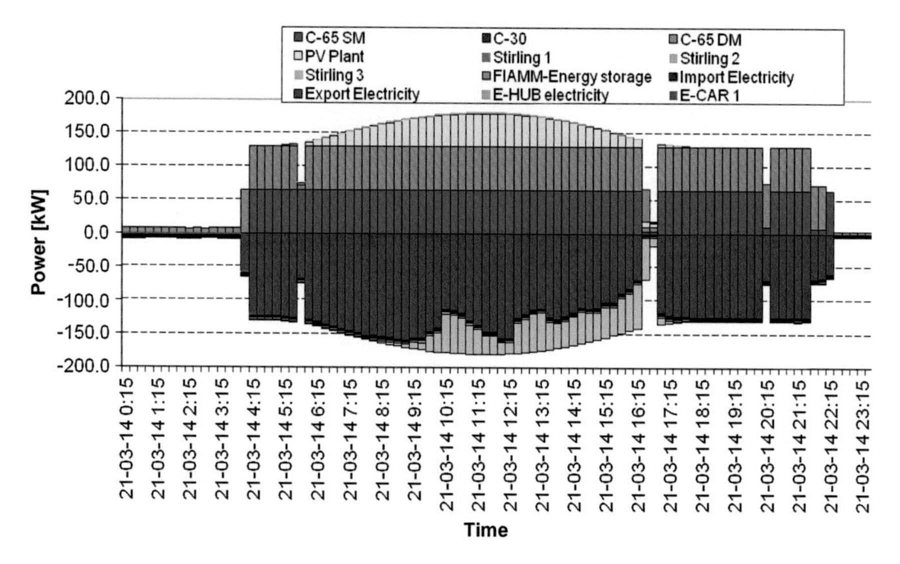

Fig. 13 The day-ahead energy resources scheduling

As stated, the main function of DEMS is the scheduling of distributed energy resources for next following day. Figure 13 shows the scheduling result of distributed energy resources for 20/03/2014. The scheduling has been carried out with a time step of 15-min.

The objective functions of DEMS for day-ahead scheduling are cost and emission minimization. The operational cost of scheduling for each period has been shown in Fig. 14.

The forecasted power of CSPs for the day-ahead scheduling is illustrated in Fig. 15.

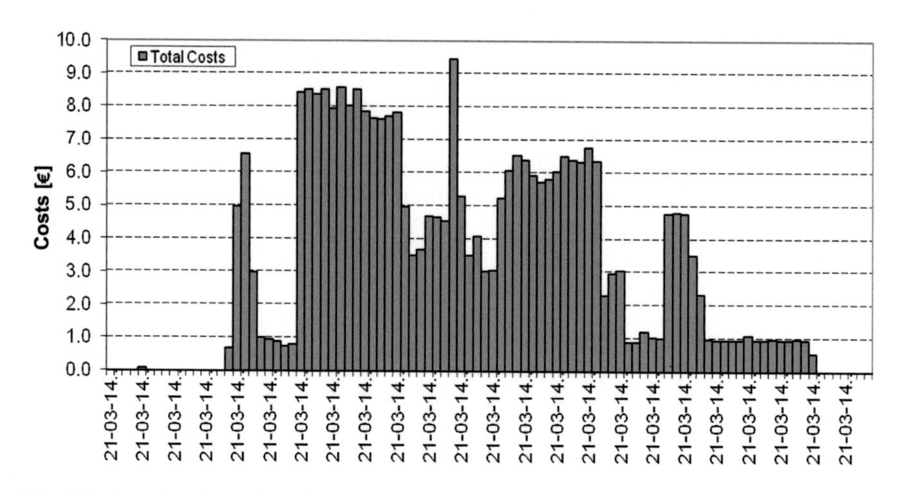

Fig. 14 Operational cost in each time-step with the computed scheduling

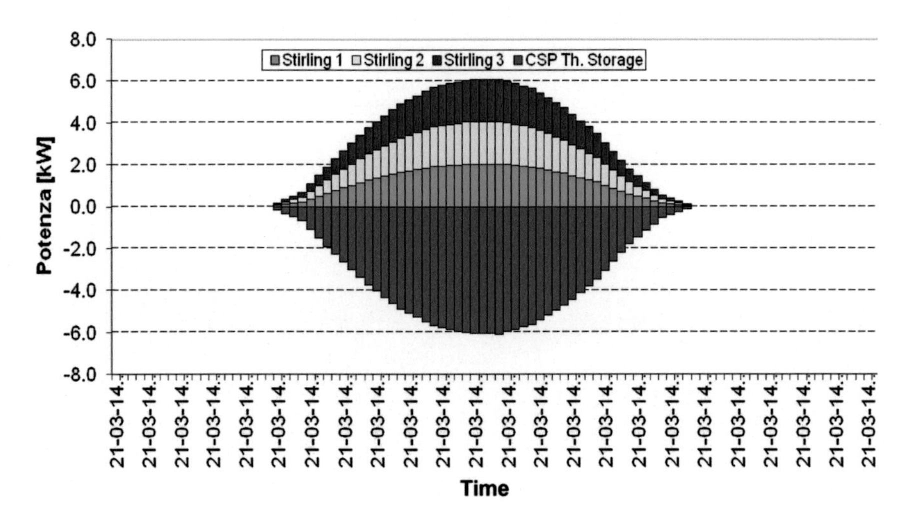

Fig. 15 CSP forecasted power for day-ahead scheduling

6 Conclusions and Future Research Lines

The performances and the functionalities of the Decentralized Energy Management System (DEMS) for a renewable-based microgrid project have been presented in this paper. DEMS provides the online set points for generation units while minimizing the operational costs and environmental targets, such as CO_2 emissions. In DEMS, the forecasting procedures of renewable resources, load and thermal consumptions are also implemented in order to provide the input data of the day-ahead scheduling. This way, it is possible to perform the day-ahead and real-time scheduling of all the distributed energy resourced in the SPM.

Work is now in progress in order to export the SPM architecture and its control logics and energy management procedures at the city level, in order to define a "smart" residential district concept. In this context, the following research issues will be addressed:

- identification of optimal strategies for both energy consumption and production;
- test and development of new products and systems in the Smart Grid sector (in cooperation with industries and Distribution Systems Operators—DSOs);
- development of new "smart" communication protocols;
- analysis and validation on the field of techniques for a seamless transition between grid-tied microgrid and islanding mode microgrid;
- analysis of protection coordination issues for islanded microgrids.

References

Bracco, S., Delfino, F., Pampararo, F., Robba, M., Rossi, M.: The University of Genoa smart polygeneration microgrid test-bed facility: the overall system, the technologies and the research challenges. Renew. Sustain. Energy Rev. **18**, 442–459 (2013)

Bracco, S., Delfino, F., Pampararo, F., Robba, M., Rossi, M.: A mathematical model for the optimal operation of the University of Genoa smart polygeneration microgrid: evaluation of technical, economic and environmental performance indicators. Energy **64**, 912–922 (2014)

Chakraborty, S., Weiss, M.D., Simoes, M.G.: Distributed intelligent energy management system for a single-phase high-frequency AC microgrid. IEEE Trans. Ind. Electron. **54**, 97–109 (2007)

Hajizadeh, A., Golkar, M.A.: Intelligent power management strategy of hybrid distributed generation system. Electr. Power Energy Syst. **29**, 783–795 (2007)

Marnay, C., Venkataramanan, G., Stadler, M.: Optimal technology selection and operation of commercial-building microgrids. IEEE Trans. Power Syst. **23**, 975–982 (2008)

Palma-Behnke, R., Benavides, C., Lanas, F., Severino, B., Reyes, L., Llanos, J., Sáez, D.A.: Microgrid energy management system based on the rolling horizon strategy. IEEE Trans. Smart Grid **4**(2), 996–1006 (2013)

Teleke, S., Baran, M., Bhattacharya, S., Huang, A.: Rule-base control of battery energy storage for dispatching intermittent renewable sources. IEEE Trans. Sustain. Energy **1**(3), 117–124 (2010)

Tsikalakis, A.G., Hazizargyriou, N.D.: Centralized control for optimizing microgrids operation. IEEE Trans. Energy Convers. **23**, 241–248 (2008)

Westermann, D., John, A.: Demand matching wind power generation with wide-area measurement and demand-side management. IEEE Trans. Energy Convers. **22**(1), 145–149 (2007)

List of Resources

Consortium for Electric Reliability Technology Solutions: Retrieved 13 Jan 2015, from http://certs.lbl.gov/certs-der-micro.html (n.d.)

Dufresne: Energy storage world forum. Retrieved 13 Jan 2015, from http://www.energystorageforum.com (n.d.)

Dufresne: Retrieved 13 Jan 2015, from http://www.dufresneresearch.com/ (n.d.)

Homer Energy: Retrieved 13 Jan 2015, from http://www.homerenergy.com/index.html (n.d.)

Siemens: Ten reasons for a smart grid with DEMS. Retrieved 13 Jan 2015, from https://www.swe.siemens.com/italy/web/IC/SG/EA/applicazioni/Gestione%20di%20Microgrid%20e%20Virtual%20Power%20Plant/Documents/DEMS_en.pdf (n.d.)

Smart Grid Observer: Microgrid world forum. Retrieved 13 Jan 2015, from http://www.microgridworldforum.com/ (n.d.)

University of Genoa, (Smart Energy Team): The Savona Campus test-bed facilities for the optimal planning and control of Smart grids. Retrieved 13 Jan 2015, from https://www.youtube.com/watch?v=RHzV7_YjZtM

U.S Department of Energy: Smart grid. Retrieved 13 Jan 2015, from https://www.smartgrid.gov/ (n.d.)

Transit Journaling and Traffic Sensitive Routing for a Mixed Mode Public Transportation System

Joshua Balagapo, Jerome Sabidong and Jaime Caro

Abstract In this paper, we propose *transit journaling*, a crowdsourcing solution for public transit data collection, and we describe CommYouTer, an Android app for this purpose. CommYouTer enables the user to (1) document his public transit trips with automated transfer detection, (2) participate in crowdsourcing real-time traffic conditions and other relevant data, and (3) query for efficient commuting directions via a traffic-sensitive routing algorithm. For transit journaling, the app offers a recording feature that uses the smartphone's GPS antenna and accelerometer to track the user's location and activity. Activity detection is applied to the mobile phone's accelerometer data to differentiate between two user states (walking vs. non-walking), which are then used to determine vehicle transfers along the journey. We also implement our modification of RAPTOR, an existing round-based public transit routing algorithm. Our modification allows the system to account for real-time crowdsourced traffic conditions. We test our system in Metro Manila, Philippines, where public transit is primarily headway-based (non-scheduled).

1 Introduction

Several routing and traffic update apps exist for motorists on private vehicles, since only street data is typically required in these systems. Public transit information services, on the other hand, depend on the amount and quality of data provided by

J. Balagapo (✉) · J. Sabidong · J. Caro
Service Science and Software Engineering Laboratory,
Department of Computer Science, University of the Philippines Diliman,
1101 Quezon City, Philippines
e-mail: joshuabalagapo@gmail.com

J. Sabidong
e-mail: jrsabidong@upd.edu.ph

J. Caro
e-mail: jdlcaro@up.edu.ph

G.A. Tsihrintzis et al. (eds.), *Intelligent Computing Systems*,
Studies in Computational Intelligence 627, DOI 10.1007/978-3-662-49179-9_11

the governing agencies. In some cities, public transit is abundant but relevant information is very limited, since data gathering can be a costly task.

Ching (2012) indentified 4 categories of bus data (space, time, conditions and perception). Spatial/geographical data is used to map transit routes, usually on top of existing road maps. Time data can be used to estimate average traffic flow in different places at different times, while condition and perception data can help transit providers and government agencies assess the quality and efficiency of public transit services.

In the Philippines, there has been a successful undertaking by the Department of Transportation and Communication (DOTC) to map out the Metro Manila transit routes (Republic of the Philippines and Department of Transportation and Communications 2013). The data, which is publicly available in Google Transit Feed Specification (GTFS) format, can be used to build useful applications such as trip planners for commuters. However, we highlight significant shortcomings in the data that we wish to tackle.

1.1 Limited Scope of Data

The GTFS data from DOTC currently includes only public transit routes in Metro Manila. Furthermore, the only modes of transit included are rail, bus, and jeepneys. There is currently no information regarding other commonly used public utility vehicles (PUVs), such as *UV Express Service* vehicles and tricycles. We wish to have a system that will facilitate the collaborative mapping of these modes of transportation. Moreover, we wish to include routes in areas outside of Metro Manila.

1.2 Formal Route Names Versus Informal Headsigns

PUVs in the Philippines have officially licensed routes, but drivers use informal headsigns for identification. The official route names in the GTFS data indicate the endpoint locations of the route, and the main road traversed. On the other hand, the informal headsigns only hold popular landmarks and roads, and a PUV driver usually uses one or more of these headsigns to inform passengers which landmarks his route will pass. Thus, using the formal route names may be a source of confusion for users of a trip planner, because the informal headsigns are more familiar in everyday use. Our system also aims to record these headsigns from the contributions of commuters.

1.3 Insufficient Stop Descriptions

Stops are primarily identified by their GPS coordinates. In the GTFS data, the naming convention used is station name (for rail), or intersection of roads (for other vehicles). Stops without nearby intersections, however, are simply named after the street they are on (thus, many stops currently share the same name).

Commuters may have different ways of identifying particular stops (e.g., when there are many landmarks near one stop). As with the previous item, we suggest that these identifiers should be stored and linked to the corresponding stops, so that each stop will be identified uniquely by the most popular identifiers (or aliases). Popular/commercial landmarks can be obtained from providers such as Google Maps, but in less-populated areas it may be helpful to get user-contributed identifiers (*e.g. "pink waiting shed beside a large tree"*).

1.4 Traffic Sensitivity in Routing/Trip Planning

Existing trip planning solutions usually use travel time, fare, and number of transfers as optimization criteria. Since most of public transit in Metro Manila is not scheduled, expected travel times may vary greatly throughout different times of the day and different days of the week. Thus we aim to implement trip planning that is sensitive to traffic conditions, as traffic jams can be severe enough to prompt commuters to take alternate routes. Since our system will rely heavily on crowd-sourcing, we expect users to contribute traffic conditions, which we can use to come up with more realistic trip planning results.

2 Related Work

2.1 Crowdsourced Mapping and Real-time Tracking

Crowdsourcing is becoming a common approach to information-gathering tasks in public transit systems. A method called *flocksourcing*, or guided crowdsourcing, was used to build Dhaka's first map of bus routes (Ching 2012). The "flock" also used the GPS-enabled smartphones to trace the paths of the routes, and the data gathering was completed by 1000 volunteer commuters over a period of eight weeks. Thus, crowdsourcing is a viable option in obtaining data in areas where public transit data is incomplete or not available at all. We set GPS recording as one of the main features of CommYouTer.

Another application of transit crowdsourcing is real-time tracking, where commuters allow their movement to be tracked via smartphone GPS sensors, and contribute the data for the benefit of others. A popular application is Waze (2013), which offers real-time crowdsourced traffic data but which is geared towards motorists on private vehicles (no public transit routing feature). Public transit counterparts are Moovit (2013) and Tiramisu (2013), but they are only usable in officially supported cities.

2.2 *Activity Detection*

Reddy et al. (2010) described a method for determining a person's transportation mode using the sensors on a mobile phone, namely the GPS receiver and the accelerometer. Their work explored several classification systems and found that a two-stage system involving a Decision Tree and a Discrete Hidden Markov Model achieved an accuracy level of 93.6 % when classifying between stationary, walking, running, biking, or in motorized transport. Thiagarajan et al. (2010) did classification between walk and non-walk with Decision Tree only, while taking into account the *peak power* feature, which achieved up to 97.5 % accuracy. Our goal is to employ activity detection to give our app the ability to automatically determine when vehicle transfers take place while the app is recording GPS coordinates. At the end of the recording process, the user needs only to identify the route names/headsigns of the vehicles taken, and not where the transfers took place.

2.3 *Trip Planning/Routing*

In our review of related literature, we explore routing algorithms both for road networks and public transit networks. Road network routing can be considered the general case, where the typical assumption is that all roads can be utilized. In public transit networking, we consider specific routes along these roads, and usually, specific times when those routes are being served. In headway-based public transit systems, however, the problem is similar to plain road network routing, so we may apply variations of shortest-path-problems to a pre-built graph.

2.3.1 Dijkstra's Algorithm

Graph-based solutions are usually derived from Dijkstra's algorithm. This algorithm solves the shortest-path problem for a graph with path costs by calculating tentative distances as the search expands outward (Dijkstra 1959). To search schedule-based transit networks, an augmented variant of the algorithm can be used to scan vertices with increasing arrival time, and evaluate each edge $e = (u, v)$ at time $dist(u)$ (the arrival time at u) (Delling et al. 2012). Several speedups exist for Dijkstra's algorithm in large road networks (Geisberger et al. 2008; Sanders 2005; Bast et al. 2006).

2.3.2 A* Search

A* Search is an extension of Dijkstra's algorithm that achieves better time performance using a heuristic function (Hart et al. 1968). It begins from the start node, working its way to the target node, while maintaining an open set of nodes to be traversed. It chooses the minimum-cost node at every iteration of the search, using the function $f(x) = g(x) + h(x)$, where $g(x)$ is the known cost from the origin to the node being evaluated, and $h(x)$ is the heuristic function estimating the cost from the node to the origin (usually the distance between the goal and the node in question). The process is repeated until the goal node becomes the minimum-cost node in the open set (the nodes yet to be traversed). At the end of the search, the algorithm does a backward trace of minimum-cost nodes from the target node to the starting node, and returns the optimal path. Goldberg et al. gave an improvement to the basic A* search, which is basically a lower-bounding technique using landmarks and the triangle inequality (Goldberg and Harrelson 2004). Delling et al. (2007) showed that a combination of ALT and highway hierarchies yields a significant speedup over highway hierarchies alone.

2.3.3 Raptor

RAPTOR is a new round-based trip-planning algorithm designed for the bicriteria problem (minimizing both arrival time an number of transfers) (Delling et al. 2012). Instead of a graph, it operates on a timetable $(\Pi, \mathcal{S}, \mathcal{T}, \mathcal{R}, \mathcal{F})$ where $\Pi \subset \mathbb{N}_0$ is the period of operation, $\mathcal{S}$ is a set of stops, $\mathcal{T}$ a set of trips, $\mathcal{R}$ a set of routes and $\mathcal{F}$ a set of transfers (or foot-paths).

The algorithm works in rounds. Round k computes the fastest way of getting to every stop with at most k − 1 transfers (i.e., by taking at most k trips). That is, the algorithm associates with each stop p a multi-label $(\tau_0(p), \tau_1(p), \ldots, \tau_K(p))$, where $\tau_i(p)$ represents the earliest known arrival time at p with up to i trips. The goal of round k is to compute $\tau_k(p)$ for all p. It does so in three stages:

The first stage of round k sets $\tau_k(p) = \tau_{k-1}(p)$ for all stops p: this sets an upper bound on the earliest arrival time at p with at most k trips. The second stage then processes each route in the timetable exactly once. This is done by visiting each stop of each route until we find a stop p_i such that $et(r, p_i)$, the earliest trip in route r that one can catch at stop p_i, is defined. For each subsequent stop p_j, we update $\tau_k(p_j)$ using this trip, and set the parent pointer to the stop at which t was boarded. Finally, the third stage of round k considers foot-paths. For each foot-path $(p_i, p_j) \in \mathcal{F}$, it sets $\tau_k(p_j) = min(\tau_k(p_j), \tau_k(p_i) + walking\_time(p_i, p_j))$. The algorithm can be stopped after round k, if no label $\tau_k(p)$ was improved. The total running time of the algorithm is linear per round. In total, it takes $\mathcal{O}(K(\sum_{r \in \mathcal{R}} |r| + |\mathcal{T}| + |\mathcal{F}|))$, where K is the number of rounds.

The advantage of timetable-based algorithms over graph-based is that they directly exploit the fact that public transit vehicles operate on predefined lines

(Delling et al. 2012). In general, graph-based approaches work well for road networks but not for public transportation networks (Bast 2009). Thus, we use RAPTOR for our app.

2.4 Trip Planning with Real-time Data

Jariyasunant et al. (2009) demonstrated that travel time estimate accuracy can be improved for schedule-based transit systems by taking real-time data into account. Their implementation runs a K-shortest paths algorithm on a graph, while interfacing to a third party bus arrival prediction system for better travel time estimates. Our goal is to provide a similar solution, but instead using RAPTOR and traffic flow indications at stops.

3 Methodology/Design

We propose transit journaling, a crowdsourcing activity for public transit data collection, and we describe CommYouTer, an Android application for this purpose. It functions as a journal with which the user can record the details of his entire *journey*, with minimal effort. That is, CommYouTer records time-stamped GPS locations, tracing the user's movement along his commute. These GPS locations are to be partitioned automatically into trips, presumably of different modes of transportation.

Analogous to transit journaling is the usage of the social networking service Foursquare (2013). Foursquare lets a user check-into places they visit and have this activity broadcasted on his social media accounts (e.g. Facebook, Twitter, etc.). User check-ins award points and badges in a game-like reward system. Foursquare also features crowdsourced tips, which serve as suggestions for things to do, see or eat at the location. In transit journaling, a user will be able to document various trips (e.g. going to work, or visiting a venue for the first time) and share this information online for the benefit of the general public.

Prior to the digitization of public transit information, commuting directions in Metro Manila were shared by word of mouth. If a person wishes to know the commuting directions from one place to another, he must directly ask someone who might know, whether it is a peer, bystander, or driver of a public utility vehicle encountered along the way. The same goes for qualitative descriptions of commuting experiences—it is often the case that an experienced commuter will recommend to his peers which public transit routes he thinks are convenient, or warn others about how crowded the vehicles are at particular times of the day. The eventual proliferation of online information on cities allowed these exchanges to take place in personal blogs and social media sites, and even gave rise to crowdsourced trip planning services in Metro Manila such as ParaSaTabi.com (2013).

Transit journaling aims to take this development a step further by encouraging people to talk about their commuting experiences in social media, and have the important data automatically generated from the contributions. Some of the useful things that can be done with the crowdsourced data are: (1) automated mapping of new routes from recorded journeys (as in Dhaka's bus map) and (2) aggregation of traffic conditions to come up with a predictive model (Padmanaban et al. 2010). People involved in transit journaling will also be encouraged to discuss traveling strategies, or identify problems in particular places and transit agencies. Our system, CommYouTer, is primarily intended to bridge existing transit information to crowdsourced data by transit journaling. The result is a dynamic information system where users help grow the database (which includes route as well as traffic data), and at the same time benefit from a traffic-sensitive routing service.

3.1 *The Server/Back-End*

Our server application stores and manages the crowdsourced contributions, and runs instances of the routing algorithm upon users' requests. We implement it as a Java Servlet deployable in PAAS environments such as the Google App Engine, which takes care of web app scaling.

3.1.1 GTFS Data Pre-processing

Upon initialization, the server parses the uploaded GTFS text files to build a StaticData object, which is an efficient structuring of public transit data in memory.

Footpaths are not included in the GTFS data provided, so we generate them in a pre-processing stage by calculating the distance between every stop pair and considering all pairs that are within *walking_distance* apart. We set *walking_distance* as 0.3 (300 m). For efficiency's sake, we do not keep an actual list of footpaths, but directly attach each (*stop, walking_time*) tuple to the walkable stop and vice versa. In doing so, we only need to look at each tuple attached to each stop being scanned at Stage 3 of RAPTOR. *walking_time* is estimated as

$$walking\_time = walking\_distance / walking\_speed \tag{1}$$

with *walking_speed* set to 5 km/h.

3.1.2 Server Design

We implement our server as a Java Servlet deployable in PAAS environments such as the Google App Engine, which takes care of web app scaling for CommYouTer. Upon initialization, the server parses the (most recent) text files uploaded with it to

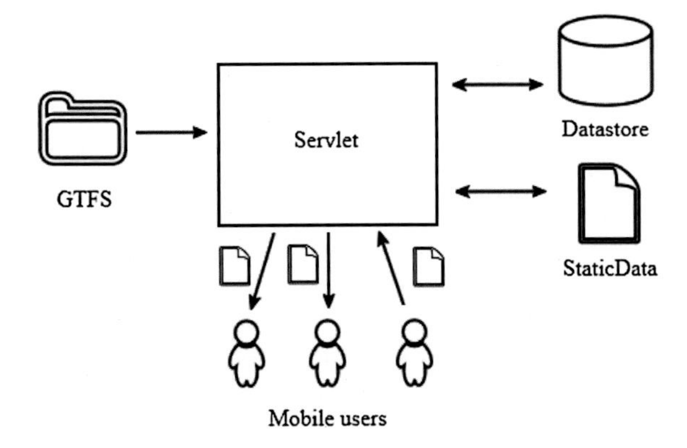

Fig. 1 CommYouTer system design

build a StaticData object in memory, which is an efficient structuring of non-modifiable public transit data. Thus, subsequent searches will access the same StaticData, while modifiable values such as traffic conditions will be kept in a (slower) separate datastore provided by the Google App Engine. If the GTFS data is updated (e.g., by uploading the new text files to the server), the servlet must be re-initialized. The interaction is described by the following diagram (Fig. 1):

The Google App Engine enables persistent data through the Java Persistence API, with which we store and manage our crowdsourced data.

3.1.3 The Modified RAPTOR Search Algorithm

To implement traffic sensitive routing, we use our modification of the Round-based public transit routing algorithm RAPTOR. Instead of using scheduled trips, we compute travel times for a headway-based public transit system at runtime. For every pair of connected stops, we use the travel time estimate

$$travel\_time = f * distance/average\_speed \tag{2}$$

where f is a scale factor determined by the traffic flow at two adjacent stops:

$$f = \frac{average\_speed}{mean(traffic\_flow\_1, traffic\_flow\_2)} \tag{3}$$

and *distance* is the Haversine distance (Robusto 1957) between the two stops, expressed in km. To illustrate the computation of the factor f (Fig. 2):

To keep traffic flow records, we keep a *Traffic Reports* table in the datastore. Every new traffic report made at the same stop replaces the old one. At the start of each search, the algorithm fetches all the traffic reports and loads it to a hashmap for faster execution. At each travel time computation, the routing algorithm checks the

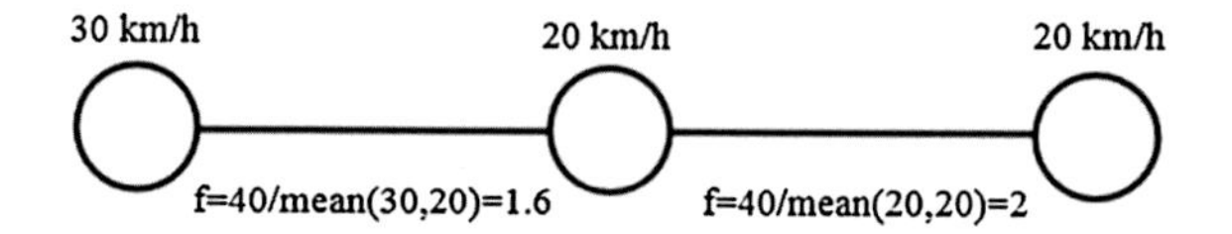

Fig. 2 Sample instance of f computation for two road segments, average_speed = 40

hashmap if there are entries for each of the two stops in question. If none is found, the algorithm will use the default *average_speed* value for that stop.

3.2 *The Mobile App*

In the following section, we describe our app by its Android activities and their functions:

3.2.1 Search

The default activity is the Search function, where the UI consists of two autocomplete text fields for origin and destination, respectively, and a "Get Directions" button. At startup, the Activity checks the registration status of the program and proceeds to the Signup activity if the app has not been registered yet. If there is a working data connection and the user/device is already registered, the Activity fetches the text autocomplete data from the server in real time, and loads it on to memory (Fig. 3).

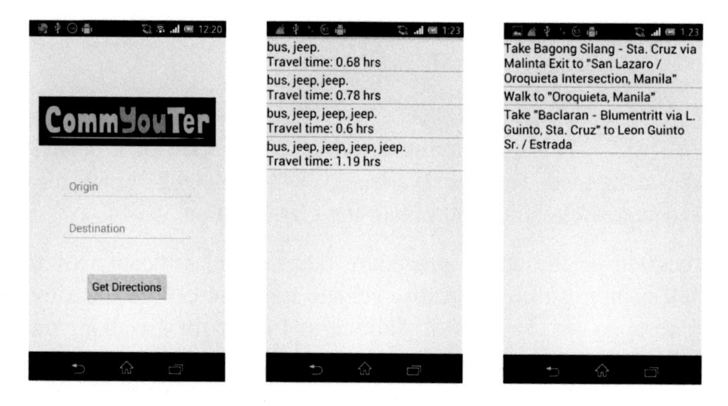

Fig. 3 Search, results, journey display

3.2.2 Results/Journey Displays

The results display is a list view of journey summaries returned by the search query. Each journey summary contains the basic sequence of trips (*e.g., WALK, BUS, JEEPNEY*) along with important details such as total travel time or fare cost. The journey display shows a single journey, which is a list of trips (rides or walks). Each list item is expandable to show the intermediate stops, which are marked one by one if they have been reached by the user.

3.2.3 Recording

Journey recording is achieved with the App running two concurrent threads; one for activity detection and another for polling the GPS receiver. Activity detection is closely patterned after the implementation in Thiagarajan et al. (2010). For CommYouTer, variance and the 1–3 Hz DFT bins (typical walking periodicity) are used as features. *JTransforms* (2013), a free Java library, was used to compute the DFT values. *Weka*, a collection of free Machine Learning tools written in Java (Hall et al. 2009), was used for the Decision Tree Classifier. Weka can automoatically generate DT classifier code from a training set, which is basically a text file containing samples of the corresponding activity. The training set for CommYouTer contained a total of 15 min of walk and non-walk instances, and the smartphone position on the user was varied (on hand, in pocket) in the walk instances.

At runtime, The following procedure is periodically executed at 20 Hz:\

(1) Take the raw accelerometer values (x, y, z).\
(2) Compute the $\mathcal{L}^2$ norm (magnitude).\
(3) For a sliding window w, compute the variance, and\
(4) compute the Discrete Fourier Transform

$$M_k = \sum_{n=0}^{|w|-1} m_n e^{-\frac{2\pi i}{|w|}kn} \tag{4}$$

where $k = 0, \ldots, N - 1$. Take the values in the 1–3 Hz DFT bins.\
(5) Run the computed features (variance, 1–3 Hz DFT bins) through the Weka-generated classifier and obtain the classification.

Each execution of the above procedure returns a classification of the user's activity (walk or non-walk). Typically, vehicle transfers can be done by walking only a few steps, so we set 4.5 s of walking as trigger for a walking/transferring segment. Activity detector output determines whether an interval between two GPS points was walked or traveled by vehicle, by the following cases:\

Case 1: points are too close to each other; no movement occurred.\
Case 2: points are significantly distant from each other, activity was "walk" for at least 3 consecutive times (4.5 s) within the segment: path was walked or user got off the vehicle.\
Case 3: points are significantly distant from each other, no 3 consecutive "walk" activities: path was traveled by vehicle.

Below is a diagram illustrating the process (Fig. 4):

The recorded trip, along with its details, is then sent to the server where route building is performed. The server attempts to find a *match* for each coordinate in the trip. Two stops are said to match if they are significantly near each other (i.e., the distance between them is less than a threshold, which we set to 0.2 km). Note that we use the words 'stops' and 'points' interchangeably to describe locations along the route that roughly describe its shape. This is because buses and jeepneys in Metro Manila do not necessarily stop on predefined bus stops, and instead do so nearly anywhere along the route. In areas where formal bus stops are specified, the other stops can be marked as invalid so that they are ignored during search.

In route building, we tackle the problem of consolidating fragments of recorded trips into whole routes. The real-life scenario is that the routes in $\mathcal{R}$ do not represent complete routes mapped by a careful data-gathering process, but rather disjoint segments from user contributions. Thus, the task is to attempt to combine segments of the same route name every time a new one is added to the database.

Same-name routes can be merged if they overlap. However, it is important to note that segments must share at least one stop in order to be eligible for merging. Since recorded GPS points may have wide gaps in between, it is possible to have overlapping segments which cannot be merged because their stops do not coincide. We refrain from assuming the actual path or sequence of recorded stops along the route in the absence of a common stop. Based on these considerations, we introduce the following algorithm for route consolidation:

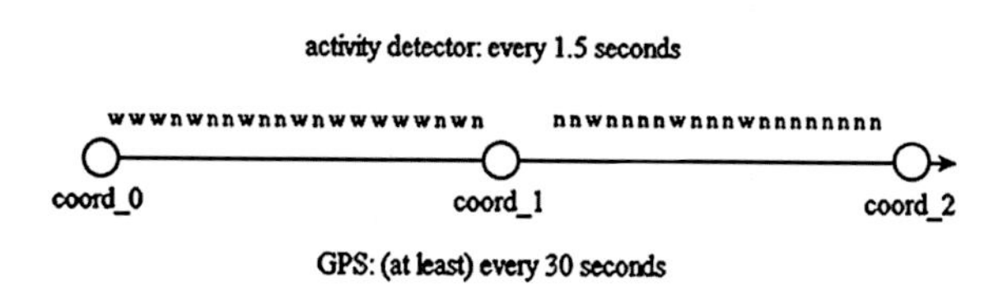

Fig. 4 Sample run of recording feature. Segment from coord_0 to coord_1 is considered walked (there are 3 consecutive w's, indicating intermittent walking), while segment from coord_1 to coord_2 is non-walked, since it contains only spikes of w's

Let $r_1, r_2, r_3, \ldots, r_i$ be existing route segments sharing the same name r; r_{rec} the recently recorded segment; and r_{temp} a new empty segment. The algorithm works by using the most recently recorded segment to make attempted merges. Its goal is to maximize the number of connected stops that can be constructed by combining segments within r_{rec} and an overlapping segment.

A variable *max_length* is used to record the construction with the most number of stops. It is initialized with the length of r_{rec}. For each stop in r_{rec}, the algorithm searches throughout the existing segments for a matching (i.e., significantly nearby) stop.

If one is found, the matching stop becomes a pivot point for the different combinations of the two overlapping segments: $r_{rec} - r_m$ and $r_m - r_{rec}$. We also consider the case that r_m is the maximum-length segment (r_{rec} is already set as the default). Only the maximum-length construction is retained throughout the process, and routes which contained matching stops (including r_{rec}) are discarded (Fig. 5).

If no matching stop is found at an iteration (i.e., the algorithm has cycled through all the segments without finding a match), we move on to the next stop in r_{rec} without updating *max_length*. The maximum-length construction is stored in r_{temp}, which is added as a new route r_{i+1} at the last step (Fig. 6).

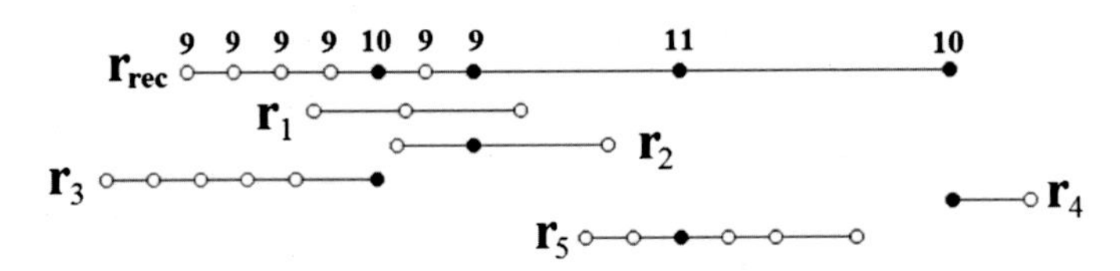

Fig. 5 Sample run of the route-building algorithm. Vertical adjacency represents correspondence of stops (i.e., *solid dots* match). Numbers on top of r_{rec} show maximum length at each iteration

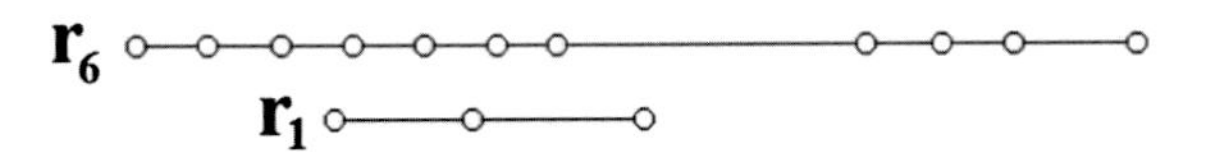

Fig. 6 The result: r_6 is constructed out of r_{rec} and r_5, with a total of 11 stops. Previous routes containing matches have been deleted

The algorithm is described formally as follows:

```
k ← 1
m ← 1
n ← 1
r_temp ← r_rec
max_length ← length(r_rec)
for each stop r_rec(k) do
    searching ← true
    searchstart_m ← m
    searchstart_n ← n
    while searching do
        if r_m(n) matches r_rec(k) then
            searching ← false
            //compare 3 different segment constructions
            r_m_r_rec_length ← n+(length(r_rec)-k)
            r_rec_r_m_length ← k+(length(r_m)-n)
            r_m_length ← length(r_m)
            //use the longest construction found
            //then update max length
            if r_m_r_rec_length > max_length then
                r_temp ← concatenate(r_m(1...n),
                r_rec(k+1...length(r_rec))
                max_length = r_m_r_rec_length
                //delete the used segment,
                //point to the next one
                delete r_m
                n ← 1
                m++
            else if r_rec_r_m_length > max_length then
                r_temp ← concatenate(r_rec(1...k),
                r_m(n+1...length(r_m))
                max_length ← r_rec_r_m_length
                //delete the used segment,
                //point to the next one
                delete r_m
                n ← 1
                m++
            else if r_m_length > max_length then
                max_length ← r_m_length
            end
        end
        //no match has been found
        if searchstart_m == m and
        searchstart_n == n then
            searching ← false
        end
        //increment stop index such that it points to next
        //segment upon reaching end of the current one
        if n ==length(r_m) then
            n ← 1
            if m == i then
                m ← 1
            else
                m++
            end
        else
            n++
        end
    end
end
if length(r_temp) > 0 then
    r_{i+1} = r_temp
end
```

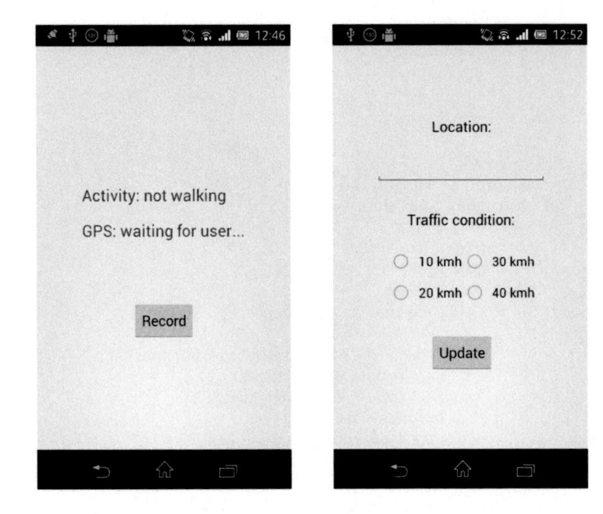

Fig. 7 Recording, traffic report

It is important to clarify that the length being maximized pertains to the number of stops in the sequence, not the actual distance covered by the routes. The tradeoff to maximizing the number of connected stops is that geographically distant/sparsely distributed stops may be discarded in favor of better-connected ones.

The UI consists of a text describing the status of the app (waiting for user/recording, walking/not walking) and the record button. Upon completing a single Record-Stop sequence, the Journey Display Activity is brought up, with the blank fields pertaining to other journey details ready to be filled out by the user (Fig. 7).

3.2.4 Traffic Report

The traffic report activity contains an autocomplete field similar to the ones in the Main Activity, which allows the user to select which at which stop he wishes to make the traffic report. This defaults to the stop nearest to the user, which is obtained by automatically taking the user's current location, and then querying the server's database for the nearest stop. After selecting a stop, the user can proceed to input the traffic flow (in km/h) by choosing from one of the predefined values.

3.2.5 Results Display

The results display is a list view of journey summaries returned by the search query. Each journey summary contains the basic sequence of trips (*e.g., WALK, BUS, JEEPNEY*) along with important details such as total time and fare cost. For the user's convenience, the results are ordered such that the following are placed at the top of the list: Fastest, Most comfortable, Least Transfers, and Cheapest. Any result can take more than one of those labels.

3.2.6 Journey Display

The journey display shows a single journey, which is a list of trips (rides or walks). It is modifiable when brought up after recording (since fields remain to be filled out by the user), and non-modifiable when used to view search results from the server. Each list item is expandable to show the intermediate stops, which are marked one by one if they have been reached by the user.

3.2.7 Journal

The journal displays the personal recorded journeys stored in the smartphone's database. This allows the user to review journeys taken in the past, and re-share them as needed.

3.2.8 Stop Editor

The stop editor contains an autocomplete stop selector, and the fields pertaining to the modifiable properties of a stop.

3.2.9 Route Editor

The stop editor contains an autocomplete route selector, and the fields pertaining to the modifiable properties of a route.

4 Tests and Results

4.1 Basic Routing Capacity

Due to few number of users at the time of writing, we initially assume no traffic reports (i.e., $f = 1$ at all stops) and test the routing algorithm's default ability to produce results similar to actual human commuting preferences. Our approach is to perform a qualitative evaluation. This is because we are not measuring the efficiency of the algorithm itself, but the usefulness of our implementation's results to commuters in real world situations. Respondents were asked to use CommYouTer to generate directions for an origin-destination pair that they frequently traverse. 36 respondents were asked to test the app's search feature by comparing its results with directions that they know of.

4.1.1 Survey

Each respondent was given a survey containing questions to get the following data:

- How often they commute in a week
- How many transfers they take on their longest commute
- How they ask for directions from one place to another
- How relevant/informative the application's results were
- Whether the application was able to generate the directions that they knew of
- Whether those directions were preferred by the respondent
- Whether they would recommend the application to other commuters.

4.1.2 Demographics

- 88 % of the respondents commute moderately to most frequent (5–10 on a scale of 1–10)
- The top choice for asking for directions is via word of mouth, the internet comes next
- The number of transfers range from 0-5. Thus it is safe to assume that users can arrive to their destination by taking up to 5 trips.

4.1.3 Algorithm Evaluation

- 86 % of the respondents were able to get directions that they know of
- 56 % said that the app generated their preferred route
- 92 % of the respondents said that the results were useful
- 78 % of the respondents would recommend the application to others.

4.2 Traffic Sensitivity

To demonstrate how traffic sensitivity affects the routing algorithm's results, we highlight a practical example. Consider the origin-destination pair:

Quezon Avenue/EDSA intersection (LTFRB_1398) to Senator Gil Puyat Ave/Makati Avenue Intersection, Makati City, Manila (LTFRB_1977)

There are 2 highway-based paths to that can be taken from the point of origin: one via Quezon Avenue and one via EDSA. The routing algorithm generates one for each, shown in the following map (Fig. 8):

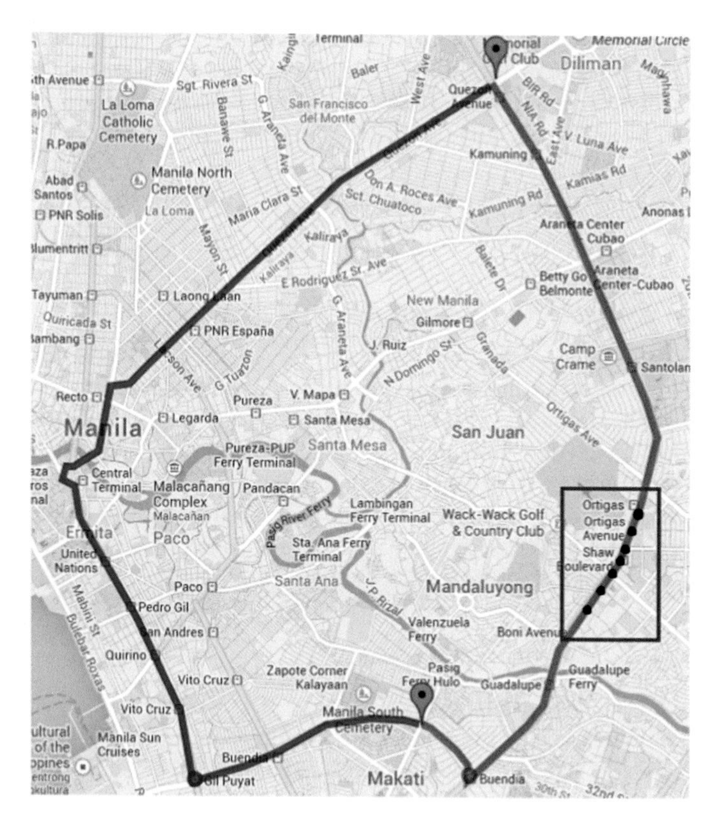

Fig. 8 Two journeys from LTFRB_1398 (*top*) to LTFRB_1977 (*bottom*): journey A (via Quezon avenue, *left*) and journey B (via EDSA, *right*). Traffic reports are represented by *black dots* along Journey B

Journey A is approximately 17.03 km in length, while Journey B is 13.71 km. At default traffic flow, Journey A takes approximately 0.4258 h (26 min), while Journey B takes 0.3428 h (21 min). Note that these travel time estimates are purely based on distance and constant travelling speed (no traffic lights, etc.) We simulate heavy traffic at EDSA (Ortigas area) simply by reporting traffic flow of 10 km/h for the following stops: LTFRB_2055, LTFRB_2056, LTFRB_2057, LTFRB_4233, LTFRB_4234, LTFRB_4235, and LTFRB_4236. This instantly changes Journey B's travel time to 0.4670 h (28 min), which will cause a shift in the algorithm's preferred route.

4.3 Journey Recorder

The Journey Recorder was tested by attempting to record 10 different journeys. In 8 of these, CommYouTer recorded and partitioned the GPS points correctly. The factors affecting the accuracy of the Journey Recorder are (1) GPS hardware reliablity, (2) accessibility of satellites and (3) ambiguous user movement. (1) and (2) are highly dependent on the quality of the smartphone's GPS unit. Some models cannot get a GPS fix at all while under the roof of a vehicle. Obstructions such as buildings and trees also affect the precision of the GPS points.

5 Future Work

5.1 Base Estimate Correction

It is sometimes unclear which of the smaller roads the route traverses, since the spacing of GPS locations can be wide. This can cause our straight line distance-based travel time estimate to be highly inaccurate. Furthermore, traffic lights and other driving considerations are currently not taken into account. We might be able to apply correction to our base estimates by recording actual point-to-point travel times, if enough data can be obtained by transit journaling.

5.2 Preference-Weighing System

We can use a more detailed weighing system to implement a hierarchy of vehicle preferences. In real situations, people prefer certain modes of transportation over others (e.g., train over buses).

5.3 Traffic Flow Prediction

Since traffic reporting is not as frequent and timely as we want it to be, we might try prediction methods similar to those used in Padmanaban et al. (2010), by aggregating traffic reports and/or point-to-point travel times.

5.4 Further Evaluation of Mapping Ability

A more detailed evaluation of the system's mapping ability (specifically, the route building method) can be carried out if there are several users on a non-mapped city.

6 Conclusion

We have successfully implemented CommYouTer, a mobile app and server that facilitates transit journaling, and provides traffic sensitive routing. The system can be used anywhere the traffic conditions are similar to those in Metro Manila, specifically, if public transit is primarily headway-based, and traffic conditions can drastically alter travel times on a regular basis.

References

Bast, H.: Car or public transport two worlds (2009)

Bast, H., Funke, S., Matijevic, D.: TRANSIT: ultrafast shortest-path queries with linear-time processing. In: 9th DIMACS Implementation Challenge Shortest Path, DIMACS, 01/2006 (2006)

Ching, Albert M. L.: A User-flocksourced bus experiment in Dhaka. Massachusetts Institute of Technology, Department of Urban Studies and Planning (2012)

Delling, D., Pajor, T., Werneck, R.: Round-based public transit routing. In: Proceedings of the 14th Meeting on Algorithm Engineering and Experiments (2012)

Delling, D., Sanders, P., Schultes, D., Wagner, D.: Highway hierarchies star. In: 9th DIMACS Challenge on Shortest Paths (2007)

Dijkstra, E.W.: A Note on two problems in connexion with graphs. Numer. Math. **1**(1), 269–271 (1959)

Foursqare: Retrieved 20 Oct 2013 from https://foursquare.com

Geisberger, R., Sanders, P., Schultes, D., Delling, D.: Contraction hierarchies: faster and simpler hierarchical routing in road networks. In: WEA 08 Proceedings of the 7th International Conference on Experimental Algorithms (2008)

Robusto, C.: The cosine-haversine formula. Am. Math. Mon. **64**, 38–40 (1957)

Goldberg, A., Harrelson, C.: Computing the shortest path: A* search meets graph theory (2004)

Hall, M., Frank, E., Holmes, G., Pfahringer, B., Reutemann, P., Witten, I.: The WEKA data mining software: an update. ACM SIGKDD Explorations Newsletter (2009)

Hart, P.E., Nilsson, N.J., Raphael, B.A.: A formal basis for the heuristic determination of minimum cost paths. Syst. Sci. Cybern. IEEE Trans. **4**(2) (1968)

Jariyasunant, J., Work, D., Kerkez, B., Sengupta, R., Glaser, S., Bayen, A.: Mobile transit trip planning with realtime data (2009)

JTransforms: Retrieved 20 Oct 2013 from https://sites.google.com/site/piotrwendykier/software/jtransforms

Moovit: Retrieved 20 Oct 2013 from http://m.moovitapp.com/English/index.html

ParaSaTabi: Retrieved 20 Oct 2013 from http://www.parasatabi.com

Padmanaban, R., Divakar, K., Vanajakshi, L., Subramanian, S.: Development of a real-time bus arrival prediction system for Indian traffic conditions. University of California Transportation Center (2010)

Reddy, S., Mun, M., Burke, J., Estrin, D., Hansen, M., Srivastava, M.: Using mobile phones to determine transportation modes. ACM Trans. Sensor Networks. **6**(2), Article 13, p. 27 (2010)

Republic of the Philippines, Department of Transportation and Communications: Philippine Transit Information Service (2013). Retrieved 20 Oct 2013 from http://www.dotc.gov.ph/index.php

Sanders, P., Schultes, D.: Highway hierarchies hasten exact shortest path queries. In: ESA 2005, LNCS 3669, pp. 568–579 (2005)

Thiagarajan, A., Biagioni, J., Gerlich, T., Eriksson, J.: Cooperative transit tracking using smart-phones. In: SenSys 10, 3–5 Nov 2010, Switzerland (2010)
Tiramisu: The real-time bus tracker. Retrieved 20 Oct 2013 from http://www.tiramisutransit.com
Waze (Community-Based Traffic and Navigation Application): Retrieved 20 Oct 2013 from https://www.waze.com

Adaptation of Automatic Information Extraction Method for Environmental Heatmaps to U-Matrices of Self Organising Maps

Urszula Markowska-Kaczmar, Agnieszka Szymanska and Lukasz Culer

Abstract The paper is focused on adaptation of the information extraction method for environmental heatmaps to U-matrices of Self Organising Maps—SOM neural networks. Our method bases on OCR, image processing and image recognition techniques. The approach was designed to be as much as possible general but information acquired from a heatmap and the form of a heatmap defer depending on the heatmap type. In the paper we introduce some dedicated processing steps while trying to minimize the number of changes in the previously proposed method. The results for U-matrices of SOM neural network are evaluated in the experimental study and compared with efficiency of the method for environmental maps.

1 Introduction

Heatmap is a diagram type that can show different dependencies using color as an indicator of third dimension value, where two other dimensions are usually shown on the axes. To bind color values to some numeric values, heatmaps have legend with proper descriptions. Typically blue or purple color is used to show low values, but it is not necessary. Visualising information on heatmaps is used in many disciplines. The way of showing feature values on the maps can differ depending on the application.

U. Markowska-Kaczmar (✉)
Department of Computational Intelligence, Wroclaw University of Technology, Wyb. Wyspianskiego 27, 50-370 Wroclaw, Poland
e-mail: urszula.markowska-kaczmar@pwr.edu.pl

A. Szymanska · L. Culer
Department of Technical Informatics, Wroclaw University of Technology, Wyb. Wyspianskiego 27, 50-370 Wroclaw, Poland
e-mail: a.szymanska@pwr.edu.pl

L. Culer
e-mail: lukasz.culer@pwr.edu.pl

G.A. Tsihrintzis et al. (eds.), *Intelligent Computing Systems*,
Studies in Computational Intelligence 627, DOI 10.1007/978-3-662-49179-9_12

Heatmaps representing genetic data matrices usually consist of different colored rectangles and descriptions on both axes. They enable the exploration and detection of patterns, correlations (Verhaak et al. 2006) and finding sub-patterns inside a subset of genes and conditions based on biclustering or clustering data rows and columns separately (Sun and Li 2013).

In medicine a trend to use thermographic heatmaps for diagnosing diseases becomes more and more visible. Thermography can be used as a non-invasive way of detecting breast cancer (Ng 2009). To make this process easier algorithms for automatic data extraction were presented. Some of them base for example on k- and fuzzy c-means (EtehadTavakol et al. 2010), texture features (Acharya et al. 2012), fractal dimension (EtehadTavakol et al. 2010) or independent components analysis (Boquete et al. 2012). Also other medical symptoms such as lungs diseases (Klosowicz et al. 2001) can be observed on thermographic heatmaps.

Other heatmaps type which is widely known from everyday life are environmental heatmaps (Fig. 1). They often use gradients to visualise such parameters as temperature, air pressure or humidity. On both axes of such maps occur geographic coordinates. There were some approaches to get environmental information from heatmap semi-automatically. They based on annotation tool and OCR techniques (Moumtzidou et al. 2012, 2013).

Heatmaps are also used to visualise Kohonen's Self Organising Maps (SOM). This neural network creates a mapping from a N-dimensional space of input vectors into a two-dimensional plane, which shows relations between patterns. These relations are visualised by heatmap called U-matrix (Fig. 1).

The chapter starts with problem formulation that is described in Sect. 2. Then, because we want to take as much as possible from our developed HInEx method (Markowska-Kaczmar et al. 2014), its original form is presented in Sect. 3. This section shows also original application of the method to environmental heatmaps. Next section is devoted to adaptation of HInEx to SOM U-Matrix maps. The

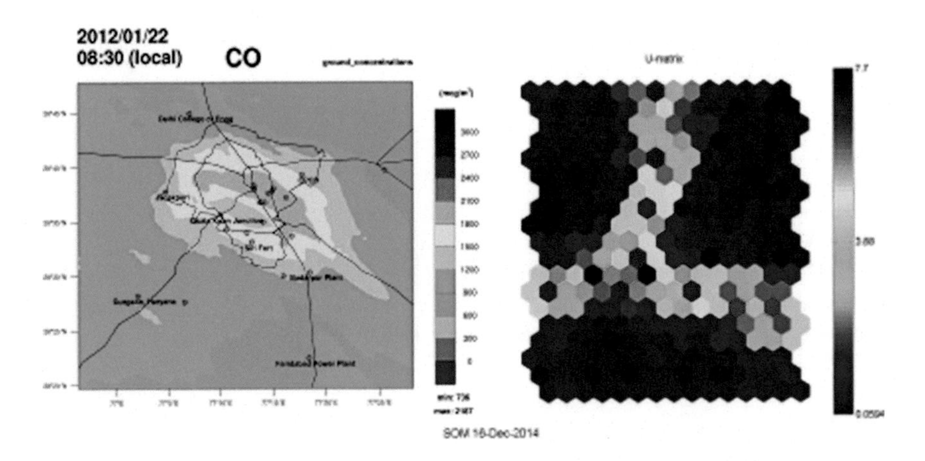

Fig. 1 Examples of heatmaps: *left* environmental heatmap and *right* SOM U-matrix

following section describes the Matlab-based SOM generator, that was used to generate our test set. The experimental study (Sect. 6) contains information about proceeded tests and shows obtained results, which are interpreted in Sect. 7.

2 Problem Formulation

In the paper (Markowska-Kaczmar et al. 2014) we have presented the method named HInEx for automatic information extraction from heatmaps but we were concentrated our research on environmental maps. Its example is shown in Fig. 1. Although generally, heatmaps regardless of the application area have some common ideas, they differ in the way of information visualization and consequently the way of information extraction must be adjusted.

Our aim was to develop a general method of automatic information extraction from heatmaps that could be easy adapted to more specific applications with low effort of work. For environmental maps we got automatically the value of the parameter visualised on the map, for example a temperature for the city in a query given by the user. To achieve it we recognize the latitude and longitude values and a color value using legend detection and then merge it with the information about the city's location from a google maps module. In this chapter we are interested in adaptation of the HInEx method to extract information from SOM neural network shown in Fig. 2 represented by U-matrix. A kind of heatmaps that is presented in Fig. 1.

The network consists of two layers: input and SOM. The input vector $x = [x_1, x_2, \ldots, x_n]$ corresponds to a given pattern. SOM layer is a competitive structure of neurons—a winning neuron is determined for the input vector based on the similarity between input vector and neuron's weight vector. During training, the SOM

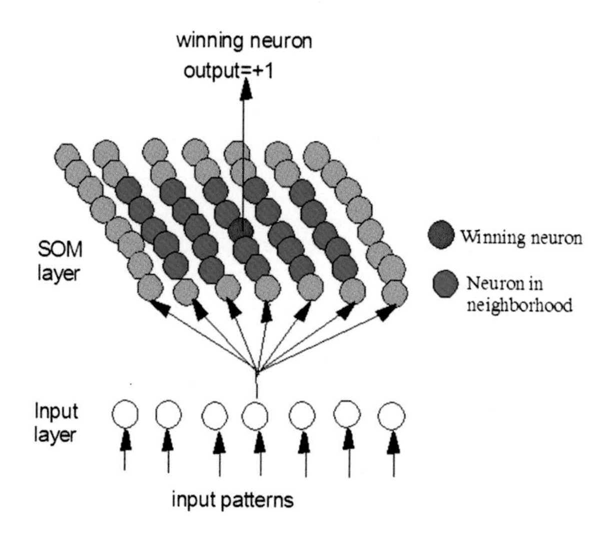

Fig. 2 SOM neural network architecture

finds the neuron that has weights with the least distance from the training pattern. It then changes the neuron's weights to increase the similarity to the input vector. It also influences the weights of neighboring nodes. Different input vectors activate different winning neurones that influence different neighbors. The overall effect of this process is to move the weights associated with the output neurones such that they map the distribution of the input vectors. Consequently, after training is ended, the weights of each output neuron model the features that characterize a cluster in the input data. In effect, SOM creates through an unsupervised learning process a mapping from multidimensional space of the input data to a two-dimensional plane of neurons. This mapping shows topological relations of the multidimensional space.

To visualize the structure of SOM layer the "Unified distance matrix" (U-matrix) technique was developed. It is used to show similarities between neurons weights and weights of their neighbours and therefore also similarities in the data of the input space. U-matrix is a heatmap, where small distances between weights are depicted by cells with blue colors while red color indicates more widely separated weights. Thus, groups of cells with blue color in U-matrix can be considered as clusters, and the dark red parts as the boundaries between the clusters. This representation can help to visualize the clusters in the high-dimensional spaces.

In the case of research presented in this chapter, HInEx will be adapted in order to find the number of clusters in SOM on the basis of U-matrix.

3 HInEx—Heatmap Information Extraction

As we mentioned, the *HInEx* method (from *H*eatmap *IN*formation *EX*traction) was originally developed for retrieving information from environmental heatmaps. Created interface gives the possibility to get automatically the value of the characteristics for the location given by the user in the form of geographical coordinates or a city name.

3.1 The Idea

The method consists of several steps showed in Fig. 1. A heatmap image is taken as the algorithm input. Initially, our aim was to separate a heatmap area from the legend and the rest of the image elements, such as title and other descriptions. Because heatmaps contain large number of colors, especially when they do not have quantized color scale, we decided to reduce the amount of information by describing the heatmap as a tree using hierarchical clustering. Leaves of this tree are created by splitting the original image into parts until it is possible to describe a given part with one dominant color. Dominant colors are then bounded to the key descriptions and axes coordinates (Fig. 3).

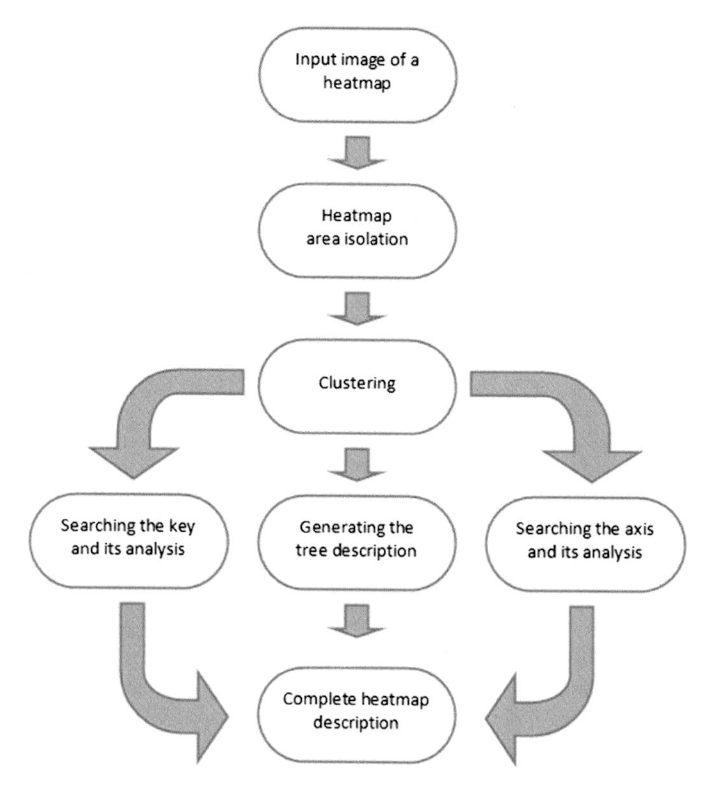

Fig. 3 The general idea of the HInEx method of automatic information extraction from heatmaps

3.2 Heatmap Area Isolation

The input for this step is the original image of a heatmap. The aim of this phase is removing all additional elements existing in the heatmap image (the title, the key). Usually, all elements are surrounded by the common background, so an image is a collection of elements separated by the background color pixels. It is worth noticing that in a heatmap image the map area is usually the biggest one, i.e. it has the biggest surface. In order to extract the heatmap area from the original image, an operation similar to thresholding is performed.

As a background color the averaged color of corner pixels of the heatmap image is considered. To find all pixels from the background the similarity of pixels is considered which is measured by the distance between their colors. We use RGB color model, so the color is represented by 3 numbers from the range of 0–255. Each of them corresponds to one of the color component, respectively red, green and blue. The distance is measured on the basis of the Manhattan norm. Colors were considered as similar, if the distance between them was lesser than a threshold

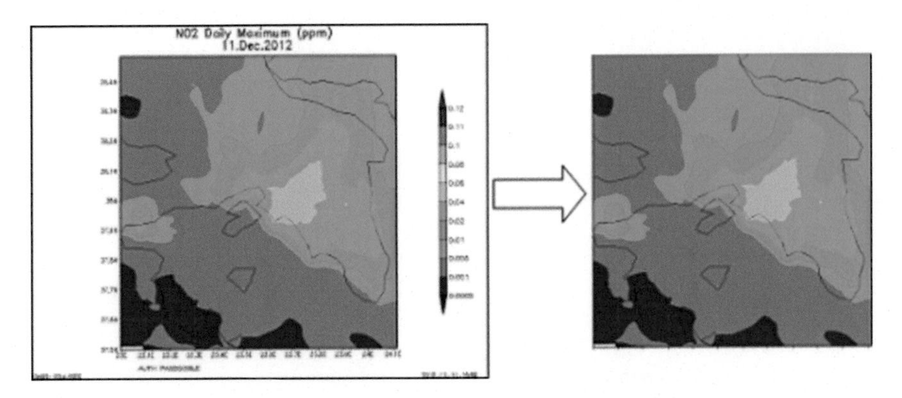

Fig. 4 The extracted heatmap area from a heatmap image

value, which is defined as BACKGROUND MAX DISTANCE and on the basis of preliminary experiments is set to 5.

After separation of background pixels, they are changed to the white color. Other pixels become black. Next, a uniform area, defined as a set of black pixels having in the nearest neighborhood (pixels with x-coordinate or y-coordinate greater or lesser by 1 are considered pixels' neighbors) black pixels from their own uniform area are selected. Then, the area with the biggest amount of pixels is isolated as the heatmap area. This step provides new a smaller image, extracted from the original one, limited by borders of the smallest rectangle containing the selected area. The exemplary result is shown in Fig. 4. The whole procedure in the pseudocode is given as Algorithm 1.

Algorithm 1: The extraction of the heatmap area

Determine background color;
Using the background color proceed thresholding-like operation;
Split pixels into uniform black-colored areas;
Find GUBA - the Greatest Uniform Black colored Area;
Find the smallest rectangle containing GUBA (its upper-left and bottom -right corners define the extracted map);

3.3 *Clustering Image Pixels Based on Colors*

The agglomerative hierarchical clustering method (Agarwal et al. 2010; Day and Elelsbrunner 1984) is used in this step of the *HInEx* method. Hierarchical clustering is a general family of clustering algorithms that build nested clusters by merging them successively. Its aim in the *HInEx* method is to decrease the number of colors or discretize the values corresponding to the environmental feature expressed in the heatmap. This step uses an image acquired in the previous step. It characterizes by a wide range of colors, as in the original heatmap. The output of this step is the heatmap with a reduced number of colors.

Let us remind that colors are represented in 3-dimentional space, given by the RGB color model. The idea of the applied clustering algorithm is to merge clusters until the stop condition when all possible unions will be achieved. First, the clusters with unique colors are joined (pixels with the same color are merged together, independently of their position in the map). In each iteration a merge attempt for two closest clusters is proceeded. If they could not be connected because of the stop condition, the next closest pair of clusters is considered. If there is no possibility to merge any of the cluster pairs, this part of the algorithm is finished. Then, all colors contained in a cluster are replaced by the color of the cluster center.

The cluster center is considered as an average color of all colors included in the cluster. In the cluster description the radius is used. It is the greatest distance between the center of a given cluster and a color contained in the cluster. The distance between colors is measured using euclidean norm. The distance between clusters is the lesser of the two values: the distance between both cluster centers decremented by the radius of the first cluster and the distance between both cluster centers decremented by the radius of the second cluster. The stop condition is given by the maximum radius of cluster. Before connecting two clusters their common center and the radius is determined. If the radius for the new cluster would be greater than `MAX CLUSTER RADIUS` = 60, the clusters cannot be merged. The effect of clustering is shown in Fig. 5. The algorithm in pseudocode is presented in Algorithm 2.

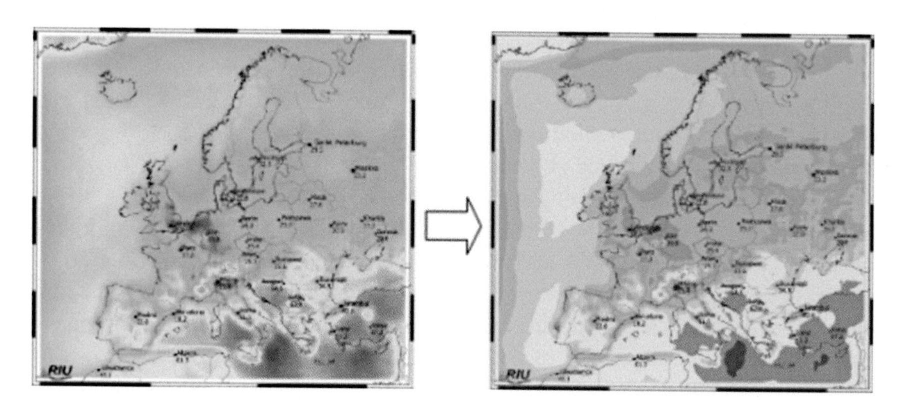

Fig. 5 The heatmap image before and after clustering

Algorithm 2: The clustering method

```
Find the clusters with unique colors in the heatmap area;
Store the clusters in the cluster base;
Set stop condition unsatisfied;
repeat
    Determine the distance between each cluster pair;
    Find the pair of clusters with the smallest distance that could be merged;
    if the pair was found then
        Remove it from the cluster base;
        Create the merged cluster and put it into the cluster base;
    else
        set stop condition;
    end
until stop condition satisfied;
Replace all colors contained in each cluster with its color;
```

3.4 Generating Tree Description

In order to store the map in a more compact form and ready for quick response to a query referring to the environmental feature value visualized in a given heatmap location, the heatmap is represented in the form of a tree and then labeled. The input for this step of the *HInEx* method is the clustered heatmap image from the previous step.

Each node in the tree corresponds to a part of a given heatmap image. Its root represents the whole heatmap. Next, nodes are created as follows. Starting from the root, the node is checked whether the image assigned to it contains uniform color area. If it is true it becomes a leaf. Otherwise, its image is divided into 4 parts. Next, every part is stored in new child nodes. The procedure is repeated for every child node. Construction of the tree is completed if the nodes cannot be split anymore. There are two conditions turning a node into a leaf. The first one refers to the dominating image colors of the leaf. If the percentage value of the number of pixels of the most common color in a leaf image is greater than `MIN CONTAIN` = 80 % then the node becomes a leaf. The second one is about maximum depth of the tree limited by `MAX DEPTH` = 5 parameter. This condition limits the image fragmentation. The recursive algorithm of the tree generation in pseudocode is shown in Algorithms 3 and 4. At the end of this step, the input heatmap is transformed to the output heatmap image. For each leaf its area in the input image is filled by a leaf color, i.e. color dominating in the images corresponding to it. The example of the output heatmap provided in this step is shown in Fig. 6.

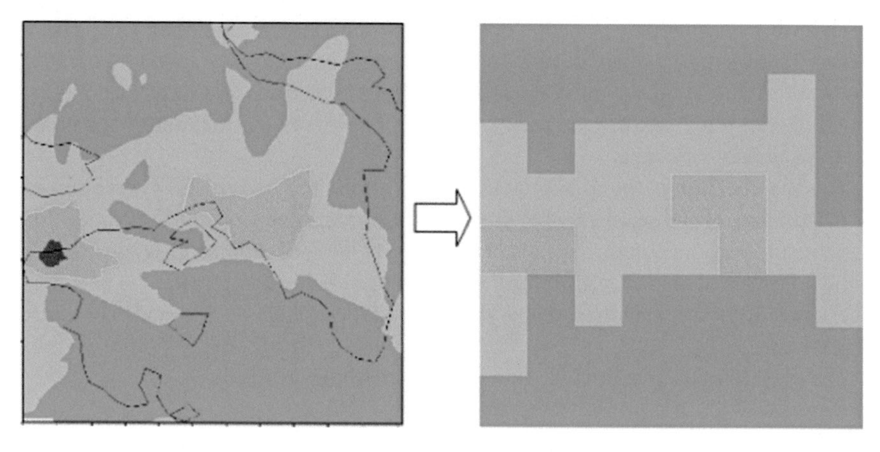

Fig. 6 On the *right side* the heatmap image clustered, on the *left side* the example of the heatmap after filling it by leaf colors

Algorithm 3: The tree generation method

Create the root node from the input image;
Execute Process Node algorithm for the root node;
Fill areas in the input image with the leaf color corresponding to them;

Algorithm 4: Process Node (input node)

if *the input node cannot be converted into leaf* **then**
  Split the input node image into 4 parts;
  Create nodes from each of them;
  for *every created node* **do**
    Execute Process Node;
  end
else
  Convert the node into the leaf;
end

3.5 The Key Search and Its Analysis

The heatmap key caries a lot of information. Therefore it is searched and then analysed in this step of our method. As an input the heatmap image in the original form is considered, as in Fig. 1. The procedure applied in this step uses information about a heatmap area location in the whole image. Also the list of colors occurring in the map is needed. The output of this step provides values of an environmental feature acquired from the key.

First, to search the key, for each color used in the extracted heatmap area, pixels of this color are searched outside of this area. Then minimum and maximum positions of these pixels must be found in both directions (vertically and horizontally). For a given color, on the basis of the minimum and maximum positions, the medians are calculated.

The key location is found and its direction (horizontal or vertical) is recognized by calculation of the pixel number from the minimum and maximum position list that are similar to the medians. Labels for color values are defined as labels nearest to the left pixel of an upper limit for the occurrence of this color inside the key. The labels positions are searched using the method described in Sas and Zolnierek (2013).

The algorithm in pseudocode is presented in Algorithm 5.

Algorithm 5: The key search method

for *all colors from the heatmap area* **do**
 Find all pixels of this color outside the heat map area;
 Find pixels in this set with maximal and minimal coordinates on at least one axis;
end
For all colors find median of minimal coordinates and median of maximal coordinates for each axis;
Count how many pixel positions from the minimal and maximal position list are similar to the median for each axis;
if *there are more pixels similar to the median of horizontal axis coordinates* **then**
 the key is vertical;
else
 the key is horizontal;
end
if *the key is vertical* **then**
 Set the boundaries of the key on the X axis as the median of the minimal and the median of the maximal positions;
 Set the boundaries of the key on the Y as the smallest minimum and the highest maximum from the list of all colors positions;
else
 Perform similar operations for the horizontal key;
end
Find label description for each color as the label nearest to the left pixel of an upper limit for the occurrence of this color inside the key;

3.6 The Axis Search and Their Analysis

This step can be proceeded simultaneously with previous two steps. Its aim is extraction of longitude and latitude values of the locations presented in the geographical map visualised in a heatmap image. As an input in this step the method gets the original heatmap image and the clustered heatmap area. The output provides parameters for two linear functions. These functions are used to obtain geographical coordinates for pixel coordinates in the clustered map area.

First, to generate the function parameters, strings in the original heatmap image, close to the heatmap area bounds are searched. Strings corresponding to the geographical coordinates are usually located near the extracted heatmap area bounds. Then, the strings are separated into four groups—left border group, right border group, top border group and bottom border group. The left and right groups are the vertical groups, respectively the top and bottom border groups are horizontal. Vertically located strings describe the latitude, and horizontally the longitude. Usually, descriptions can be found only on one side of a heatmap, therefore strings on the other side are removed because they are incorrectly recognized. On the other hand, there are heatmaps having the same description on both sides. Therefore the side with the higher number of strings than in corresponding one, seems to be more promising for further recognition. Additional points will provide more data to determine linear parameters, therefore the group with the lesser number of strings is removed.

The next operation is parsing both groups of strings and storing them in the uniform form. For instance both strings: *12.345W* and *−12.345* will be stored as −12.345. Strings that cannot be parsed are removed. Then the appropriate coordinates are linked with strings. For horizontal strings, it will be the x-coordinate value, for vertical strings the y-coordinate value of the string center relative to y-coordinate value of the top map border. In consequence, a set of points with two values and a parsed value from a string and the determined coordinate value are given for vertical and horizontal groups.

Next, the best linear transformation describing these points is searched. Classical interpolation methods did not give good results. It is caused by errors in the string recognition. In the *HInEx* method to find this transformation, for each pair of points the horizontal line passing through two points is determined. Then for each line, other points are tested whether they match the line. Finally, the parameters of the line with the highest number of matched points, are selected. This method of searching linear parameters is efficient even for sets with a big amount of incorrectly recognized strings. The linear parameters for the vertical line are defined in the same way. The details of this procedure is shown in pseudocode in Algorithm. 6. It is worth mentioning that during checking the line parameters the decision about matching is based on the difference between the values obtained from considered linear function and the value obtained from the parsed string. If the distance is greater than `MAX RELATIVE AXIS ERROR` = 10 % then it does not match the line. The algorithm in pseudocode is presented in Algorithm 6.

Algorithm 6: The axis search method

Recognize strings in the input image;
Separate the strings into top-border, bottom-border, left-border, and right-border strings;
From the horizontal strings (top-border and bottom border strings) choose one with the greatest number of strings;
Perform similar steps for the vertical string group;
Parse strings to obtain geographical coordinates;
Remove the invalid ones;
Extract strings coordinates relative to the top and left map border;
for *every pair of points in the group* **do**
  Determine the line passing through them;
  Count the number of other points that match it;
end
Store the linear parameters of the line, which has the greatest number of matched points;
Proceed the last 2 steps for the second group;

3.7 Complete Heatmap Description

This is the final step of the algorithm. Here all the data are combined. As an input, this step takes the tree generated in previous steps, as well as the key and axis description. The output of this step is a fully described heatmap. In this step description of trees is supplemented by real parameters obtained from the key and axis description. First, the algorithm checks every leaf boundary position on the extracted heatmap. Then, using linear functions obtained from axis description, the geographical coordinates are set. Next, the leaf real value is obtained from the key color label.

4 SOM Cluster Number Extraction Based on U-Matrix

Using U-matrices allows to determine the number of clusters (groups) existing in the training data. The groups naturally split the data based on their similarity measured by the distance between them (Ultsch 1993). This approach is based on fact, that neurons representing single group are "close" to each other—distances between them are small. The distances between neurons from different clusters are usually big—neurons are "far" from each other, visually separating groups from each other. Usually, humans can very easily notice those borders and separate groups of neurons and consequently obtain a number of groups. A cluster membership for a given pattern is assigned on the basis of the neuron position in the

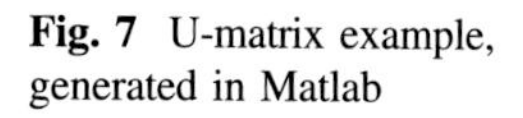
Fig. 7 U-matrix example, generated in Matlab

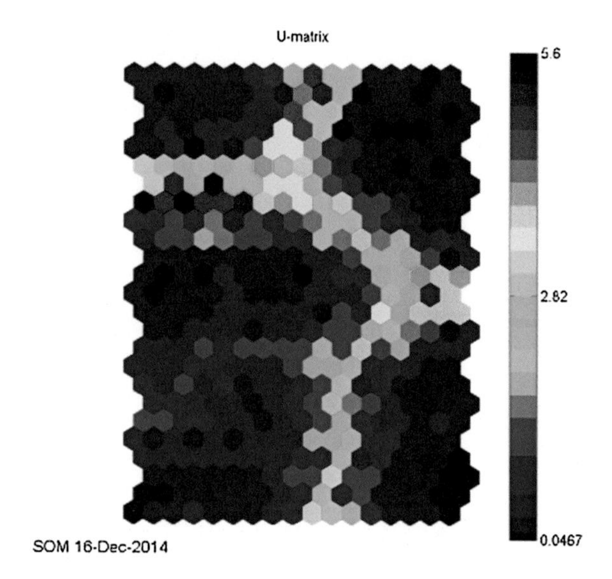

SOM network that responded for that pattern. Regions of neurons create clusters. The number of clusters is essential because it influences on the quality of clustering. Various clustering methods need to set the number of clusters in advance (for instance K-means). In this paper we tried to use *HInEx* method to find the number of clusters in SOM network visualised by U-matrix, Fig. 7.

4.1 The Idea of HInEx Application to SOM U-Matrix

First steps are common with environmental maps application. Firstly, the proper heatmap area is isolated from original image, to separate it from other heatmaps' elements and make further processing easier. Next step is the use of developed clustering algorithm on isolated heatmap area to significantly reduce colors amount on the map. Then, the key area is determined, based on colors from previously clustered map, and those colors are linked with textual key description. The next steps are unique for SOM U-matrices. Firstly, for further use single distance representation area between neurons is obtained from image (e.g. hexagon or square). Secondly, relationships between each color based on the key are determined—this step gives the information which color represents lower and which higher values. The color representing minimal distance is obtained and, based on that, specific threshold-like operation is processed to separate minimal distance areas from others. At this point heatmap area is represented by irregular white and black image. The next step are morphological dilatation- and erosion-like operations. This results with much more regular image, which is used at the final step—the uniform areas

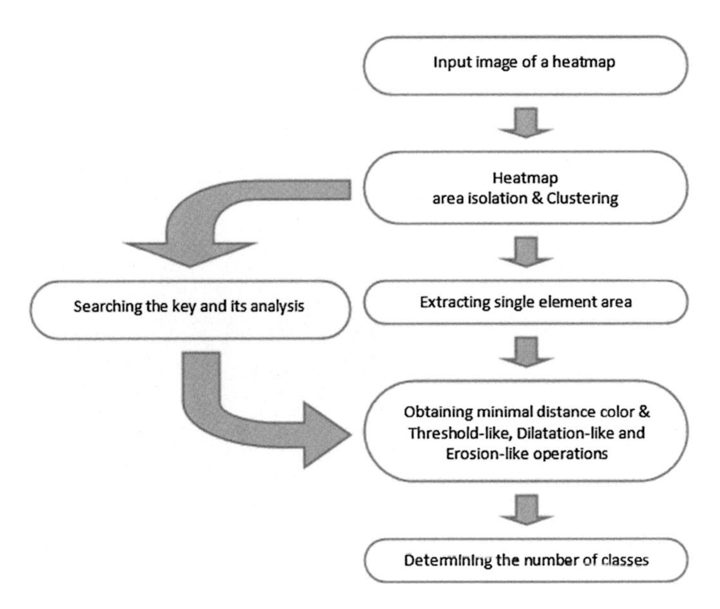

Fig. 8 The idea of HInEx application to SOM U-matrix

are extracted and the number of white uniform areas are considered as the number of heatmap classes. All steps are visualized on diagram in Fig. 8

In the next sections we will focus on parts of method that differ from the procedures used for the environmental maps. Exemplary result of Heatmap area isolation is presented in Fig. 9.

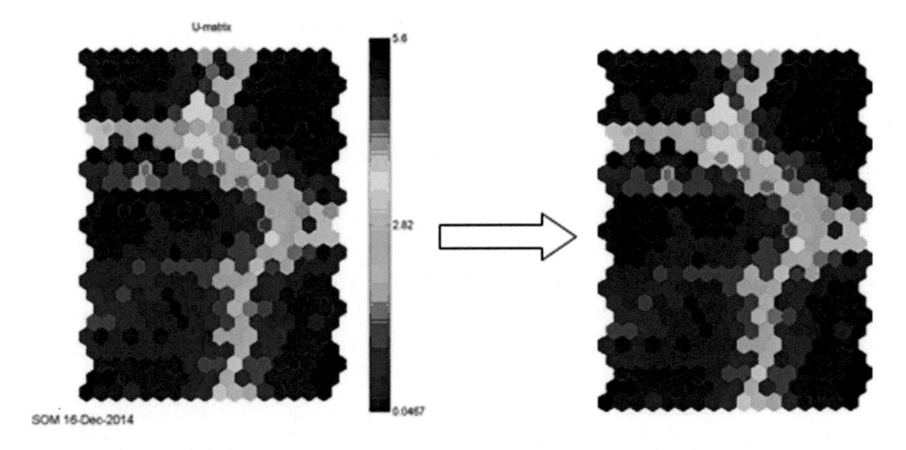

Fig. 9 Heatmap area isolation

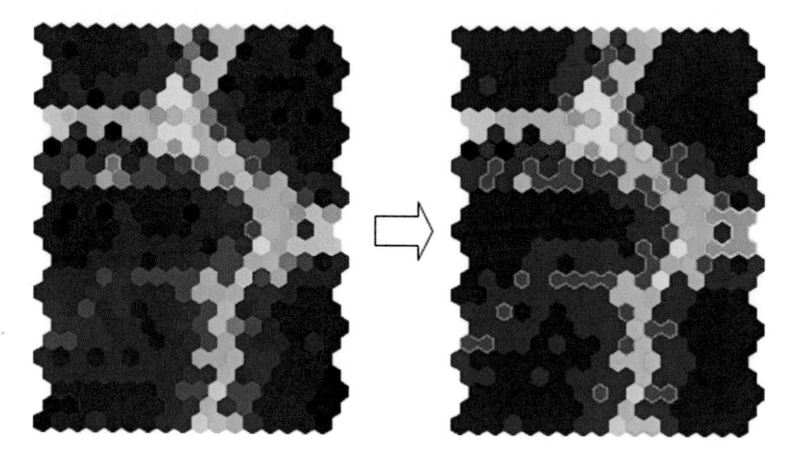

Fig. 10 Clustering operation—unclustered image on the *left* and clustered on the *right*

4.2 Clustering

The step is performed analogously as for environmental maps, but in this case this operation has double goal. At first, as previously, it allows to pack greater number of colors (thus the values) under single label, what will be used to extract minimal distance areas. At second, clustering reduces small distance values fluctuations, which could lead to irregularities in searching minimal distance areas, connected with single classes. Example of image before and after clustering is in Fig. 10.

4.3 Extracting a U-Matrix Cell Corresponding to a Single Distance Between Neurons

The task of this step is to extract a cell being a uniform area which represents single distance between two neurons. It is based on the fact that they are the smallest elements of U-matrix image, therefore areas representing them should be also the smallest areas on a heamap area. The algorithm that is shown in Algorithm 7 checks every uniform area and chooses the smallest one to be a single distance between neurons representation area.

Algorithm 7: Extracting single element area creating a U-matrix cell

```
Obtain all uniform areas from clustered heatmap;
Set minimalSingleElementArea to ∞ ;
Set currentSingleElementArea to null;
for every uniform area on clustered heatmap do
    Find smallest rectangle containing current uniform area;
    Determine its area;
    if current uniform area area is smaller than minimalSingleElementArea
    then
        Set minimalSingleElementArea to current uniform area area;
        Set curretSingleElemetArea to current uniform area;
    end
end
return currentSingleElementArea;
```

4.4 Searching a Color Representing the Minimal Neuron Distance in SOM

It is based on a previously extracted area of the heatmap key and labeled colors. The key usually contains information about values of distances between neurons. However obtaining minimal distance color is impossible at this moment, because many colors can be labeled with the same value during key search and analysis step. To solve this problem the idea is to find direction of growing values in the heatmaps key and sort colors from clustered heatmap by their positions on the heatmaps key. In this way accurate values will not be acquired, but it gives information about relations between those colors, especially which one refers to the biggest and to the lowest value.

The algorithm starts with picking colors labeled with the lowest and the highest labeled values and separating them into two groups. Each color has connected with its position on the key. For vertical keys it will be y-coordinate, and for horizontal keys x-coordinate. The procedure describing how to find a color representing the minimal distance between neurons is presented in Algorithm 8. Inside those groups, average position of points with those labels is calculated. Then, for example, if for vertical key colors from a group labeled with the lowest value an average y-coordinate is lower than average y-coordinate of colors from group labeled with highest value it means that values on heatmaps key increases with y-coordinate. The ideal projection is not necessary—relationships between colors are sufficient. On this basis, output order of colors is determined and the color connected with the lowest value is obtained. The detailed procedure is presented in Algorithm 8.

Algorithm 8: Searching a color representing the minimal neuron distance in SOM

```
Find key label containing minimal value;
Find key label containing maximal value;
Create minimalValueGroup;
Create maximalValueGroup;
for every labeled color from clustered heatmap do
    if current color label equals minimal value label then
        Add current color to minimalValueGroup;
    else if current color label equals maxinal value label then
        Add current color to maximalValueGroup;
end
if current key type is VERTICAL then
    Set getPositionOnKey(color) function to getY-coordinateOnKey(color);
else
    Set getPositionOnKey(color) function to getX-coordinateOnKey(color);
end
Count averageMinimalPosition using getPositionOnKey(color);
Count averageMaximalPosition using getPositionOnKey(color);
if current key type is VERTICAL then
    if averageMinimalPosition < averageMaximalPosition then
        Set keyGrowingType to VERTICAL VALUE ASCENDING WITH
        POSITION;
    else
        Set keyGrowingType to VERTICAL VALUE DESCENDING WITH
        POSITION;
    end
else
    if averageMinimalPosition < averageMaximalPosition then
        Set keyGrowingType to HORIZONTAL VALUE ASCENDING
        WITH POSITION;
    else
        Set keyGrowingType to HORIZONTAL VALUE DESCENDING
        WITH POSITION;
    end
end
return sorted labeled colors from the clustered heatmap by keyGrowingType
and getPosition(color);
```

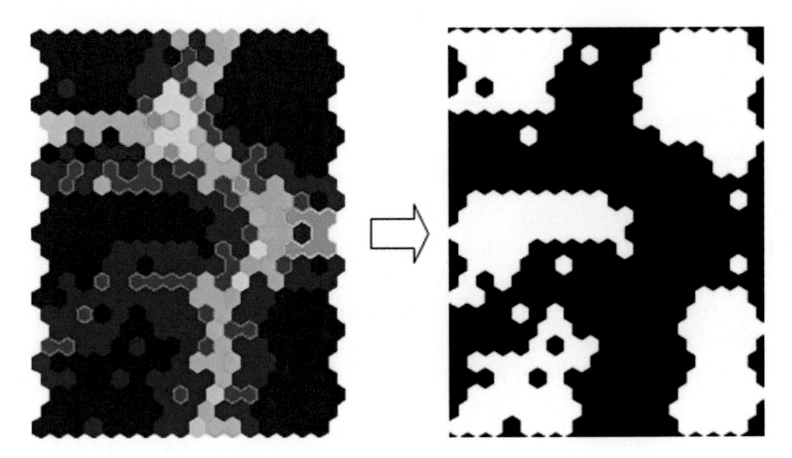

Fig. 11 Threshold-like operation

4.5 Threshold-like Operation

The algorithm used in this phase is shown in Algorithm 9. It uses the color representing the minimal distance obtained in the previous step. It simply replaces minimal distance color by white and other colors by black, what results in black-white image, similar to those after standard threshold operation. It allows to separate minimal-value areas, from others. The white areas correspond to clusters.

As it can be seen in Fig. 11, the output image is irregular—it contains many areas connected with single distances between neurons, defined by color that differs from its neighborhood. Therefore the next step is dedicated to removing them.

Algorithm 9: Threshold-like operation

```
for every pixel on clustered heatmap do
    if current pixel color equals minimal distance color then
        Set current pixel color to white;
    else
        Set current pixel color to black;
    end
end
```

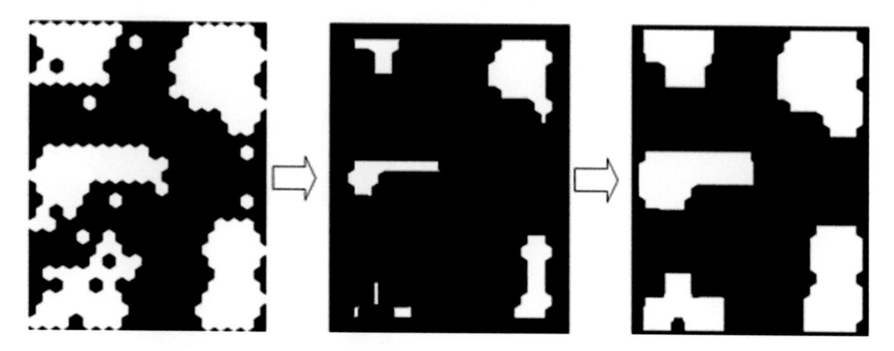

Fig. 12 Dilatation and erosion-like operations

4.6 Dilatation and Erosion-like Operations

Dilatation-like operation is inspired by morphological image processing dilatation. It uses previously extracted single element area. In the algorithm that in pseudocode is shown in Algorithm 10, each black pixel is replaced with the smallest black rectangle, that contains this single element area. It allows to eliminate small white areas, that shouldn't be considered lately as separate groups.

Erosion is an identical operation to the dilatation, but it is processed for white pixels. Proceeding those two operations results with reduction of white areas number and making remaining white areas more uniform.

Exemplary sequence of dilatation and erosion-like operations is presented in Fig. 12.

Algorithm 10: Dilatation-like operation

Obtain smallestRectangleContainingSingleElementAreaSide from SingleElementArea;
for *every pixel on clustered heatmap* **do**
    if *current pixel color equals black* **then**
        for *every existing on image pixel, which x-coordinate ∈ <currentPixel x-coorinate - smallestRectangleContainingSingleElementAreaSide / 2 - 1; currentPixel x-coorinate + smallestRectangleContainingSingleElementAreaSide / 2 + 1> and y-coordinate ∈ <currentPixel y-coorinate - smallestRectangleContainingSingleElementAreaSide / 2 - 1; currentPixel y-coorinate + smallestRectangleContainingSingleElementAreaSide / 2 + 1>* **do**
            Set color to black;
        end
    end
end

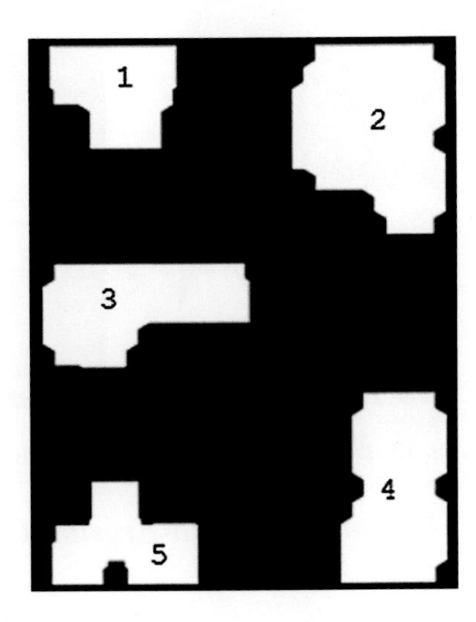

Fig. 13 Determining the number of clusters

4.7 *Searching for the Number of Groups in SOM*

This is the final step of algorithm. During this step each uniform area on a processed image is checked and if its color is white—it is considered as a single cluster. All those areas are counted and its number is amount of total clusters distinguished by Self Organized Map. Example of heatmap with tagged clusters is shown in Fig. 13 and the procedure in pseudocode in Algorithm 11.

Algorithm 11: Determining number of classes

```
for every uniform area on processed heatmap do
    Set currentlyFoundClasses to 0;
    if current uniform area is white then
        Increase currentlyFoundClasses by 1;
    end
end
return currentlyFoundClasses;
```

5 SOM Generator Description

To generate a test set of SOM U-matrices a random generator was created using MATLAB SOM Toolbox. It generates U-matrices with a random cluster number. Every cluster contains vectors of patterns which are similar to each other—the difference between their elements cannot be greater than 1. The number of vectors in clusters is also random. This algorithm in pseudo code is shown in Algorithm 12.

Algorithm 12: Random SOM U-matrix generator

```
Set class number to integer random number between MIN CL NUMBER and
MAX CL NUMBER;
Set feature number to integer random number between MIN FEAT NUMBER
and MAX FEAT NUMBER;
Generate base classes vectors:
for I = 1 to class number do
    for J = 1 to feature number do
        SOM Matrix [I, J] = random number between 0 and MAX MATRIX
        VAL;
    end
end
Generate other classes members which should be similar to base classes
vectors:
for I = 1 to class number do
    Set class members count to integer random number between MIN CL
    MEM and MAX CL MEM;
    for J = 1 to class members count do
        SOM Matrix extension [1, J] = SOM Matrix [I, J] + random number
        from MIN THRESHOLD to MAX THRESHOLD;
    end
    SOM Matrix = [SOM Matrix; SOM Matrix extension];
end
Generate SOM Data Struct from SOM Matrix using MATLAB SOM Toolbox;
Make SOM using SOM Data Struct;
Choose randomly hexagonal or rectangular neighborhood model;
Use random initialization;
```

In our tests we have chosen the following values of parameters:

MIN CL NUMBER = 2
MAX CL NUMBER = 8
MAX MATRIX VAL = 10
MIN CL MEM = 3
MAX CL MEM = 8
MIN THRESHOLD = −0.5
MAX THRESHOLD = +0.5

6 Experimental Study

Experimental study performed for environmental heatmaps was designed to test performance of each step of the method. It is presented in Markowska-Kaczmar et al. (2014). The experiment described in this chapter was performed to test the method as a whole. Tests were proceeded with 74 environmental heatmaps. Each heatmap was manually tagged with one place located on a heatmap and value assigned by the user for this place. Then the heatmap was processed with the algorithm. Tagged places coordinates were defined using Google Maps API and, based on them, the value from clustered map was obtained and compared with the tagged value. The relative difference between both values is an outcome. The results of the experiment are shown in Table 1. Final results depend on every step of algorithm—invalid isolation of heatmap area, heatmap key or axis detection.

Environmental maps with detected places are maps, on which tagged place was found. For those places values from map was obtained and compared to the tagged value. Relative errors greater than 100 % were considered as invalid value. For valid values, an average relative error was determined.

Experimental study for SOMs was design to evaluate the quality of the method that should return the number of clusters. To do this, 100 SOMs were generated using created generator. Each SOM was tagged with the number of groups that was used to create it. As an error we consider a difference between the recognized and the real number of groups in the image. Results are given in Table 2.

The average absolute error is an average difference between the recognized and the real number of clusters. The average relative error presents the previous value related to the average number of real clusters. The last parameter shows the number of SOMs, that our algorithm processed faultless.

Table 1 Results of environmental heatmaps information extraction

Tested feature	Result (%)
Environmental maps with detected place	59.0
Places with valid value obtained	79.5
Average relative error for valid values	16.9

Table 2 Results of SOM U-matrix group number extraction

Tested feature	Result
Average absolute error	0.59
Average relative error	12.74 %
Error-free SOMs	69 %

7 Conclusion

For environmental maps, only 59 % tagged places was found—it was caused mainly by invalid axis detection, as its performance was on 45.5 % (Markowska-Kaczmar et al. 2014). As we observed, the majority of huge differences between tagged and detected values was caused by invalid label parsing, so we decided to remove them from final results. For the remaining results, the average relative error was determined and equals 16.9 %. For SOMs, there were possible two kinds of sources of errors:

- incorrect dividing one cluster into two or more,
- incorrect merging two or more clusters into one.

The first kind of error was mainly caused by low number of examples in a given cluster with a high number of attributes describing each pattern—it caused a big distances between neurons inside a cluster that was incorrectly considered as a border between them.

The second error was common for clusters that were similar, which means that the number of neurons responding to the examples from both classes was high. The higher number of neurons associated with both clusters was making a border between them on U-Matrix less clear.

Future study will be focused on axes recognition improvement, which would make environmental maps information extraction more efficient. We also plan to adapt heatmap recognition method to other purposes, for example to detect building heat losses on thermographic maps.

Acknowledgements This work was partially supported by the Innovative Economy Programme project POIG.01.01.02-14-013/09- 00 and partially by the European Commission under the 7th Framework Programme, Coordination and Support Action, Grant Agreement Number 316097, ENGINE—European research centre of Network intelliGence for INnovation Enhancement. The authors also would like to thank dr Jerzy Sas for making available his application of the text detection for our research.

References

Acharya, U.R., Ng, E.Y.K., et al.: Thermography based breast cancer detection using texture features and support vector machine. J. Med. Syst. **36**(3), 1503–1510 (2012)

Agarwal, P., Alam, M., Biswas, R.: Analysing the agglomerative hierarchical clustering algorithm for categorical attributes. Int. J. Innov. Manage. Technol. **1**(2), 186–190 (2010)

Boquete, L., Ortega, S., et al.: Automated detection of breast cancer in thermal infrared images, based on independent component analysis. J. Med. Syst. **36**(1), 103–111 (2012)

Day, W., Elelsbrunner, H.: Efficient algorithms for agglomerative hierarchical clustering methds. J. Classif. **1**, 7–24 (1984)

EtehadTavakol, M., Lucas, C., et al.: Analysis of breast thermography using fractal dimension to establish possible difference between malignant and benign patterns. J. Healthc. Eng. **1**(1), 27–44 (2010a)

EtehadTavakol, M., Sadri, S., et al.: Application of K-and fuzzy c-means for color segmentation of thermal infrared breast images. J. Med. Syst. **34**(1), 35–42 (2010b)

Klosowicz, S.J., Jung, A., Zuber, J.: Liquid crystal thermography and thermovision in medical applications: pulmonological diagnostics. In: Systems of Optical Security, pp. 24–29. International Society for Optics and Photonics (Aug 2001)

Markowska-Kaczmar, U., Szymanska, A., Culer, L.: Automatic information extraction from heatmaps. In: The 5th International Conference on Information, Intelligence, Systems and Applications, IISA 2014, pp. 267–272. IEEE, NJ (2014)

Moumtzidou, A., Epitropou, V., et al.: A model for environmental data extraction from multimedia and its evaluation against various chemical weather forecasting datasets. Ecol. Inf. (2013)

Moumtzidou, A., Epitropou, V., et al.: Environmental data extraction from multimedia resources. In: Proceedings of the 1st ACM International Workshop on Multimedia Analysis for Ecological Data, pp. 13–18 (Nov 2012)

Ng, E.K.: A review of thermography as promising non-invasive detection modality for breast tumor. Int. J. Therm. Sci. **48**(5), 849–859 (2009)

Sas, J., Zolnierek, A.: Three-stage method of text region extraction from diagram raster images. In: Burduk, R. (ed.) Proceedings of the 8th International Conference on Computer Recognition Systems: CORES 2013, pp. 527–538. Springer, Berlin (2013)

Sun, X., Li, J.: Pairheatmap: comparing expression profiles of gene groups in heatmaps. Comput. Methods Programs Biomed. **112**(3), 599–606 (2013)

Ultsch, A.: Self-organizing neural networks for visualisation and classification, pp. 307–313. Springer, Berlin Heidelberg (1993)

Verhaak, R.G., Sanders, M.A., et al.: HeatMapper: powerful combined visualization of gene expression profile correlations, genotypes, phenotypes and sample characteristics. BMC Bioinf. **7**(1), 337 (2006)

Evolutionary Computing and Genetic Algorithms: Paradigm Applications in 3D Printing Process Optimization

Vassilios Canellidis, John Giannatsis and Vassilis Dedoussis

Abstract 3D printing is a relatively new group of manufacturing technologies, methods and processes that produce parts through material addition. 3D printing technologies are mainly employed for the fabrication of prototypes and physical models during product design and development; however as they continuously improve in terms of accuracy and range of raw materials they are increasingly employed in the actual manufacturing process. This puts a new emphasis on the study of some of the process planning problems and issues that are related with the cost efficient use of 3D printing systems and the quality of their products. Among the most crucial process planning problems are: (i) the selection of fabrication orientation and parameters which is by definition a multi-criteria optimization problem in which the operator seeks to achieve the optimum trade-off between cost and quality, under given fabrication constraints and requirements, and (ii) the batch selection/planning or "packing" problem, at which the selection and placement of various different parts inside the machine workspace is considered. As such, the primary goal of the chapter is to present the effective utilization of Genetic Algorithms, which are a particular class of Evolutionary Computing, as a means of optimizing the 3D printing process planning.

Keywords 3D printing · Stereolithography · Build orientation · 2D packing · Genetic algorithm · Multi-objective optimization

1 Introduction

Manufacturing technology has changed significantly during the second half of the 20th century due to the application of computer technology in programming and control of processes and machines, such as the Computer Numerical Control

V. Canellidis · J. Giannatsis (✉) · V. Dedoussis
Laboratory of Advanced Manufacturing Technologies and Testing, Department of Industrial Management and Technology, University of Piraeus, 80 Karaoli & Dimitriou str., 18534 Piraeus, Greece
e-mail: ggian@unipi.gr

G.A. Tsihrintzis et al. (eds.), *Intelligent Computing Systems*,
Studies in Computational Intelligence 627, DOI 10.1007/978-3-662-49179-9_13

(CNC) machines and robots, as well as for the integration of the various manufacturing stages through Computer-Aided Design and Computer-Integrated Manufacturing. Based on this development in computer-aided design and manufacturing, a new class of manufacturing technologies has emerged at the end of the century, namely 3D Printing (3DP) or Additive Manufacturing (AM) technologies. As the more scientifically-oriented term "Additive Manufacturing" implies, this family of technologies shares one common operational principle, namely that parts are fabricated by gradual material addition. The part is usually "built" in a bottom-up fashion through the consecutive addition of thin layers, which can be a few microns thick.

Since the late 1980s, when the first 3DP machines (or Rapid Prototyping machines, as they were then called) were introduced, several technologies and systems that operate according to the material addition principle have been developed. These technologies can be classified into three broad categories, according to the form and type of material being processed: liquid, powder and solid 3DP (Gibson et al. 2010). Examples of liquid (or semi-liquid) processing technologies are Stereolithography and Polyjet printing. Both technologies employ a light source in order to solidify and join thin layers of photocurable polymers, thereby forming the final part. Powder systems cover a more diverse range of raw materials (polymers, metals and ceramics) which are shaped in layer form either through selective heating (e.g. Selective Laser Sintering—SLS) or through the selective deposition of glue on powder, much like the inkjet process in conventional "2D" printing. Solid systems employ material in solid form, usually in filament or sheet form. In the case of FDM (Fused Deposition Modeling), which is one of the earliest and most common solid 3DP technologies, the filament material is a thermoplastic polymer that is deposited in liquid melted form through a moving extrusion head. Finally, sheet-based systems, like that based on LOM (Laminated Object technology), process sheets of material, such as paper, from which the appropriate layers are successively laser-cut and joined with previous layers.

The 3DP production process involves four main phases, the CAD or 3D modeling phase, the pre-processing (or setup) phase, the fabrication phase and the post-processing (or post-fabrication) phase. The 3D model of the part is a necessary requirement for 3DP; hence the process begins with its creation. Since there is a variety of CAD/3D design software but no standard format for the representation of 3D geometric data, the 3D model is then translated into the neutral STL format, which is the de facto standard for the 3DP sector. The STL model is the basic input for the pre-processing phase, which involves the selection of build parameters, such as layer thickness, the virtual "slicing" of the part (computation of the layers geometries) and the virtual construction of support structures, wherever they are required.[1] Pre-processing involves also the construction of "build jobs", where

[1]Support structures are aiding structures which secure part stability during fabrication. They are virtually constructed by special software and they are physically removed in post-processing phase. 3DP technologies that require support structures belong usually to the liquid and solid filament categories.

several parts are placed in machine workspace for concurrent fabrication. Following this, the actual fabrication (building) takes places in the 3D printer. Fabrication is fully automated and machine operation requires minimal time and attention. The final post-processing phase involves cleaning and finishing of the fabricated part and perhaps some further treatment for better stability and strength. Post-processing tasks depend mainly on the 3DP technology used, and are usually done manually.

3DP systems have been initially employed for the fabrication of models and prototypes for design and product development teams, because they offered significant savings in time and allowed greater geometric flexibility, compared to other manufacturing technologies. During recent years, however, a trend to apply 3DP in actual manufacturing practice is observed; especially when production quantity is small and the product's geometric complexity is relatively high (Wohlers 2011; Powley 2013). This trend can be attributed both to improvements in accuracy and speed of 3DP systems as well as to the parallel trend towards customized and sustainable manufacturing, for which 3DP seems as a potential enabling technology. In this context 3DP has also gained significant attention by the press and media, as one of the main pillars for a "third industrial revolution" and the resurgence of manufacturing in the developed countries (Markillie 2012; Hsu 2012). Exaggerated as these claims may prove to be, it is a fact that 3DP processes are increasingly employed in actual manufacturing of products and this trend is expected to continue in the near future (Wohlers Associates 2013).

Employing 3DP systems for the production of actual products poses a challenge for the developers as well as the operators of these systems, because cost and quality requirements are significantly higher in the manufacturing setting compared to that of model/prototype fabrication. In the manufacturing context it is, therefore, important to use the system as efficiently as possible, a requirement that can be also expressed as achieving the best-possible (or sufficient) product quality with minimum cost. Since the actual fabrication of the part is computer-controlled, product cost and quality are defined mainly by the operator's choices in the initial process planning phase. One of the most important parameters defined by the operator is that of part's orientation in the machine workspace (build orientation), because this choice has a direct effect on product quality and cost. Let's suppose for example that the bottle of Fig. 1 should be fabricated by Stereolithography, and that one of the two orientations presented should be selected. In this case, fabrication in orientation 'a' would require a significantly lower number of layers than in 'b' (lower height) and would, therefore, require less machine time and cost. Orientation 'b', on the other hand, can be considered better in terms of quality because less stair stepping and contact with support structures is expected. The relationship between fabrication quality and sources of inaccuracies, such as stair stepping and supports, will be further analyzed in Sect. 3.

Another very important task in the process-planning phase is that of placing as densely as possible (packing) a number of usually not-identical parts in the machine workspace, in order to make better use of machine operation time through maximizing the use of the available machine volume, as well as minimizing setup and total batch fabrication times. The two tasks can be, and usually are, interrelated

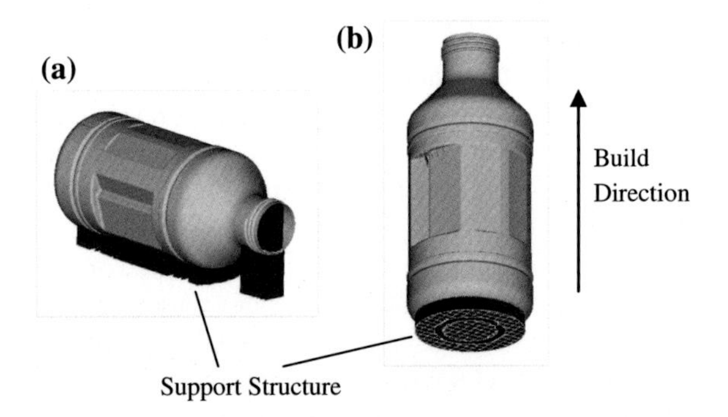

Fig. 1 Example of part fabrication in different orientations

since orientation selection defines the number of possible packing arrangements. The relative importance of each problem, orientation selection and parts packing, varies according to the 3DP system/technology examined and specific production requirements for each production batch. Efficient packing is e.g. quite important for technologies like Stereolithography and SLS, where the 'dead' time associated with the necessary layer recoating or material spreading step, is allocated in as many parts as possible, but relatively less important in the case of FDM.

Both tasks can be considered as optimization problems where a solution that minimizes the cost and inaccuracies of the process or an optimum trade-off solution between cost and quality is sought. Both problems can also be considered as multi-objective optimization problems with several local optima, for the solution of which evolutionary optimization techniques could be efficiently applied, in order to find a sufficiently good or near-optimum solution within practically acceptable time. In the present paper the application of Genetic Algorithm (GA) evolutionary techniques for addressing the orientation selection and parts packing problem is presented. Since the actual definition of the two problems varies significantly between different 3DP technologies, the proposed approaches are investigated in the context of a specific technology, namely Stereolithography (SL), which is one of the most widely spread and applied technologies in the 3DP manufacturing sector.

2 Evolutionary Optimization

Decision makers in disciplines like engineering, mathematics, economics and even social sciences are confronted every day with optimization problems of ever-growing complexity. An optimization problem can be mathematically expressed as a pair (S,f), where $S \neq \emptyset$ represents the solution space of the problem, i.e. the set of all possible points (sets of decision variables values) that satisfy the problem's constraints, and $f : S \rightarrow \mathbb{R}$ is the objective function expressing the quality of a set of decision variables values. Solving an optimization problem entails finding a set of decision variables values (solution) $i' \in S$, which satisfies the relation $f(i') \leq f(i), \forall i \in S$, assuming minimization of f is required, or $f(i') \geq f(i), \forall i \in S$ assuming maximization (Alba and Dorronsoro 2008).

In order to deal with an optimization problem, the decision maker has a wide variety of algorithms at his disposal, which can be generally classified into three categories: exact algorithms, heuristics and meta-heuristics. Exact algorithms are capable of providing the global optimum for an optimization problem, but the growth of the execution time needed is exponential to the instance size of the problem. Thus, utilizing an exact algorithm for real problems of great complexity (NP-hard problems) seems to be rather impractical. In such cases, as well as in cases where exact algorithms cannot be used approximate techniques, which focus on finding a near optimal solution within a comparatively short time frame, are employed. The most common class of approximate techniques is that of heuristic algorithms, which are capable of producing a relatively good solution in reasonable time, but there is no guarantee that the solution found represents the overall (global) optimum. Meta-heuristics are a relatively new kind of approximate techniques that offer a better balance between the quality of the obtained solution and the execution time needed. Metaheuristic algorithms can be better described as the embodiment of ad hoc heuristic methods to a higher level environment in order to boost the capabilities for exploration and exploitation of the solution space (Alba and Dorronsoro 2008).

A special branch of meta-heuristics, which is of high-interest nowadays, is that of Evolutionary Algorithms (EAs) (Drėo et al. 2006). The design and function of EAs are inspired by the idea of biological evolution of species and are loosely based on some of the associated biological processes occurring in nature. EAs should be considered as tools for efficient searching of optimal solutions in large or complex solutions spaces and not as tools for capturing the complexity and richness of nature's evolution. Thus, EAs can be better described as selective "breading" processes where populations of solutions are "evolved" by rounds of evaluation, biased selection, reproduction, variation and replacement. The goal is to discover the global optimum (extremum) among the many local optima (extrema), through a process similar to natural selection, where the survival and propagation of the most promising (fit) solutions is inherited by subsequent generations and the whole population gradually evolves towards better/fitter solutions.

Genetic Algorithms (GAs) are among the most well known and widely applied approaches to EA (Reeves and Rowe 2002; Sivanandam and Deepa 2008). In GAs, a set of possible solutions is represented by a population of individuals, each of which is described by an associated chromosome, whose genes represent various elementary component/variables of the problem. Evolution proceeds in an iterative way by successively evolving the population of individuals according to their fitness values. The fitness value of an individual expresses the corresponding degree of adaptation, with respect to the objective aimed, and defines its probability of survival and reproduction in its (stationary or changing) environment. The process begins with an initial population, which is usually generated randomly and gradually evolves through the following series of steps.

- The first step is that of evaluating the individuals and selecting the best for reproduction. First, the fitness value of each individual is computed according to the objective function. Based on these fitness values, the mating/reproduction probability of each individual is computed, employing the so-called selection operator. This probability is approximately proportional to the relative fitness value of the individual compared to others in the population.
- The reproduction step involves the matting of selected individuals (parents) and the generation of new ones (offspring) by employing the reproduction operators. Offspring are produced by crossing over genetic material from two or more solutions of the population (crossover operator) and by randomly altering their genes with a small probability following a recombination or replication event (mutation operators). The use of both operators allows the recombination of the parents' genetic material in new variations as well as the generation of offspring with new potentialities.
- Finally there is the step of eliminating part of the individuals, according to a replacement scheme, so that the population reverts to its initial size. The replacement scheme may be based on selection criteria associated with the fitness value of each individual or even be entirely random. In most cases though, extra consideration is being taken for the best few individuals (the elites) of the parent population to be protected from replacement, so that they survive across generations ensuring "non-retardation" of the best solution.

Through successive repetition of the above-mentioned steps, the population is driven towards increasing average fitness values such that, after an adequate number of generations, a (near-)optimal solution to the problem emerges from the population.

3 Determination of the Pareto-Optimal Build Orientations in Stereolithography

In 3D Printing a part is being fabricated through the successive formation and addition of layers, usually in the "bottom-top" direction. Due to the layering character of the various 3DP processes the layer addition direction, as defined by the selection of the orientation of the part with respect to the platform of machine, influences significantly the part fabrication cost, time and quality. In terms of quality, fabrication orientation is important because it implicitly defines the areas that exhibit stair stepping which is the main type of manufacturing error in 3DP. Fabrication orientation implicitly defines also the required number of layers, which is a parameter that is directly associated to manufacturing time and cost, as well as the size of the supporting structure, which is a necessary characteristic of many 3DP technologies and another parameter associated with cost, time and in some cases quality.

In its most usual form the orientation problem is considered in isolation, disregarding its interrelation with the optimum placement/packing problem. In this form the problem is defined as that of selecting the optimum orientation for a single part, out of a specific set of possible orientations and under part-specific fabrication constraints and requirements. Since the orientation selection is by definition a multi-criteria optimization problem, the optimum orientation is usually the one that gives the best compromise between the often conflicting objectives of minimizing cost/time on one hand and maximizing quality on the other. In the context of orientation selection, fabrication time and cost are quite interrelated since the basic orientation dependent cost factor is build time, i.e. the time required by the machine for the construction of the part. The proposed solutions found in literature can be summarized according to three basic characteristics: (a) the nature of the criteria, (b) the method employed for the definition of the solution space, namely which and how many orientations are evaluated, and (c) the optimization method employed.

The proposed criteria can be roughly categorized into two groups, those related with the time and cost, and those related with quality and accuracy. In terms of time and cost, the required number of layers and part height (Hur and Lee 1998; Pandey et al. 2004; Kim and Lee 2005; Alexander et al. 1998; Lan et al. 1997; Hur et al. 2001a), as well as the estimated build and/or or post-processing time and cost have been proposed (Alexander et al. 1998; Pham et al. 1999; Thrimurthulu et al. 2004; Byun and Lee 2006a, b; Ahn et al. 2007). In terms of fabrication quality and accuracy, the factor of surface roughness is mostly considered, as measured by the estimated average roughness (Pandey et al. 2004; Thrimurthulu et al. 2004) the weighted average surface roughness (Byun and Lee 2006a, b) and the total area of surfaces with estimated roughness above a certain limit (Majhi et al. 1999).

Furthermore, several criteria related with known sources of dimensional inaccuracies such as the volumetric error (Masood and Rattanawong 2002) and the process planning or stair-stepping error (Hur and Lee 1998; Majhi et al. 1999) have also been proposed. The optimization approach varies also significantly depending on the method of determining the candidate solutions (solution space), as well as the optimization method followed. A common approach found in the literature is to define a limited set of "good" candidate orientations by employing the part's (or of its convex hull) planar surfaces as bases for candidate orientations (Lan et al. 1997; Byun and Lee 2006a, b; Majhi et al. 1999), or by defining the optimum orientations of the basic part features and surfaces (Pham et al. 1999). Another proposed method is that of incremental rotation of the part around user-specified or the coordinate system axes (Hur and Lee 1998; Masood and Rattanawong 2002). In order, to further explore the solution space in a computationally efficient time, the use of evolutionary optimization techniques has also been proposed (Pandey et al. 2004; Kim and Lee 2005; Thrimurthulu et al. 2004; Ahn et al. 2007).

For the optimization of a single criterion, like the part height or of the average cusp-height, specific geometric algorithms have been proposed in previous studies (Lan et al. 1997; Ahn et al. 2007; Majhi et al. 1999). On the other hand, for the simultaneous assessment of all conflicting factors the weighted multi-objective function method (Kim and Lee 2005; Pham et al. 1999; Thrimurthulu et al. 2004;

Byun and Lee 2006a), the TOPSIS method (Byun and Lee 2006b) as well as the multi-criteria genetic algorithm method have been employed (Pandey et al. 2004). For further and more detailed discussion of the various proposed methodologies to the orientation problem the reader could refer to the study by Pandey et al. (2007).

3.1 *Orientation Selection in SL*

Although the problem of orientation is common for all 3DP processes, any practical approach at solving it should consider the specific cost and quality characteristics of the process under investigation, namely Stereolithography for the present study. The fabrication cost of a SL part can be calculated as the sum of the costs associated with its three production phases, i.e. the pre-processing (process-planning), the build (construction of the part by the machine) and the post-processing (finishing) phase. While the cost of the pre-processing phase can be consider insignificant, the cost associated with the build and post-processing phase dominates the final total cost. Respectively, the cost of the build and post-processing phase for a given part to be fabricated depends on the build time by the SL machine and on the post-processing time required for manual finishing and removal of the supports. Build time and support removal time can be estimated according to estimation models developed previously by the authors (Giannatsis and Dedoussis 2001; Canellidis et al. 2009). Thus, in order to minimize the cost of a part made via Stereolithography one has to minimize the sum (appropriate called Fabrication Time) of build time needed and post-processing time required for manual finishing and removal of the supports.

One of the major issues concerning quality and accuracy of a part fabricated with a SL system is surface roughness. Excessive surface roughness is mostly due to the stair-stepping effect that is observed in all inclined surfaces of a part, as a result of the layer nature of the building process. The intensity of stair-stepping depends on the layer thickness and the angle of the surface with respect to the build axis. Besides stair-stepping, roughness of down-facing surfaces in SL is being affected by the remains of the support structure, which is built simultaneously with the part and is removed manually in post-processing phase. Support structures are a common necessity for many 3DP technologies in order to safely attach the part on the machine platform and prevent part/layer drifting and deformations. In SL, as in many other 3DP technologies (especially metal), support structures are constructed by the same material that the part is built, a fact that makes the removal of their remains quite challenging. Both the extent of stair-stepping as well as the support-associated roughness depend on the build orientation and could, therefore, be reduced by judiciously selecting-'optimizing' the orientation of the fabrication.

As a representative measure of the overall part quality in a given orientation one can use the average surface roughness R_a over the whole surface of the part. For the evaluation of a part's average surface roughness in a given orientation a prediction model developed by Giannatsis and Dedoussis (2007) is employed. The prediction model entails profiles of surface roughness that have been measured on specifically prepared SL test parts. Based on the experimentally measured roughness and the STL part representation in a given orientation, the roughness of the various part surfaces is estimated, according to the slope of the associated facets.

According to the above analysis is obvious that the orientation problem in SL is a multi-criteria problem in which the often conflicting objectives of minimizing average surface roughness (measure of quality defects) and minimizing time (associated with cost) should be simultaneously satisfied. In this case, several optimum solutions that represent different degrees of tradeoffs between the objectives may be identified, instead of a single optimum-one, as expected in single-objective problems. These are the Pareto optimal solutions to the problem (Branke et al. 2008), also known as non-dominated, efficient or non-inferior solutions. Central for the identification of Pareto optimal solutions is the idea of dominance. A solution is said to dominate another if it is better in terms of all the criteria under consideration. Assume for example that the minimization of two criteria f_1 and f_2 is considered and that three solutions 'a', 'b' and 'c' have been identified. As can be observed in Fig. 2 solutions 'a' and 'b' both dominate solution 'c' because they obtain smaller values in both criteria compared to 'c'. The set of non-dominated solutions, such as 'a' and 'b', is also known as Pareto Front.

The Pareto front concept can be quite useful in multi-objective optimization because it helps the human decision maker in considering the multiple objectives simultaneously and assessing the possible tradeoffs. The best choice is then the Pareto optimal solution that best satisfies the decision-maker's preferences, which can be stated before (a priori) or after (a posteriori) the initialization of the optimization procedure. A priori statement usually requires the decision-maker to define the relative importance of each objective. This information is then used to construct the corresponding weighted multi-objective function that should be optimized. In this case, the optimum solution is that solution of the Pareto optimal set which minimizes (or maximizes) the weighted multi-objective function. In a posteriori statement, the Pareto optimal set is constructed first and the best solution is selected

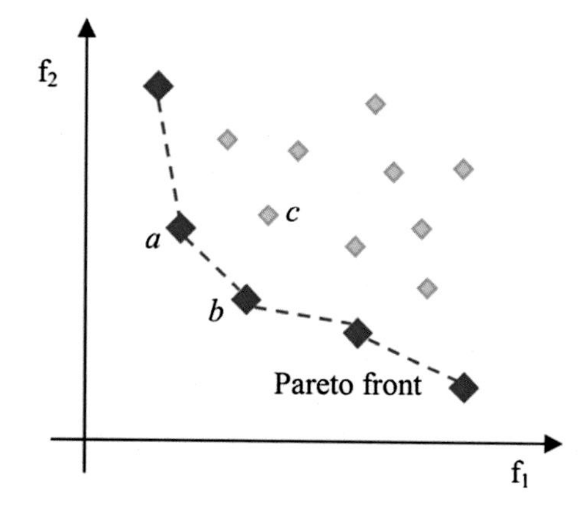

Fig. 2 Schematic representation of the Pareto front in a multi-objective optimization problem

afterwards, after examination of the set. This way the decision-maker can better assess the extent of possible tradeoffs, which can not always be known in advance.

For the fabrication orientation problem both a priori and a posteriori methods can be used. However, a priori statement of preferences presents some serious drawbacks. The first one concerns the assumption that the operator completely understands the nature and extent of tradeoffs between fabrication quality and cost for a given part and can accurately represent his preferences (based on manufacturing requirements) in the statement of the criteria weight values. This is, however, an assumption that is not easily satisfied in reality without significant experience and attention in the weight selection process. Furthermore, in actual practice, the orientation problem is not considered entirely independent of other process planning problems, such as the packing problem. As noted earlier the main objective for the packing problem is to minimize the unused machine work-space in batch fabrication, thereby enhancing the cost-efficient use of the machine. In the a priori approach a single optimum orientation for a part is chosen and this is considered fixed during the following placement/packing phase. This approach leads to good packing arrangements in reasonable computational time, but it could be reasonably argued that even better arrangements could be found if more orientations per part were considered. Thus, selecting only one optimum solution narrows the potential of the packing procedures to create even better packing layouts without compromising the quality and cost of the parts being fabricated.

3.2 Algorithm Configuration and Implementation

In an effort to alleviate the disadvantage of the single optimum solution approach both in the context of the orientation problem and the packing problem, without enlarging the optimal set to a practically infeasible size, the generation of a Pareto-optimal set of solutions has been adopted. For the specific problem under investigation the Pareto optimal set should comprise of all orientations that provide the best compromises between fabrication time/cost and quality. In order to construct this set of solutions for a given part the algorithm NSGA-II, proposed by Deb et al. (2002), has been employed. NSGA-II utilizes elitism and an explicit diversity preserving mechanism for the discovery of non-dominated solutions. The evolution process begins with a randomly generated population of N solutions/orientations, which is then reproduced in order to generate offspring. The new population, of size $2N$, is then sorted into different non-domination levels/classes. The level of an individual is defined by evaluating its performance (in terms of the selected objectives), and counting the number of other individuals that are dominated by or dominate it. The domination level acts as the fitness index for the evolving population, by keeping for breed only the best N, in terms of non-dominance, chromosomes. The algorithm employs also a specific mechanism, i.e. the crowded comparison operator, for maintaining the diversity of the population as well as for increasing the sparsity of solutions on the Pareto front. For a detailed description of the NSGA-II algorithm the reader is referred to (Deb et al. 2002).

Assuming a part in a random initial fabrication orientation, any rotation about the *X* and *Y* axis of the machine, i.e. the two axes that define its platform plane, will lead to changes to the values of the two criteria, namely fabrication time and average part roughness. On the contrary rotations about the *Z* axis, i.e. the build direction, do not affect the two criteria. Thus, the problem solution space includes all orientations that can be derived by *X*- and *Y*-wise rotations of the part with respect to the initial orientation. Each orientation-solution is represented in the NSGA-II by a chromosome comprising of two genes, the first representing the *X*-wise rotation (θ_x) and the second the *Y*-wise rotation (θ_y).

In order to successfully utilize the NSGA-II procedure, as in any metaheuristic algorithm, appropriately adapted reproduction operators must be established. In the proposed implementation of the algorithm, the reproduction operator must be able to cope with this continuous representation scheme, since the two genes are defined in floating point format. An appropriate crossover operator, proposed by Haupt and Haupt in (2004), is employed in the current implementation. The operator assigns randomly a value to a variable, which implicitly defines the crossover point, i.e. the gene (or genes) to which the crossover operation will be applied, as well as if the crossover operation will be applied on one of the two parents' genes, or on both. If the θ_x gene is selected as the crossover point, then the corresponding genes for the two children chromosomes $child_1$ and $child_2$ are created through blending of the corresponding genes of the two parents, par_1 and par_2, as follows:

$$\theta_x[child_1] = \theta_x[par_1] - b(\theta_x[par_1] - \theta_x[par_2]) \quad (1)$$

$$\theta_x[child_2] = \theta_x[par_2] + b(\theta_x[par_2] - \theta_x[par_1]) \quad (2)$$

where *b* is a single decimal random number between 0 and 1. The same procedure applies for the generation of the children θy genes. It is obvious that if *b* is zero then $child_1$ is a clone of par_1 and $child_2$ is a clone of par_2. The genes can be blended using the same *b* value for each new gene or by choosing different values. As *b* value ranges between 0 and 1, the introduction of values beyond the extremes already represented in the population is not allowed. Thus, in order to escape premature convergence of the algorithm to a local optimum an extrapolation method that will permit the creation of genes with values out of the parents' limits should be facilitated. This is achieved by providing values greater than unity for the *b* factor. Another common mechanism to reduce the probability of converging to a local optimum and add diversity to the population is the application of a mutation operation. In the proposed implementation, the mutation operator simply selects a gene with a predetermined probability and a random new value is assigned to it. For the implementation of the above described GA the commercial package Matlab R2008 has been used.

3.3 Build Orientation Case Study

In order to study and present the above mentioned methodology the case of a "real world" mechanical knuckle has been examined. The geometry of the part under examination is considerably complex, comprising of several flat, cylindrical and freeform surfaces as well as containing various geometrical features, such as holes of different sizes and several chamfer and fillet features (Fig. 3). Part dimensions are 134.01 × 110.29 × 158.14 mm and the corresponding STL model is composed of 9592 triangles.

For the specific computational run, a layer thickness of 0.15 mm was selected and no roughness tolerances for specific surfaces were assigned. The NSGA-II parameters selected were:

- a population size of 60 chromosomes,
- a crossover rate of 20 %, and
- the termination limit was defined at 200 generations.

The calculated Pareto front is presented in Fig. 4 and is comprised of 30 solutions/orientations. Each one of these orientations represents a good (non-dominated) trade-off between the two selected criteria, namely fabrication time and average roughness. Three characteristic members of the Pareto optimal set, solutions marked as 'A', 'B' and 'C' in Fig. 4, are graphically presented in Figs. 5, 6 and 7, respectively. Color-mapping is employed for presenting the estimated roughness value of each surface so as to allow a more complete evaluation of the overall quality, as well as a histogram showing the distribution of surface roughness as a percentage of the total part surface area.

As observed in Fig. 4, solution 'A' strongly favors the criterion of fabrication time (416 min) over average surface roughness, something that it is also evident by the relatively small build height of the part (i.e. dimension along the Z-axis) in the corresponding orientation (see Fig. 5). This is achieved though at the expense of

Fig. 3 Knuckle STL model in the initial orientation

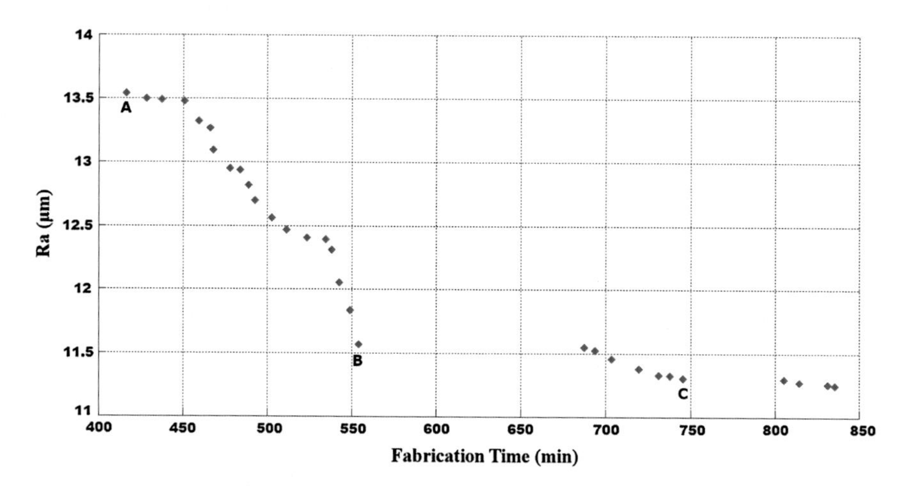

Fig. 4 The Pareto front for the knuckle STL model

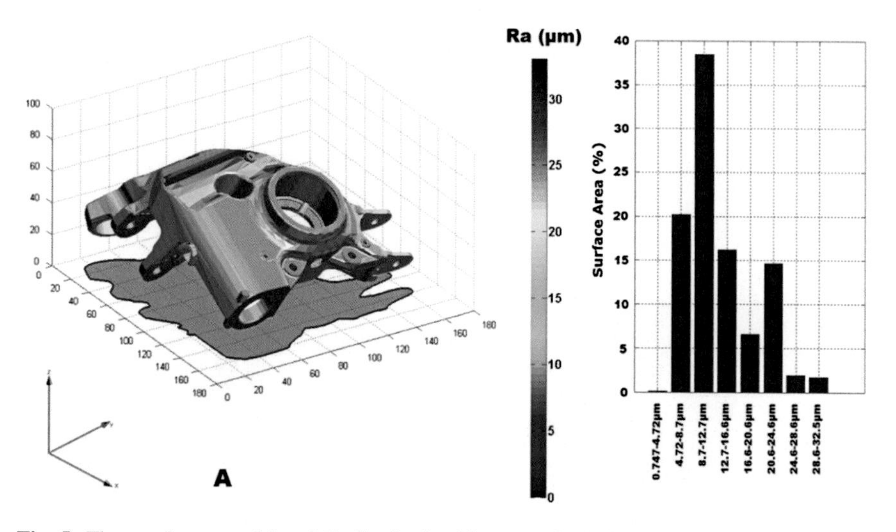

Fig. 5 The roughness model and R_a distribution histogram for build orientation A

relatively high average roughness (13.54 μm) due to stair-stepping and the relatively high amount of part surface area that is contact with the supporting structure (down-facing surfaces of the part). From the histogram of the 'A' solution it can be observed that approximately 59 % of the part's surface area exhibits roughness values in the range of (0.7–12.7 μm) while the rest 41 % exhibits roughness values bigger than 12.7 μm. Orientation 'C' (Fig. 7), on the other hand, is better in terms of average roughness (11.31 μm) with 76 % of the part's surface area exhibiting roughness values in the range of (0.7–12.7 μm) and only 24 % exhibiting roughness values greater than 12.7 μm. The corresponding time, however, is quite high

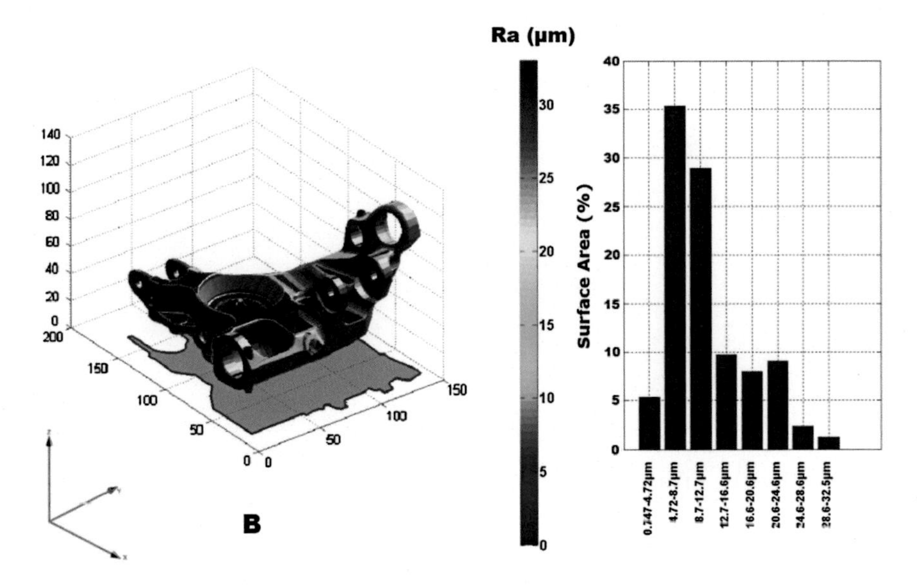

Fig. 6 The roughness model and R_a distribution histogram for build orientation B

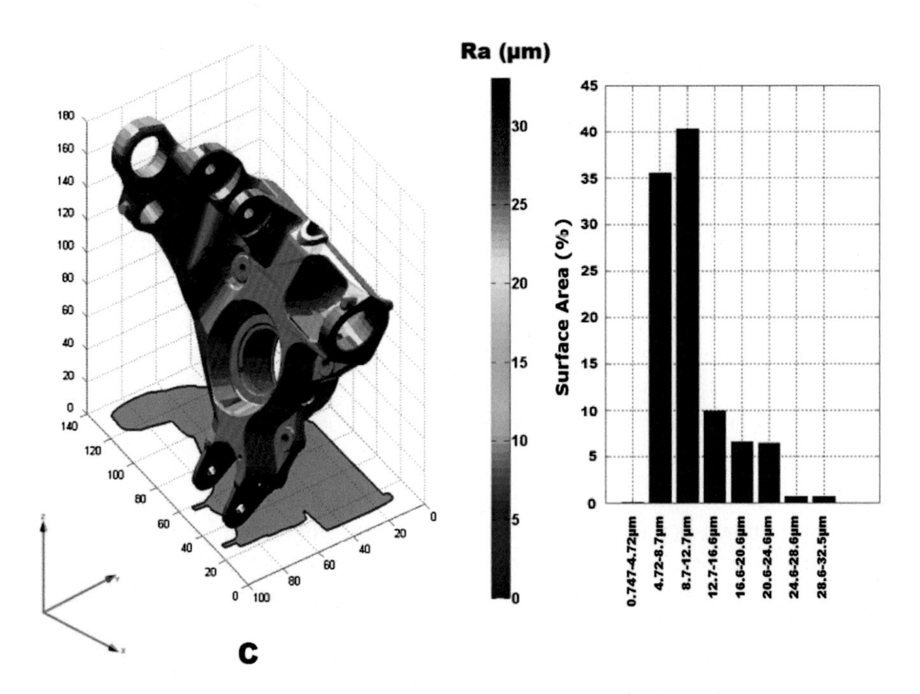

Fig. 7 The roughness model and R_a distribution histogram for build orientation C

(745.6 min) as the algorithm "oriented" the part in such a way that most of facets would have low estimated roughness, disregarding the fact that such an orientation implies relatively high build height. Finally, orientation 'B' (Fig. 6) represents a middle tradeoff alternative, in which both criteria obtain values in the middle of the observed ranges (554.4 min and 11.57 μm). In this orientation, many of the part's plane surfaces and the central cylindrical hole are oriented perpendicular to the building platform, in order to minimize stair-stepping. This scheme seems to offer a good tradeoff solution since it implies relatively low fabrication time without compromising quality, as expressed in the estimated average roughness value (11.57 μm), which is very close to the minimum found at the Pareto Front.

In order to select the final building orientation, the operator may evaluate the criteria values for the Pareto optimum orientations as well as examine their respective 3D roughness models. By an examination of the roughness model for a specific orientation the operator of the SL system is able to directly evaluate the surface roughness of critical part surfaces and features, and thus estimate the need for post processing. Moreover, the projection of each orientation to the building platform may be easily computed and visualized (see Figs. 5, 6 and 7). Thus, the operator may prefer to utilize, ultimately, an orientation that minimizes the footprint of the part on the fabrication platform or presents a more convex projection. This way a reduced and more optimized set of candidate orientations can be provided as an input to the subsequent packing and placement task.

4 Determination of the Optimum Packing Layout in Stereolithography Machine Workspace

3DP technologies excel in the simultaneous unattended production of geometrically complex and dissimilar parts in small fabrication batches/"jobs". As noted earlier, the job setup for 3DP technologies entails the selection and placement of the parts in such a way that will efficiently utilize the machine workspace and time, as well as reduce the number of the required job setups. The problem of the efficient utilization of the 3DP machine workspace has been the subject of relatively few studies (Wodziak et al. 1994; Ikonen et al. 1997; Lewis et al. 2005; Hur et al. 2001b; Gogate and Pande 2008; Zhang et al. 2002). The problem has been tackled mainly in its 3D form (instance), meaning that the researchers' optimization goal was to maximize the machine workspace volume covered by the parts. In this form, the problem belongs to the wider family of cutting and packing optimization problems (3D packing problem) that are considered to be NP complete problems. However, the size and shape variety of the parts that can be (and usually are) fabricated in a single job, makes the task of "nesting" them optimally on the machine workspace extremely challenging. As such, all the methodologies proposed entail a metaheuristic approach (i.e. Genetic Algorithm, Simulated Annealing) in order to drive the search for the optimum placement pattern.

Moreover, in order to decrease the computational burden and increase the practical applicability, simplified geometrical representations of the parts are usually employed at some step of the optimization process. Two of the proposed simplification techniques are those of employing the bounding box of a part or its voxel representation. Finally, in most of the proposed methodologies the orientation of a part is considered fixed or limited to a small set.

Though the problem, in its most general form, should be considered as a 3D problem, there are some specific characteristics and constraints of the 3DP technology under investigation, namely SL, which reduce it to two dimensions. The most basic relevant constraint, in the context of a SL technology seems to be the requirement for support structures. If a part is placed over another the support structure of the top part will interfere with the geometry of the bottom part, deteriorating the surface quality and dimensional accuracy of the latter, while at the same time increasing the associated post-processing time. Due to this, parts in SL technologies are usually placed aside each other, after a proper build orientation for each one is chosen. The geometrical interaction between the parts is, therefore, limited to the *X-Y* (platform) plane, and the resulting problem essentially becomes a two-dimensional packing problem (2D packing). The 2D packing problem is a widely-addressed subject in various manufacturing sectors such as that of metal products, textile, footwear and furniture, as well as ship building. However, even in its 2D form, the problem complexity can be still quite high. Two important methodological decisions have to be made in order to address any specific instance of the 2D packing problem, namely the selection of an optimization scheme, i.e. how to guide effectively the search in the solution space of all possible layouts/solutions, and that of a placement scheme, i.e. how a specific layout is constructed through the placement of the various parts. The characteristics and details regarding the placement and optimization schemes are discussed in the following sections.

4.1 Optimization Scheme

The 2D packing problem is considered to be an NP complete problem. It is considered also intractable, as far as computational time for obtaining the global optimum is concerned, because the associated solution tree grows exponentially with the size of the problem. Classical (exact) optimization algorithms are usually inefficient in dealing with such problems and one has to resort to metaheuristic approaches. In the present work a GA approach has been adopted. The implemented GA utilizes a permutation coding scheme for chromosomal representation, in which a solution is coded as a sequence of integers (genes) that corresponds to the placement sequence for a set of parts. An individual *i*, for example, is represented by chromosome $p_i = \{2, 4, 1, 3, 5\}$, which can be interpreted as placing part no. 2 first, then no. 4 etc. Each part is situated in a specific position following this sequence and according to prescribed placement rules that are described later.

The reproduction operators are also appropriately chosen to fit the permutation nature of the algorithm. the 2D packing problem, the most popular crossover operators, such as the PMX, SJX, etc., have been found to perform quite well (Dréo et al. 2006); thus, the SJX 'rule', introduced by Jakobs (Jakobs 1996), was adopted in the present work. The mutation is accomplished using a gene swap operator, which simply swaps the location of two genes in a chromosome.

Selection of individuals for reproduction follows the proportional selection strategy, where the expected number of times an individual is selected for reproduction is proportional to its fitness value (Holland 1992). The fitness of an individual is evaluated by the corresponding platform coverage (percent of platform area occupied by parts) that is achieved through placing the parts according to the sequence denoted by the corresponding chromosome. The area occupied by a part is defined by its projection on the platform plane; hence the value of the fitness function f for a chromosome, which contains n parts, is calculated as follows.

$$f = \frac{\sum_{i=1}^{n} Projection\_Area\_of\_the\_i_{th}\_part}{Fabrication\_Platform\_Area} \tag{3}$$

4.2 Packing Layout Construction Process

The issue of placement rules and interference checks in packing is quite important, because it influences not only the quality of the solution but also the required computational effort. A wide variety of tools concerning the geometric representation of the parts, the interference check procedures, as well as the placement and layout construction strategies have been proposed in previous studies. Some of the most commonly used techniques are the raster/pixel methods, the direct trigonometry methods, the phi-function method, the no-fit polygon method and the left-bottom placement rule.

In raster/pixel methods, the packing area is divided into discrete regions in the form of a grid, while parts shapes are also similarly represented in pixel form. The main advantage of this approach is that finding a feasible placement for a part is relatively fast. Moreover, raster representations are simple to code, and can be easily used to represent non-convex and geometrically complex pieces. However, these approaches are memory intensive and cannot always represent efficiently parts with high level of detail. In such a case, increasing the representation's accuracy by refining the size of the grid unit may lead to an unacceptably high computational effort (Bennell and Oliveira 2006; Whitwell 2004). Direct trigonometry methods deal with the geometry complexity directly by utilizing a vector representation for each part. Unlike the raster method, the feasibility of a placement is implied by direct trigonometric techniques, which are able of collision detection. These techniques deal efficiently with irregular shapes and produce high quality solutions but they are relatively computationally expensive (Hopper 2000). The Left-Bottom placement rule is a heuristic technique that guides placement of parts near the bottom-lower side of the packing area and as far to the left as possible. It is considered an easy to implement technique that provides fast solutions and is better suited for simple orthogonal geometries rather than for geometrically complex arbitrary shapes.

The No-Fit Polygon (NFP) technique has become an increasingly popular tool for conducting intersection tests between pairs of polygons. The NFP represents all the possible arrangements in which two arbitrary polygons touch, i.e. neither intersection nor distance between them is observed. Although the NFP is an excellent tool for making intersection tests between pairs of polygons, it has not been widely applied due to the difficulties in developing a robust approach capable of dealing with the large number of degenerated cases, which may arise during its construction as a direct consequence of the generality of the shapes considered (Bennell and Oliveira 2006; Whitwell 2004). The phi-function is the most recent innovation in dealing with the geometric issues that arise in nesting problems. Its purpose is to characterize the topological relationship of two shapes through mathematical expressions. Thus, the value of the phi function is greater than zero if the two objects are separated, equal to zero if their boundaries touch and less than zero if they overlap. The method has not been broadly adopted, as there isn't any algorithmic procedure for generating the phi function for arbitrary shapes (Chernov et al. 2010).

The proposed layout construction methodology employs three of the above mentioned methods: the vector form for part geometry representation, the left-bottom rule for sequential part placement and the NFP method for intersection control. The first step in the placement procedure involves the construction of the 2D parts projections on the fabrication platform. The orientation of each part is considered fixed and is selected according to the methodology described in Sect. 3. The projection area of a part is computed from the corresponding STL file and is represented as a closed series of vectors (polygon/polyline). If the level of detail in the STL file is high, which is nowadays usually the case, the corresponding vector projection may consist of a quite large number of points. In this case, appropriate care must be taken in order to alleviate the computational effort needed for geometrical calculations. This reduction in the number of points should of course retain a sufficient level and no significant loss of geometry. In the present study, the Douglas–Peucker technique (Douglas and Peucker 1973) is employed for this geometry simplification. According to this technique the part projection is being offset (enlarged) by a suitable threshold and then the number of points of the projection curve is reduced. That way a concise representation for each projection as well as significant savings in computational time is achieved.

Another important, in terms of computational time, aspect is the extent at which a projection is allowed to rotate along the building direction (i.e. the Z-axis). If projections are allowed to rotate freely, it is quite probable that quite dense packing layouts will emerge at the cost, however, of increased computational time. Some constraints in rotation freedom are, therefore, necessary in order to keep computational time within acceptable limits. In the present study, the use of the Minimum

Bounding Rectangle (MBR) approach is proposed. The MBR of a projection is the minimum rectangle that encloses it. The four sides of a projection's MBR are used as references for defining the four possible rotations of the projection, through successively placing each side parallel to the horizontal axis of the platform.

Following the construction of the MBRs the actual packing procedure is implemented as a two step procedure. The first step aims at placing a projection on the fabrication platform so that no collision with another projection occurs. This step involves the sequential placement of the projections' MBRs (defined by the solution's chromosome) according to the left-bottom rule. Employing this rule implies that a limited set of positions and rotations for an MBR are evaluated in order to place it as close to the origin (lower-left corner) of the fabrication platform as possible (see Fig. 8). For a more detailed description of the left-bottom placement process the reader could refer to (Canellidis et al. 2013).

Once a proper position for an MBR has been found, the second step of the packing algorithm tries to nest the next part/projection closer to the already packed polygons, utilizing the irregularly shaped actual projection of each part. This densification step employs the ray casting technique and the notion of NFP. As described earlier, the NFP may be used to determine all arrangements that two arbitrary polygons may assume such that the two shapes touch. When these two polygons are convex, the NFP can be constructed by tracing one polygon around the boundary of the other. Consider for example two polygons A and B, where A remains in fixed position. In order to construct the corresponding NFP (NFP_{AB}), the moving trajectory of a reference point from polygon B is recorded as B moves around A (Fig. 9 and http://www.tex.unipi.gr/anim/anim1.gif). Having constructed the NFP_{AB} it is quite simple to test whether polygon B intersects with polygon A in any possible arrangement. If polygon B is positioned so that its reference point is inside NFP_{AB} then it overlaps with polygon A. If the reference point is on the boundary of NFP_{AB} then polygon B touches polygon A. Finally, if the reference point is outside of NFP_{AB} then polygons A and B do not overlap or touch.

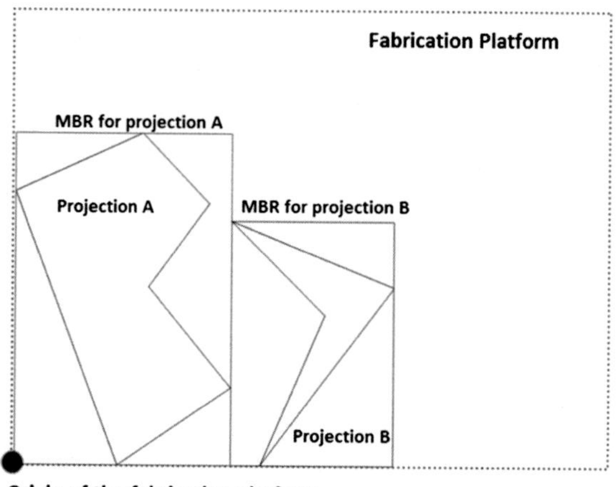

Fig. 8 Illustration of the left-bottom placement process with the MBR

The development of a procedure for deriving an analytical representation of the NFP, when it concerns convex shapes like those presented in Fig. 9, is relatively trivial. In the case of concave shapes, however, the construction of the NFP can be quite challenging and computational expensive (Bennell and Oliveira 2006; Whitwell 2004). In this case, it could be adequate to obtain a rough representation of the NFP boundaries by point sampling. Since the packing process cannot be limited to projections of convex shape, a method for obtaining an estimation of the actual NFP boundary (as a point cloud) is adopted in the present work.

The NFP construction process begins with a transformation of a possible concave polygon to convex, by covering the concave regions with dummy lines (see dummy edge E1 in Fig. 10), utilizing the Qhull algorithm. The construction of these convex approximation permits also the construction of an initial convex approximation of the NFP, through the above described process. This rough approximation already contains that part of the actual NFP which corresponds to the convex sectors of the original concave shapes. Next, the "dummy" edges of the convex NFP, i.e. every edge that does not belong on the original boundaries of the two polygons, are replaced by appropriate cloud of points so as to create a more exact approximation of the actual NFP (see replacement of dummy edge E1 in Fig. 10). The point clouds are created by sliding the reference point of one polygon on every "dummy" edge of the convex NFP with a user selected threshold and calculating the distance (vertical or horizontal according to the slope of the "dummy edge") that the polygon can move nearer the fixed polygon so as to touch but not overlap. The calculation of the distance is performed by utilizing a Ray Casting technique (Canellidis et al. 2013). When all "dummy "edges are processed an approximate point representation of the actual NFP can be obtained.

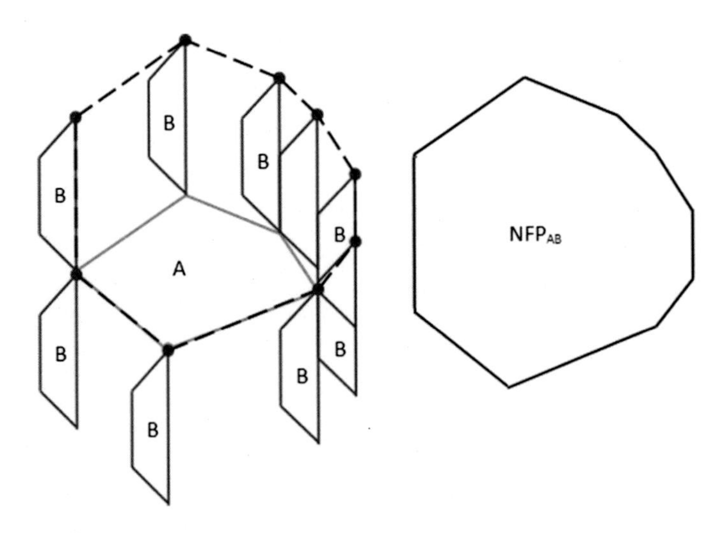

Fig. 9 Illustration of the NFP construction process for convex polygons

As noted earlier, the construction of the NFP is a necessary step in order to evaluate the distance, vertical or horizontal, that a polygon can be further moved without colliding/interfering with the already nested projections. This evaluation involves the casting of rays from the polygon's reference point towards the NFPs of its left or underneath neighbors. To better illustrate the placement process an example is presented in Fig. 11, where the incoming object, polygon B, has already been initially placed, based on the MBRs of the objects and following the

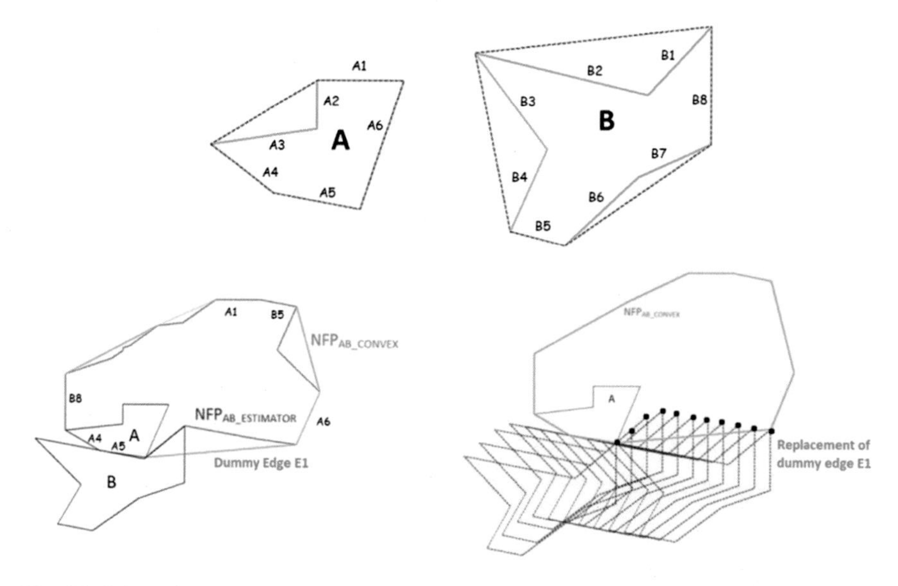

Fig. 10 Schematic representation of the NFP estimation process for non-convex polygons

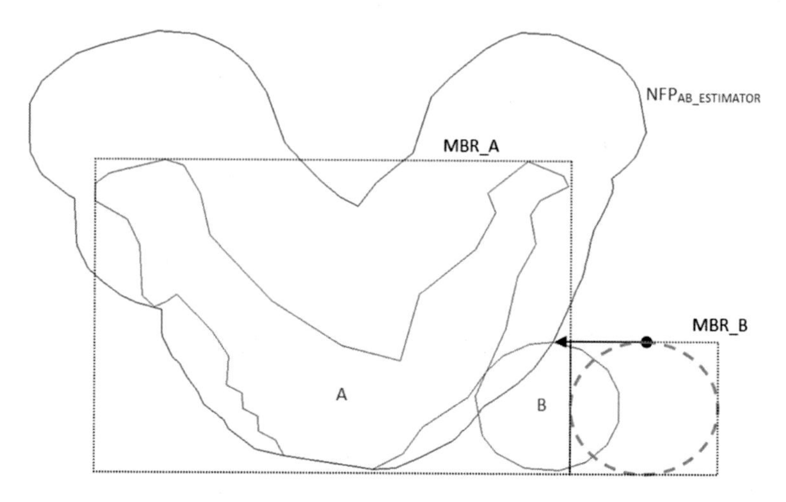

Fig. 11 Schematic representation of the placement process

left-bottom rule. In the specific arrangement there is unutilized space between the two objects that can be exploited by a leftward movement of polygon B. The corresponding travel distance is computed by casting a ray from a reference point of object B to the approximate NFP of the two objects (NFP_{AB}). Since the casting of rays from the incoming polygon may interfere with multiple other projections and NFPs, the minimum of the corresponding distances is that by which the polygon can be translated without interfering with some of its neighbors. In order to accelerate the packing process the NFPs of all pair combinations of the available objects are computed prior to the initiation of the GA heuristic process. An animated illustration of the placement procedure involving two polygons is presented in http://www.tex.unipi.gr/anim/anim2.gif

4.3 Packing Layout Case Studies

In order to investigate how the nesting methodologies are adapted to the operational characteristics of the SL technology, two representative test cases were examined. The packing methodology described has been implemented using Matlab R2008a. The test cases simulate the nesting procedure in a rectangular fabrication platform of 250 mm × 250 mm. The corresponding 'real-world' objects/parts were of various dimensions and levels of geometrical complexity (see Fig. 12) as could be expected for a representative SL build job at a prototyping/3D printing service bureau. Data regarding the corresponding STL files and the associated projections are presented in Tables 1 and 2.

As observed in Tables 1 and 2, the number of points for the most of the parts' projections is quite high due to the relatively high level of detail and complexity of the corresponding 3D STL models. In order, therefore, to keep computational time within practically reasonable limits, a reduction at the level of detail of the projected polygons has been deemed necessary. Geometry simplification has been performed by employing the Douglas–Peucker algorithm, following an initial offset of the

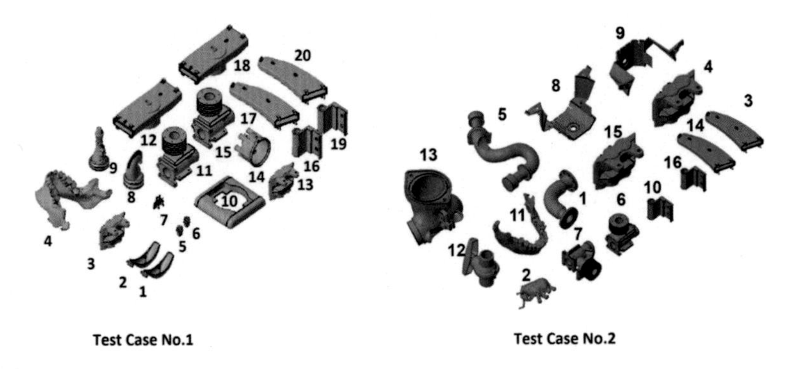

Fig. 12 Parts examined at the packing test cases

original geometry by a certain threshold, to avoid possible intersections between parts due to the simplification procedure. In Fig. 13 the initial projection of an STL model (blue line), the offset version of the initial projection (green line) as well as the simplified finally version (red line) are presented.

The parameters of the implemented GA in all test cases were: population size of 50 individuals/chromosomes, crossover rate of 100 % and mutation rate of 1.25 %. The GA terminated when the population became fully homogeneous, i.e. the mean fitness function value (mean percentage of coverage) of the population reached the corresponding maximum fitness value (maximum percentage of coverage obtained by the best packing layout in a generation). The evolution of packing arrangements

Table 1 Data of parts examined in the first test case

Part number	STL triangles	Projection's no. points	Projection's no. points after reduction
1 and 2	6552	450	14
3 and 13	49,682	894	25
4	100,948	4119	43
5 and 6	1420	256	9
7	828	231	27
8	75,934	430	16
9	30,171	439	16
10	12,300	735	13
11 and 15	4427	292	19
12 and 18	17,718	120	13
14	4028	39	7
16 and 19	4860	157	20
17 and 20	10,394	340	25

Table 2 Data of parts examined in the second test case

Part number	STL triangles	Projection's no. points	Projection's no. points after reduction
1	11,536	918	33
2	9714	2037	40
3 and 14	10,399	1535	22
4 and 15	49,564	1689	30
5	11,800	1787	54
6	4427	280	19
7	4427	845	53
8	113,827	2919	51
9	113,827	2980	60
10 and 16	4860	897	18
11	100,948	4040	43
12	64,370	1330	41
13	48,870	3098	29

is illustrated schematically in http://www.tex.unipi.gr/anim/anim3.gif, where snapshots of the best solutions for consecutive generations are presented. The packing optimization results are presented in Table 3, while the best packing layouts found are presented in Fig. 14.

As observed in Table 3, the best coverage achieved in test case no. 1 is significantly higher than in test case no. 2. This can be attributed to the less irregular (more convex) geometry of parts projections of the first test case compared to those of the second (see Fig. 12). In order to increase the amount of utilized area by placing polygons of high irregularity (consisting of large concave regions), such as mostly present in the second test case, optimum matches between concave and convex areas of neighboring polygons are required. This is not something that can be achieved employing a stochastic optimization process without significant computational cost. Due to this; the best packing pattern in the second test case has less workspace utilization rate (60.7 %) than this of the first test case (78 %). Moreover, by examining the shape of projections in the final/best solutions (Fig. 14), it can be argued that the packing algorithm in both test cases seems to favor the selection of projections of relatively large area and near convex shape. This is quite evident in the first test case, where the mandible model (no. 4) is not present in the best solution, found by the GA, since the algorithm failed to generate layout patterns that could effectively embody it. In a similar manner the parts no. 1, 5, 9 and 11 of the second test case, whose projections are highly concave, weren't selected either. It

Fig. 13 Schematic representation of the projection simplification algorithm

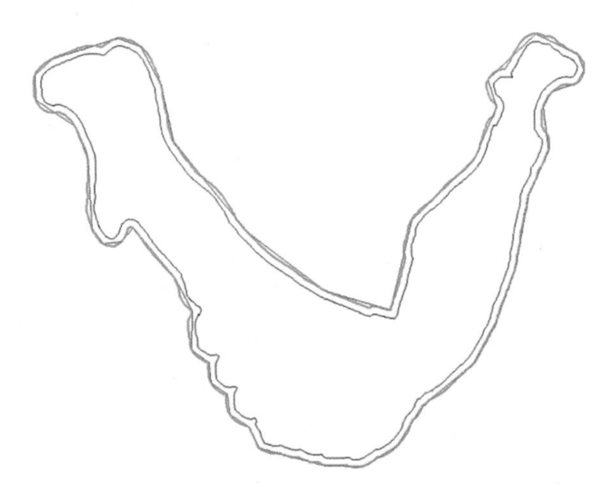

Table 3 Results of the SL test cases

Results	Test case no. 1	Test case no. 2
Platform area coverage (%)	78	60.7
No. parts packed	19/20	11/16
Mean computational time per packing solution (s)	0.65	0.57

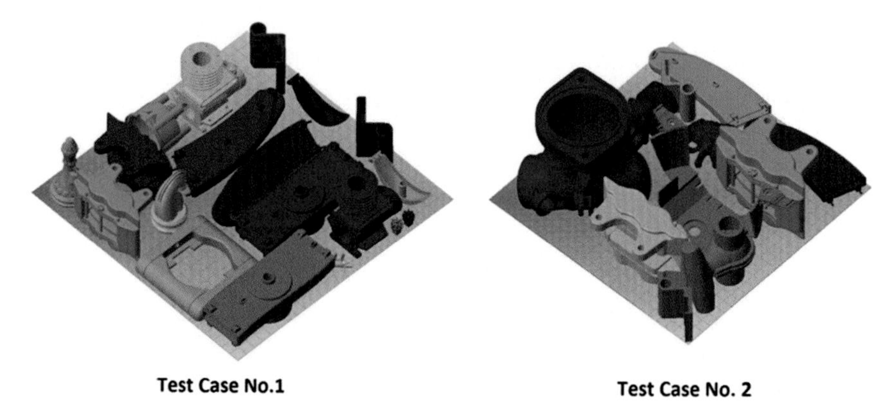

Fig. 14 “Optimum”/best packing arrangements for the packing test cases

can be reasonably, of course, argued that this phenomenon depends significantly on the configuration of each test case and that probably a repetition of the GA process could lead to better packing layouts. It seems, however, that the implemented placement strategy has some limitations when highly concave projections are present.

A possible solution to this problem could be to employ some rules for “smart” matching of convex and concave areas of different parts or to allow more freedom in rotation around the Z-axis, at the cost of course of higher computational time. Another solution could be to examine the convexity of the projection in the earlier phase of orientation selection and prioritize orientations with more convex platform projections. In this context the previously described concept of the Pareto optimal orientations per part can be employed in order to identify an orientation that both satisfies the quality/cost requirements for a certain part as well as facilitates packing procedures and the machine workspace utilization.

5 Concluding Remarks

The selection of build orientation for a number of parts and the subsequent setup of a fabrication job are the two main tasks in the process planning phase of any 3D printing process. Both tasks are quite important in terms of fabrication cost and quality and can be considered as optimization problems. The selection of build orientation is a multi-criteria problem, in which the usually conflicting objectives of minimizing cost and maximizing quality should be achieved. Job setup, on the other hand, relates to the efficient use of the machine that can be achieved through packing the highest possible number of parts and minimizing the unutilized machine workspace. In the present paper an integrated approach for addressing these problems, in the case of Stereolithography, is presented. The proposed approach employs Genetic Algorithms techniques, first for the identification of a set of Pareto-optimal build orientations for a part and then for the construction of an optimum packing scheme for a collection of parts.

Two software tools, one for each problem, have been developed according to this approach. Testing results of the build orientation tool, in the case of a relatively complex mechanical part, show that the proposed approach is efficient in the identification of a set of good (Pareto-efficient) orientations, among which the operator may select the one that best suits fabrication requirements and constraints. Case study results also show that a representative set of good build orientations can be identified relatively early in the evolutionary process, implying that the operator may choose to terminate the process after a relatively small number of generations (<50) without significant loss in the shape of the corresponding Pareto front. A similar observation can be made also for the packing layout construction tool, which has been tested in two case studies. Due to the nature of the Stereolithography process, the implemented packing approach was reduced to two dimensions and the parts' projections of the platform were considered. According to the cases studies' results, the tool can identify sufficiently good packing layouts, especially for a set of parts with relatively convex projections.

References

Ahn, D., Kim, H., Lee, S.: Fabrication direction optimization to minimize post-machining in layered manufacturing. Int. J. Mach. Tool. Manu. **47**, 593–606 (2007)

Alba, E., Dorronsoro, B.: Cellular Genetic Algorithms. Springer, New York (2008)

Alexander, P., Allen, S., Dutta, D.: Part orientation and build cost determination in layered manufacturing. Comput. Aided Design **30**, 343–356 (1998)

Bennell, J., Oliveira, J.F.: The geometry of nesting problems: a tutorial. Eur. J. Oper. Res. **184**, 397–415 (2006)

Branke, J., Deb, K., Miettinen, K., Słowiński, R.: Multiobjective Optimization: Interactive and Evolutionary Approaches. Springer, Berlin (2008)

Byun, H.S., Lee, K.H.: Determination of optimal build direction in rapid prototyping with variable slicing. Int. J. Adv. Manuf. Tech. **28**, 307–313 (2006a)

Byun, H.S., Lee, K.H.: Determination of the optimal build direction for different rapid prototyping processes using multi-criterion decision making. Robot. CIM-Int. Manuf. **22**(1), 69–80 (2006b)

Canellidis, V., Giannatsis, J., Dedoussis, V.: Genetic algorithm based multi-objective optimization of the build orientation in stereolithography. Int. J. Adv. Manuf. Tech. **4**(7-8), 714–730 (2009)

Canellidis, V., Giannatsis, J., Dedoussis, V.: Efficient parts nesting schemes for improving stereolithography utilization. Comput. Aided Design **45**(5), 875–886 (2013)

Chernov, N., Stoyan, Yu., Romanova, T.: Mathematical model and efficient algorithms for object packing problem. Comput. Geom. **43**(5), 535–553 (2010)

Deb, K., Pratap, A., Agarwal, S., Meyarivan, T.: A fast and elitist multiobjective genetic algorithm: NSGA-II. IEEE T. Evol. Comput. **6**(2), 182–197 (2002)

Douglas, D., Peucker, T.: Algorithms for the reduction of the number of points required to represent a digitized line or its caricature. Can Cartographer **10**(2), 112–122 (1973)

Drėo, J., Pėtrowski, A., Siarry, P., Taillard, E.: Metaheuristics for Hard Optimization. Springer, Berlin (2006)

Giannatsis, J., Dedoussis, V.: A study of the build-time estimation problem for stereolithography systems. Robot. CIM-Int. Manuf. **17**(4), 295–304 (2001)
Giannatsis, J., Dedoussis, V.: Decision support tool for selecting fabrication parameters in stereolithography. Int. J. Adv. Manuf. Tech. **33**, 706–718 (2007)
Gibson, I., Rosen, D.W., Stucker, B.: Additive Manufacturing Technologies. Springer, Berlin (2010)
Gogate, S., Pande, S.S.: Intelligent layout planning for rapid prototyping. Int. J. Prod. Res. **46**(20), 5607–5631 (2008)
Haupt, R.L., Haupt, S.E.: Practical genetic algorithms. Wiley, New York (2004)
Holland, J.H.: Adaptation in natural and artificial systems. MIT Press, Cambridge (1992)
Hopper, E.: Two dimensional packing utilising evolutionary algorithms and other meta-heuristic methods. Ph.D. Thesis, School of Engineering, University of Wales (2000)
Hsu, J.: Why 3-D Printing Matters for Made in USA. Scientific American. http://www.scientificamerican.com/article/why-3d-printing-matters/ (2012)
Hur, J., Lee, K.: The development of a CAD environment to determine the preferred build-up direction for layered manufacturing. Int. J. Adv. Manuf. Tech. **14**(4), 247–254 (1998)
Hur, S.M., Choi, K.H., Lee, S.H., Chang, P.K.: Determination of fabricating orientation and packing in SLS process. J. Mater. Process Tech. **112**(2-3), 236–243 (2001a)
Hur, S.M., Choi, K.H., Lee, S.H., Chang, P.K.: Determination of fabricating orientation and packing in SLS process. J. Mater. Process. Tech. **112**, 236–243 (2001b)
Ikonen, I., Biles, W., Kumar, A., Ragade, R.K., Wissel, J.C.: A genetic algorithm for packing three-dimensional non-convex objects having cavities and holes, In: Proceedings of 7th International Conference on Genetic Algorithms, Michigan, pp. 591–598 (1997)
Jakobs, S.: On genetic algorithms for the packing of polygons. Eur. J. Oper. Res. **88**, 165–181 (1996)
Kim, H.C., Lee, S.H.: Reduction of post-processing for stereolithography systems by fabrication-direction optimization. Comput. Aided Design **37**(7), 711–725 (2005)
Lan, P.-T., Chou, S.-Y., Chen, L.-L., Gemmill, D.: Determining fabrication orientations for rapid prototyping with stereolithography apparatus. Comput. Aided Design **29**, 53–62 (1997)
Lewis, J.E., Ragade, R.K., Kumar, A., Biles, W.E.: A distributed chromosome genetic algorithm for bin-packing. Robot. CIM-Int. Manuf. **21**(4-5), 486–495 (2005)
Majhi, J., Janardan, R., Smid, M., Gupta, P.: On some geometric optimization problems in layered manufacturing. Comput. Geom. **12**(3-4), 219–239 (1999)
Markillie, P.A: Third industrial revolution. The Economist, Spec. Special report: Manufacturing and innovation, Apr 21st (2012)
Masood, S.H., Rattanawong, W.: A generic part orientation system based on volumetric error in rapid prototyping. Int. J. Adv. Manuf. Tech. **19**(3), 209–216 (2002)
Pandey, P.M., Thrimurthulu, K., Reddy, N.V.: Optimal part deposition orientation in FDM by using a multicriteria genetic algorithm. Int. J. Prod. Res. **42**(19), 4069–4089 (2004)
Pandey, P.M., Reddy, N.V., Dhande, S.G.: Part deposition orientation studies in layered manufacturing. J. Mater. Process. Tech. **185**, 125–131 (2007)
Pham, D.T., Dimov, S.S., Gault, R.S.: Part orientation in Stereolithography. Int. J. Adv. Manuf. Tech. **15**(9), 674–682 (1999)
Powley, T.: 3D printing reshapes factory floor. Financial Times. http://www.ft.com/intl/cms/s/0/1de6deba-6897-11e3-bb3e-00144feabdc0.html?siteedition=intl#slide0 (2013)
Reeves, C.R., Rowe, J.E.: Genetic Algorithms: Principles and Perspectives: A Guide to GA Theory. Kluwer Academic Publishers, Dordrecht (2002)
Sivanandam, S.N., Deepa, S.N.: Introduction to Genetic Algorithms. Springer, Berlin (2008)
Thrimurthulu, K., Pandey, P.M., Reddy, N.V.: Optimum part deposition orientation in fused deposition modeling. Int. J. Mach. Tool. Manu. **44**, 585–594 (2004)
Whitwell, G.: Novel heuristic and metaheuristic approaches to cutting and packing. Ph.D. Thesis, School of Computer Science and Information Technology, University of Nottingham (2004)

Wodziak, J.R., Fadel, G.M., Kirschman, C.: A genetic algorithm for optimizing multiple part placement to reduce build time. In: Proceedings of the 5th International Conference on Rapid Prototyping, Dayton, Ohio, pp. 201–210 (1994)

Wohlers Associates: The Use of 3D Printing for Final Part Production Continues: Impressive 10-Year Growth Trend. Press release, November 18 (2013)

Wohlers, T.: Will Additive Manufacturing Change Manufacturing? Time Compression Technologies, May/June issue (2011)

Zhang, X., Zhou, B., Zeng, Y., Gu, P.: Model layout optimization for solid ground curing rapid prototyping processes. Robot. CIM-Int. Manuf. **18**, 41–51 (2002)

Car-Like Mobile Robot Navigation: A Survey

Sotirios Spanogianopoulos and Konstantinos Sirlantzis

Abstract Car-like mobile robot navigation has been an active and challenging field both in academic research an in industry over the last few decades, and it has opened the way to build and test (recently) autonomously driven robotic cars which can negotiate the complexity and uncertainties introduced by real urban and sub-urban environments. In this chapter, we review the basic principles and discuss the corresponding categories in which current methods and associated algorithms for car-like vehicle autonomous navigation belong. They are used especially for out-door activities and they have to be able to account for the constraints imposed by the non-holonomic type of movement allowable for car-like mobile robots. In addition, we present a number of projects from various application areas in the industry that are using these technologies. Our review starts with a description of a very popular and successful family of algorithms, namely the Rapidly-exploring Random Tree (RRT) planning method. After discussing the great variety and modifications proposed for the basic RRT algorithm, we turn our focus to versions which can address highly dynamic environments, especially those which become increasingly uncertain due to limited accuracy of the sensors used. We, subsequently, explore methods which use Fuzzy Logic to address the uncertainty and methods which consider navigation solutions within the holistic approach of a Simultaneous Localization and Mapping (SLAM) framework. Finally, we conclude with some remarks and thoughts about the current state of research and possible future developments.

Keywords Rapidly-exploring random trees (RRT) · Simultaneous localization and mapping (SLAM) · Sensor-based methods · Fuzzy logic · Path planning

S. Spanogianopoulos (✉) · K. Sirlantzis
School of Engineering and Digital Arts, University of Kent, Canterbury, UK
e-mail: ss976@kent.ac.uk

K. Sirlantzis
e-mail: K.Sirlantzis@kent.ac.uk

G.A. Tsihrintzis et al. (eds.), *Intelligent Computing Systems*,
Studies in Computational Intelligence 627, DOI 10.1007/978-3-662-49179-9_14

1 Introduction

During the past few years there has been significant progress in navigation applied to outdoor robots with several industrial applications in well defined environments. At the same time there is still a need of fundamental breakthroughs in autonomous systems to make them reliable in much less structured environments.

Once a reasonable representation of the environment is obtained, the vehicle needs to be controlled to perform a certain path. Path following has three main stages: navigation, path planning and guidance. The navigation module is usually responsible for the localization of the vehicle within a given map. The path planning module deals with defining global as well as local paths and the guidance module is responsible for keeping the car on the defined path within acceptable errors. The application of such a techniques has many applications in areas such as robotics, manufacturing, pharmaceutical drug design, computational biology and computer graphics.

2 RRT-Based Methods

2.1 *Unsafe Path Planning*

The real time operation of a car-like vehicle in a large unknown/uncertain environment is still a very challenging problem. In Pepy et al. (2006), in order to face unsafe path planning during the navigation process, authors are using a five degrees of freedom dynamic car model which leads to better placed estimated points, considering skidding and sliding. Using a vehicle model identical to the car that will have to follow the planned path, they use a Rapidly-exploring Random Tree (RRT) planner to quickly explore the whole configuration space. The algorithm focuses also to the computation of the unexplored parts of the space by breaking the large Voronoi areas (Fig. 1).

An interesting feature of this algorithm for path planning is that a path planned with RRT does not need local planner to find a way from a configuration to another. It also allows the RRT to rapidly explore in the beginning, and then converge to a uniform coverage of the space.

2.2 *Safe Path Planning*

In Pepy and Lambert (2006) authors address the problem of safe path planning in an uncertain-configuration space. Considering the case of a car-like robot moving in an indoor environment (three-dimensional space), they used the Extended Kalman Filter (EKF) to localize such a robot and to estimate its configuration uncertainty

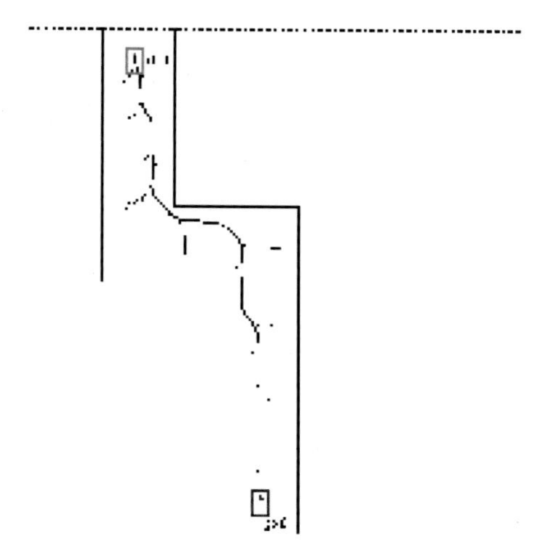

Fig. 1 Obstacle avoidance using kinematic model (Pepy et al. 2006)

during navigation. More specifically, their car-like robot involves non-holonomic constraints and their planner uses an ideal indoor 2D world map, where obstacles are represented by polygonal lines. Then, they used the Rapidly-exploring Random Trees (RRT) method, which is an incremental method to quickly explore the whole configuration space, in order to visit the unexplored parts of the space by breaking the large Voronoi areas.

But since RRT is often slow to plan a path as it randomly reaches a defined position, so in order to speed up the algorithm, they biased it towards the goal by modifying a specific function of the system (goalbias and goalzoom modifications). In the proposed method, authors argue that it is also possible to use two RRTs to plan paths faster, and using this bidirectional RRTs, the path is planned when these two trees meet each other. But since the current configuration of the robot is always uncertain and has limited accuracy, authors used also a probabilistic model to represent the uncertainty, while for the localization process, they divided it into two phases, the prediction and the estimation phase.

2.3 *Rapidly Exploring Random Tree Algorithm on Rough Terrains (RRT-RT)*

In Tahirovic and Magnani (2001) a different approach is presented, where a novel Rapidly exploring random tree algorithm on rough terrains (RRT-RT) has been developed for the purpose of outdoor mobile robot navigation. While other RRTs, which have been adopted for rough terrains, have been used to find a nearest neighbor from a new random state within the tree based on Euclidian distance, the presented algorithm in this work uses a roughness based metric. The metric is defined by the help of the roughness based navigation function, RbNF, which is a numerical function that provides the cost-to-go values (roughness-to-go) for each terrain location. The roughness-to-go value of a terrain location represents an approximate value of the remaining cost of the terrain traversability toward the goal position, or differently, to compute a numerical function RbNF which provides estimated roughness-to-go values for each patch of the given terrain.

This function is used either as a navigation function or as a cost-to-go term within the MPC optimization guiding the vehicle toward the more traversable areas while approaching the goal position. The RbNF has been also used for the purpose of navigation planning based on the model predictive (MPC) paradigm, which repeatedly performs the optimization within a region around the current vehicle position to generate an appropriate path toward the goal. In this approach, the fundamental difference comparing to the state of the art RRTs is comprised in a function, where, while a classical RRT algorithm uses a metric based on Euclidian distance to find the nearest vertex of the tree to a new random state, the algorithm proposed in Tahirovic and Magnani (2001) uses a measure based on the terrain roughness. As such, the simulation results show that the RRT-RT planner explores the terrain in an efficient manner, and even generates final paths that slightly deviate from paths obtained by the Dijkstra's algorithm.

2.4 RRT Motion Planning Subsystem

The work presented in Kuwata et al. (2008) provides a detailed analysis of a motion planning subsystem, which is based on the Rapidly-exploring Random Trees (RRT) algorithm and it is used to present the numerous extensions made to the standard RRT algorithm that enable the on-line use of RRT on robotic vehicles. It actually provides a path and a speed command to the controller that can track the path, such that the vehicle is able to avoid obstacles and stay in lane boundaries under the different conditions of urban driving. The proposed RRT algorithm samples the input to the controller instead of the input to the vehicle, and obtains the dynamically feasible trajectory by running forward a simulation of the closed-loop system, consisting of the vehicle model and the controller.

By using this stable closed-loop system, the method in Kuwata et al. (2008) has the advantage of enabling the efficient use of RRT algorithms on vehicles with unstable open-loop dynamics. Similar to the standard RRT, the proposed algorithm performs sampling, node selection, expansion and constraint check, while the planner keeps providing the command to the controller at a fixed rate. Finally, the best trajectory is selected and sent to the controller, and the tree expansion is resumed after updating the vehicle states and the situational awareness. To sum up the main extensions from the standard RRT, we can say that this approach firstly improves the computational efficiency and secondly the input to the closed-loop system is sampled, which also enables RRT to handle complex/unstable dynamics of the vehicle. Third, the lazy check enables RRT to focus on the tree expansion even with the constantly changing situational awareness, while forth, the uncertainty in the environment is captured in the form of a risk penalty in the tree. The

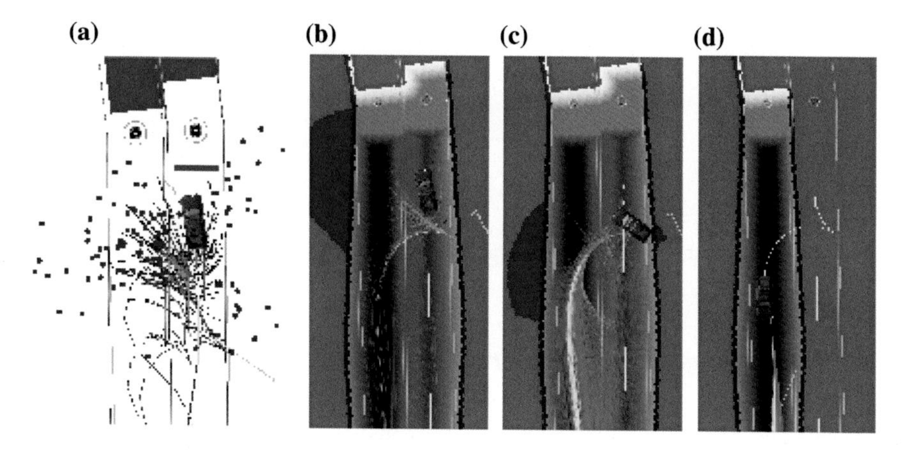

Fig. 2 Planning of a U-turn maneuver. Figures **b**, **c**, and **d** show different evolutions of the tree as the vehicle executes a U-turn. Notice in Figure **d** the trace of the path followed by the vehicle (shown in *yellow*) (Kuwata et al. 2008)

sampling using the environmental structure also significantly reduced the time to find trajectories for various maneuvers and the safety of the vehicle is guaranteed by requiring that the trajectory sent to the controller end in a stopping state (Fig. 2).

2.5 *Partial Motion Planning*

The work presented in Petti and Fraichard (2005) tries to solve the problem of computing a complete motion to the goal within a limited time. More specifically, planning in a changing environment implies to plan under real time constraints and a robotic system cannot safely remain passive, since it might be collided by a moving obstacle (decision constraint). For this reason, authors used a Partial Motion Planning (PMP) approach in order the algorithm to operate until the last state of the planned trajectory reaches a neighbourhood of the goal state. But since PMP has no control over the duration of the partial trajectory that is computed, they considered a selected milestone of a point mass robot with non zero velocity moving to the right, and depending upon its state there is a region of states for which, even though it is not in collision, it will not have the time to brake and avoid the collision with the obstacle [Inevitable Collision State (ICS)]. In order to avoid this state, they used a property which firstly proves that a trajectory is continuously safe while the states safety is verified discretely only, and secondly, it permits a practical computation of safe trajectories by integrating a dynamic collision detection module within the Rapidly-Exploring Random Tree (RRT) (Fig. 3).

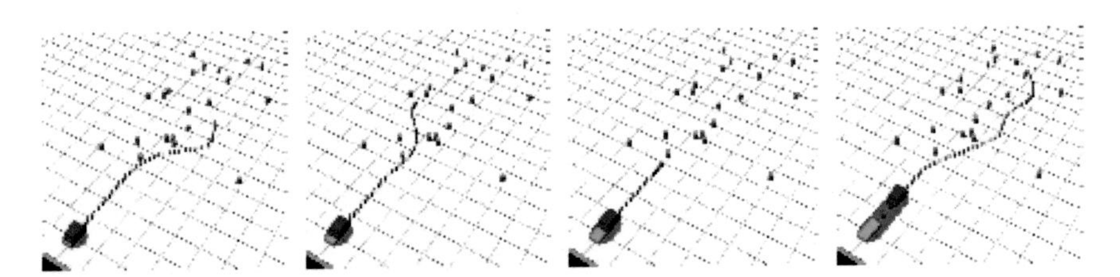

Fig. 3 Navigation within an environment cluttered with moving pedestrians (Petti and Fraichard 2005)

2.6 *Sensor-Based Random Tree (SRT)*

In Espinoza et al. (2006) a new method is presented for sensor-based exploration of unknown environments, which proceeds by building a data structure called Sensor-based Random Tree (SRT). The SRT structure represents a roadmap of the explored area with an associated safe region, and estimates the free space as perceived by the robot during the exploration. The technique which is used for this case is called SRT-Radial and deals with non-holonomic constraints using two alternative planners. More specifically, the method builds a data structure through random generation of configurations. Then, the SRT represents a roadmap of the explored area with an associated Safe Region, an estimate of the free space as perceived by the robot during the exploration. Depending on the shape of the Local Safe Region, the general method results in different exploration strategies and the idea is to increase the exploration efficiency by biasing the randomized generation of configurations towards unexplored areas. In order to do this, authors used the SRT-Radial strategy, which takes advantage of the information reported by the sensors in all directions, to generate and validate configurations candidates through reduced spaces.

The paper in Fulgenzi et al. (2008) describes a navigation algorithm for dynamic and uncertain environment, based on the assumption that moving obstacles are supposed to move on typical patterns which can be pre-learned and represented by Gaussian processes (GP). More specifically, they use a robot which is equipped with a distance sensor and models the static environment in an occupancy grid. The moving obstacles they used follow typical patterns with some amount of uncertainty, and these patterns are a priori known and represented with GP. Then, the moving obstacles are detected and tracked on-line and the prediction of their future position is computed on the base of the known typical paths. To update previously explored states with the on-line estimation, authors integrated the likelihood of obstacle paths and the probability of collision. The planning algorithm is based on an extension of the Rapidly-exploring Random Tree algorithm (RRT), where the likelihood of the obstacles trajectory and the probability of collision is explicitly taken into account. The algorithm is used in a partial motion planner, and the probability of collision is updated in real-time according to the most recent estimation.

2.7 RRT* Algorithm

In Karaman and Frazzoli (2013) a different approach is presented, where authors extend the RRT algorithm to handle a large class of non-holonomic dynamical systems. As it is known, sampling-based motion planning algorithms, such as the Probabilistic RoadMap (PRM) and the Rapidly-exploring Random Tree (RRT), guarantee asymptotic optimality, providing almost-sure convergence towards optimal solutions. In order to face the drawbacks of these methods (such as computational complexity, quality of the motion, etc.) and improve their results, authors modified them in the following way. First, they seek connections within boxes that are substantially larger in some dimensions than others, in order to ensure both asymptotic optimality and computational efficiency. Then, they computed the shape and orientation of these boxes for a large class of dynamical systems, based on differential geometry, which is dictated by the ball-box theorem of sub-Riemannian geometry. They then compute the optimal shape and orientation of these probabilistic trajectory boxes using the ball-box theorem of differential geometry, which is a class of sub-Riemannian problem where shapes can be constructed and joined in a "fuzzy" manner, i.e. without definitive boundary constraints.

2.8 Voronoi Fast Marching (VFM) and Fast Marching (FM2)

In Garrido et al. (2009) authors present a new method in order to improve the trajectories based in the Voronoi Fast Marching Method (VFM). The proposed method is suitable for improving the smoothness and the length of the trajectories calculated with probabilistic methods with bad quality trajectories, such as RRT or PRM. More specifically, in order to calculate the trajectories with the best properties, authors used a hybrid Path Planning method split into two parts: in the first one, the RRT method is used to obtain a first trajectory and in the second one, this trajectory is improved using the Voronoi Fast Marching Method. In order to apply VFM method, they calculated a tube around the previous trajectory. This tube is intersected with the walls and obstacles map and then the VFM method is applied. This way, authors take advantage of the best properties of the two methods, i.e. the possibility of working in many dimensions of RRT and the smoothness and the quality of the trajectories of VFM. In particular, the smoothing is calculated on a vectorial field that has the same goal point and it is repelled by obstacles and walls and admits the non-holonomic constraints. Finally, the VFM method uses the propagation of a wave (Fast Marching) operating on simple grid-based world model, to determine a motion plan over a slowness map (similar to the refraction index in Optics) extracted from the updated grid-map model.

In Garrido et al. (2011) authors present the application of Voronoi Fast Marching (VFM) and FM2 methods to non-holonomic mobile robot path planning. More

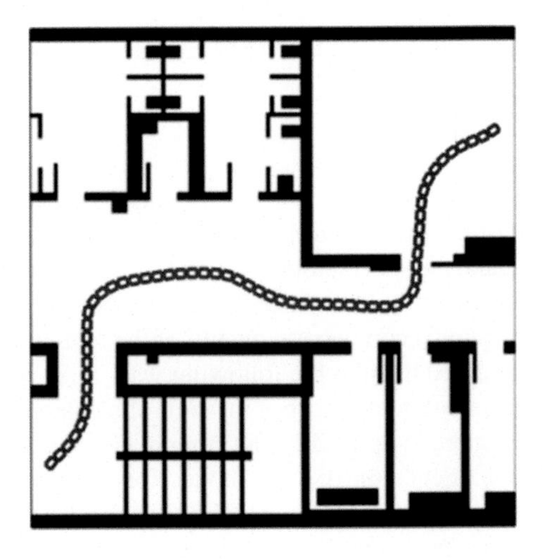

Fig. 4 Non holonomic version of the proposed method (Garrido et al. 2011)

specifically, the VFM and FM2 methods use the propagation of a wave (Fast Marching) operating on the world model (i.e. from the current position of the robot to the goal), to determine a motion plan over a slowness map (or refractive indexes, or the inverse of velocities) similar to the repulsive electrical potential of walls and obstacles, and the calculation of the path by using the gradient method from the goal to the current position point. This algorithm starts with the calculation of the Logarithm of the inverse of the Extended Voronoi Transform of the 2D updated map, in order to obtain a potential proportional to the distance to the nearest obstacles to each cell. Then they apply the Fermat's least time principle (i.e. the Fundamental Equation of the Geometrical Optics) for light propagation in order to get the refractive index and then they eliminate the poses that are not feasible. In the next step, they apply the Voronoi Extended Transform to the configuration space, along with the Fast Marching Method, and finally they calculate the trajectory from the initial pose to the goal by using the vectorial field of the expansion wave (Fig. 4).

2.9 SBL Algorithm

Authors in Balakirsky and Dimitrov (2010) presented several enhancements that improve the quality of the generated path in comparison with the simple adaptation of the Single-query, Bi-directional, Lazy roadmap (SBL) algorithm, which successfully builds upon the traditional Probabilistic Roadmaps (PRM), solving also the planning problem in the context of non-holonomic constraints of car-like robots. More specifically, the adaptations they made can be summarized as such: first, they limit the connections between milestones to a single constant curvature arc, in order to restrict the milestone's neighborhood to a relatively small subset of the geometric region to the front and rear of the robot. Secondly, they restricted the early creation

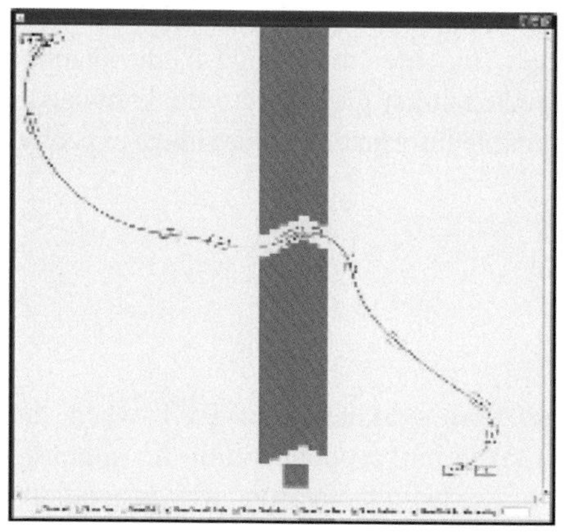
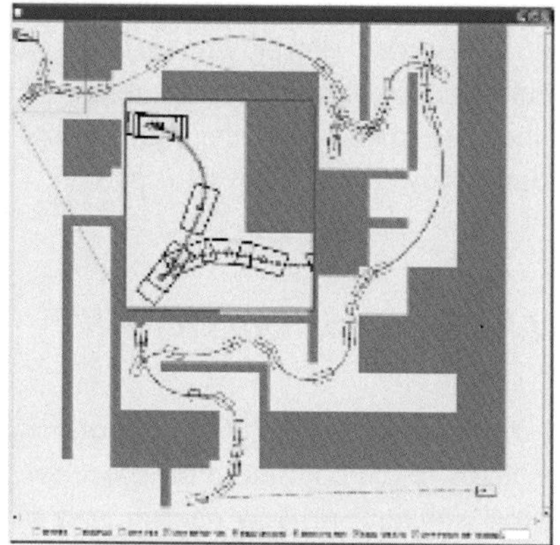

Fig. 5 PRM-AS path solution in the "corridor (maze)" environment. The magnified area (*green box*) shows the maneuver (*red curved line*) undertaken to exit the tight space (*bottom right of green box*) while positioning the vehicle for forward movement (Balakirsky and Dimitrov 2010). The maneuver (*red curved line*) also indicates the vehicle moving in reverse

of speed profiles by adding a "speed" dof to each milestone and allowing connections only between milestones with close speed values. Thirdly, they used the arc length as a metric to utilize during tree growing, tree connection, and collision checking. Forth, in order to prevent frequent switching between forward and backward motion, they used the PRM-AS with the higher probability, which generates either a forward or a backward arc according to the direction of the milestone. In addition, they defined that the modified version favors forward movement when expanding the init tree and backward movement when expanding the goal tree. They also introduced three configurable parameters, which provide a significant degree of control over the generated path by the robot, while they implemented a path smoother, in order to test each milestone for a potential connection with each of the subsequent milestones on the path (Fig. 5).

2.10 Single-Query Motion Planning

The paper in Burns and Brock (2007) presents a utility-guided algorithm for the online adaptation of the random tree expansion strategy. It is evident that the randomly expanding trees are very effective in exploring high-dimensional spaces, but as the dimensionality of the configuration space increases, the performance of the tree-based planners that use uniform expansion degrades. The proposed algorithm is based mainly on RRT and guides the expansion towards regions of maximum utility based on local characteristics of state space. More specifically, the planner guides the ongoing tree-based exploration of state space using information about state space obtained from previous tree expansions. The planner incrementally learns how to adjust the parameters of tree-based exploration based on the structure of state space that is revealed during the planning process. In order to manage it, the planner identifies expansion steps with maximal expected utility given its current knowledge of the state space. Thus, the planner uses available information to maximize expected progress towards a successful path.

2.11 Dynamic-Domain RRT

In Yershova et al. (2005) authors analyze the weaknesses of RRT when the obstacles in the configuration space are not taken into account and/or the sampling region is inappropriately chosen, and then they explain the reasons why adaptations and extensions of RRTs are generally proposed in bibliography. As a further step, they propose a general framework for minimizing the effect of some of these weaknesses by considering a new sampling strategy based on the visibility region of the nodes in the tree. More specifically, they have developed and implemented a simple new planner, which defines a boundary domain for a boundary point as the intersection of the Voronoi region of that point and an n-dimensional sphere centered at that point. In addition, authors defined also the dynamic domain of radius R for a set of points, which is the boundary domains of the boundary points combined with the Voronoi regions of all other points. They call this uniform distribution over this domain the dynamic domain distribution, and they argue that using this method, the performance of the system can become at some cases orders of magnitude better (Fig. 6).

Authors in Jaillet et al. (2005) analyze the influence of a parameter introduced in Yershova et al. (2005), which relies on a new sampling scheme that improves the performance of the RRT approach, and propose a new variant of the dynamic-domain RRT, which iteratively adapts the sampling domain for the

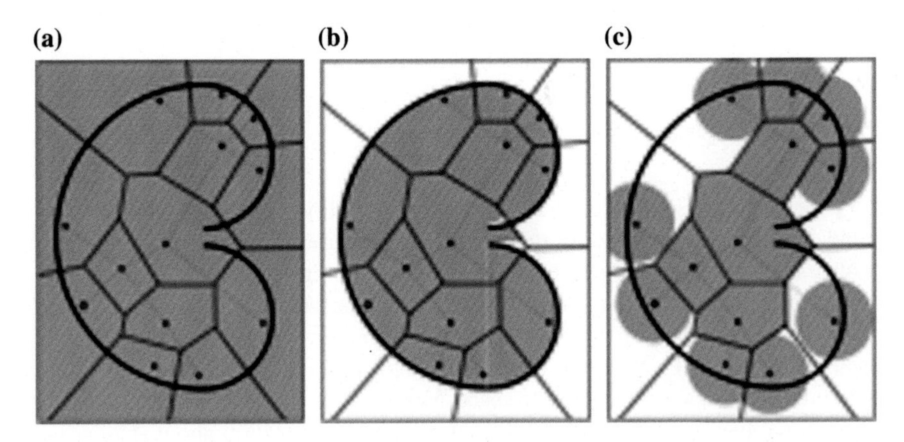

Fig. 6 For a set of points inside a bug trap different sampling domains are shown: **a** regular RRTs sampling domain, **b** visibility Voronoi region, **c** dynamic domain (Yershova et al. 2005)

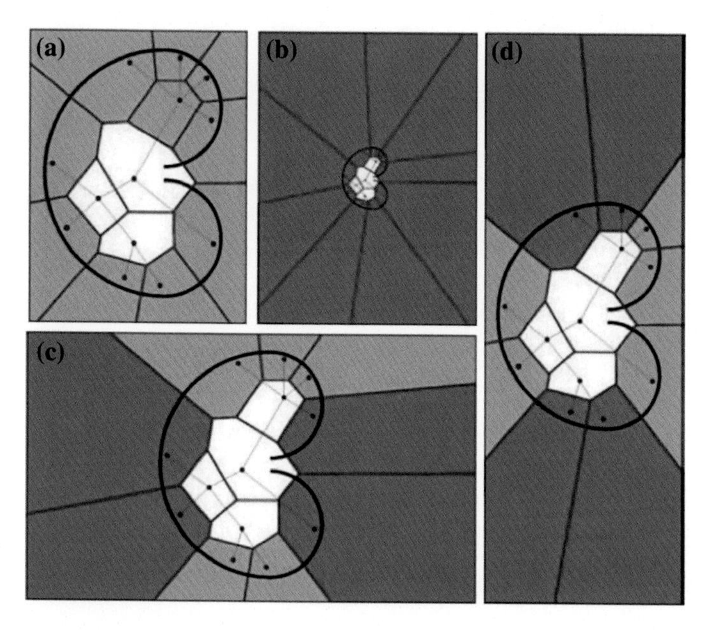

Fig. 7 For the RRT algorithm, the Voronoi region of the frontier nodes is growing together with the size of the configuration space. Therefore, depending on the boundaries of the space, the bias toward unexplored regions can be small (**a**), strong (**b**) or biased toward only some of the parts of the space (**c** and **d**) (Jaillet et al. 2005)

Voronoi region of each node during the search process. In particular, authors propose to adapt the boundary domain of a given node (i.e. its associated radius) as a function of the number of expansion attempts and failures from this node. In addition, in order to keep the probabilistic completeness of the algorithm, they ensure that the possibility for a node to be extended by putting a lower bound on the possible radius values of the nodes (Fig. 7).

2.12 Transition-Based RRT

In Jaillet et al. (2012) a new method called Transition-based RRT (T-RRT) for path planning in continuous cost spaces is presented, which combines the exploration strength of the RRT algorithm that rapidly grow random trees toward unexplored regions of the space, with the efficiency of stochastic optimization methods that use transition tests to accept or to reject a new potential state. More specifically, the proposed planner relies mainly on the notion of minimal work path that gives a quantitative way to compare path costs. The minimal work path is based on the notion of mechanical work, and the resulting loss of "energy" due to this mechanical work is the criterion that authors try to minimize. The algorithm also uses the exploration strategy of RRT resulting from the exploration bias toward

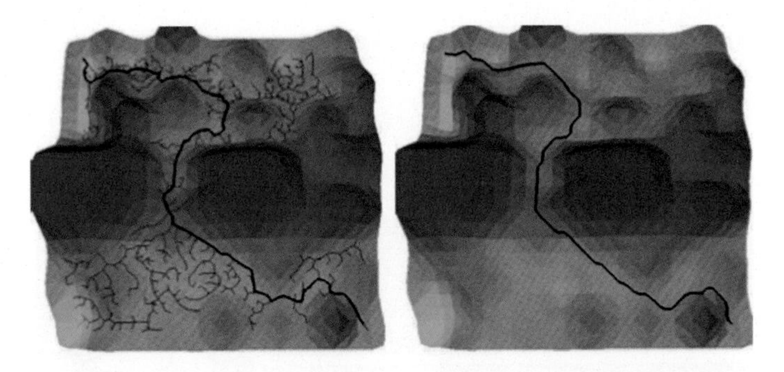

Fig. 8 Optimal paths based on the minimal work criterion (Jaillet et al. 2012)

large Voronoi regions of the space and combines it with a stochastic optimization method developed for computing global minima in complex spaces (such as Monte Carlo optimization or simulated annealing), in order to introduce it to a transition test and classify the new potential state as accepted or rejected (Fig. 8).

2.13 *Parallelizing Rapidly-Exploring Random Tree (RRT) Algorithm on Large-Scale Distributed-Memory Architectures*

In Devaurs et al. (2013) authors address the problem of parallelizing the Rapidly-exploring Random Tree (RRT) algorithm on large-scale distributed-memory architectures, using the message passing interface. The parallelization schemes they compared are the OR parallel RRT, the distributed RRT, and the manager–worker RRT. In the first case, they used the OR parallel paradigm for its implementation, where each process computes its own RRT and the first to reach a stopping condition broadcasts a termination message. The second and the third one belong to the category of collaborative RRTs, where all processes collaborate to build a single RRT. Parallelization is then achieved by partitioning the building task into subtasks assigned to the various processes. So, in order to achieve this, they rely on classical decomposition techniques and more specifically in the following two ways. First, since the construction of an RRT consists of exploring a search space, they used an exploratory decomposition, where each process performs its own sampling of the search space, without any space partitioning involved, and maintains its own copy of the tree, exchanging with the others the newly constructed nodes. This leads to a distributed (or decentralized) scheme where no task scheduling is required, aside from a termination detection mechanism. Secondly, they perform a functional decomposition of the task, leading to the choice of a manager–worker (or master–slave) scheme as the dynamic and centralized task-scheduling strategy, where the manager maintains the tree, and the workers have no access to it. The proposed methods can be used in mainly two cases. First, problems whose variability in sequential runtime is high can benefit from the OR parallel RRT, while problems for which the computational cost of an RRT expansion is high can benefit from the distributed RRT and manager–worker RRT.

In Rodriguez et al. (2006), a variant of the Rapidly-Exploring Random Tree (RRT) path planning algorithm is presented, which is able to explore narrow passages or difficult areas more effectively. More specifically, authors in Rodriguez et al. (2006) used some obstacle hints for directions to grow the tree for path planning in order to find difficult areas of configuration space (C-space). The planner they used uses obstacle vectors obtained from the obstacle and it focuses on how the tree decides to grow. They presented nine possible ways to expand a tree, in which the orientations to grow are either the same as the source configuration or random orientations. They also proposed a modification to a greedy algorithm for calculating the planner's path, such that it would take as big a step length as possible, as long as it is less than some maximum step length specified. Based on these modifications, authors argue that they could result in a path planning technique that solves motion planning problems more quickly and efficiently than other techniques (Fig. 9).

The paper in Phillips and C.S. Draper Laboratories (2004) presents a path planning algorithm for handling systems with constraints on controls or the need for relatively straight paths for real-time actions. The initial phase of the algorithm finds an efficient path using guided Expansive Spaces Trees (guided ESTs) and focuses on a randomized search on the low cost region while expanding a tree. It generates also new waypoints by probabilistically branching off of existing waypoints and weighting each waypoint based on, not only the number of close waypoints, but also on the estimated total cost of going through that waypoint on a path to the goal. The second phase of the algorithm refines the existing path according to a cost function by following the gradient of the path. This technique does not enforce elastic properties of the path and it is able to take more robust precautions in repelling from obstacles.

In Kim et al. (2005) authors try to address the problem of testing complex reactive control systems and validating the effectiveness of multi-agent controllers. More

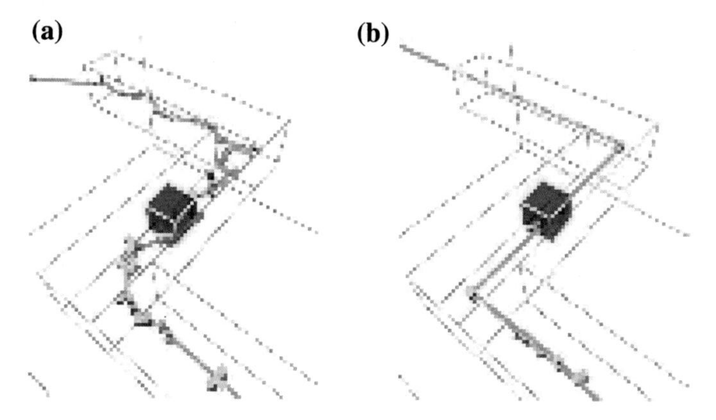

Fig. 9 Differences when growing a tree: **a** basic expansion and **b** using obstacle information (Rodriguez et al. 2006)

specifically, they consider the application of the Rapidly-exploring Random Tree (RRT) algorithm to the testing and validation problem and they propose three modifications in order to improve its results. First, they introduce a new distance function which encodes local information about the system's dynamic constraints with a first order approximation (i.e. dynamics-based selection of proximal node). Secondly, because the reachable state space is generally a small fraction of the total state space, they developed a weighting factor to penalize nodes which are repeatedly selected but fail to extend (i.e. history-based selection of proximal node), and finally, they proposed a scheme for adaptively modifying the sampling probability distribution between the traditional uniform distribution and heavily biased toward the specification set based on tree growth (i.e. adaptively biased sample generation).

2.14 Obstacle Sensitive Cost Function for Navigating Car-Like Robots

In Ziegler and Werling (2008) authors propose a new method for navigating a car-like vehicle within an unstructured environment, using a path planning technique which is posed as a graph search problem. Actually they define an implicit graph that is expanded on the fly by an A* search algorithm. In this algorithm, the search graph is set up in a way that implies derivation of a feed forward term for a downstream closed loop controller. More specifically, authors have added a feed forward term, which makes the controller react more quickly and accurate, since reaction of the vehicle to steering input is modelled separately from controller offset introduced by noise. Then, an informed search algorithm is used, which is guided by a heuristic cost function that accounts for both kinematic constraints of the vehicle and the topology of the vehicle's free space. This cost function gives expected cost-to-go for each node of the search graph, so if the cost function underestimates the actual distance to the goal, A* is guaranteed to find the least-cost path. If the error of the cost function is big, A* quickly degenerates to an exponential time algorithm. The configuration space obstacles are then computed from an obstacle map acquired from a high definition laser range scanner and search is restricted to the collision free subset of the configuration space. Authors suggest that this algorithm is suitable for solving all of the following problems: precise parking maneuvers, narrow turns and long distance navigation.

3 Methods Based on Fuzzy Logic

In Baturone and Gersnoviez (2007) a novel work is proposed, which combines some neuro-fuzzy techniques with geometric analysis in order to get a good trade-off between purely heuristics and purely physical approaches. Specifically, the

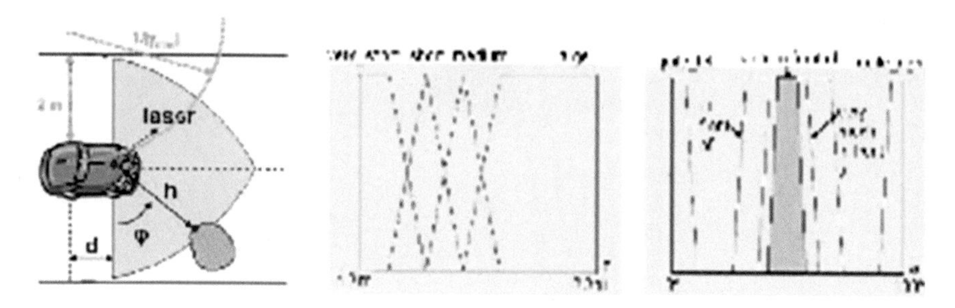

Fig. 10 Area of unavoidable and very close obstacles (Baturone and Gersnoviez 2007)

controller follows a reactive technique and generates trajectories of near-minimal lengths when no obstacles are detected, while in presence of obstacles, it generates minimum deviations from them. The fine structure of the controller constituent modules has been obtained by applying supervised learning with numerical data, satisfying the non-holonomic constraints. In particular, it consists of several neuro-fuzzy modules, whose fuzzy classifiers have two inputs and one output. For close objects, authors used two approaches to define a fuzzy classifier, a monolithic system and a hierarchical one, while for obstacle avoidance, the controller evaluates an angular aperture and determines the sign of the curvature to avoid the obstacle. Without obstacles, the robot can navigate towards the goal by the shortest path, using again a monolithic or a hierarchical fuzzy system. Experimental results indicate that if no obstacles are detected, the robot goes to the goal by a near-minimal length path, while if obstacles are close, they are avoided by minimum deviations from the quasi-optimal path (Fig. 10).

3.1 Distributed Active-Vision Network-Space System

In Hwang and Shih (2009) a navigation scheme that contains complex pattern, non-uniform illumination, and strong reflection based on a distributed active-vision network-space system (DAVNSS), is presented. This system is subject to three fuzzy variable-structure decentralized controls (FVSDCs), which includes trajectory tracking and obstacle avoidance. Two distributed wireless charge-coupled-device (CCD) cameras individually driven by two stepping motors are constructed to capture the dynamic pose of the car-like wheeled robot (CLWR) and the obstacle. The control system includes quad processors with multiple sampling rates, while a personal computer (PC) is employed to receive the image of the CLWR or obstacle by a wireless transmitter and then to plan three reference commands for the CLWR and the cameras. Next, a six-step image-processing routine and the calibration between the world coordinate and the image plane coordinate using multilayer perceptrons (MLPs) are established, while in the final step the radial distortion of ACCD is reduced for better localization and tracking (Fig. 11).

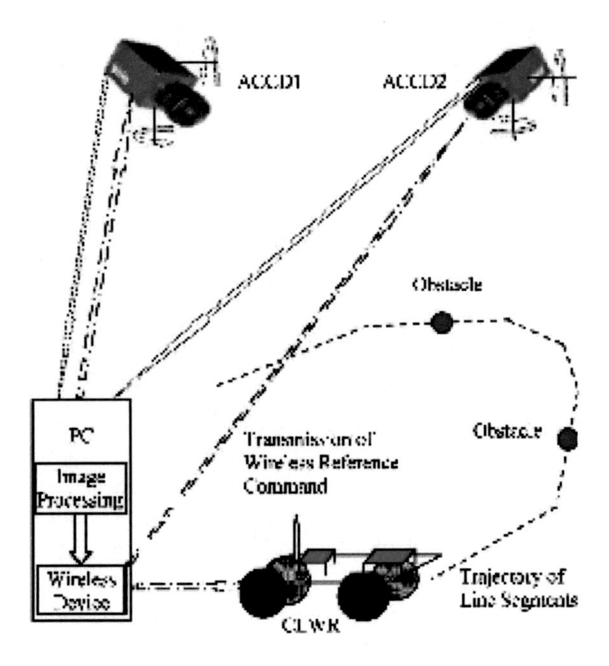

Fig. 11 Diagram of the overall control system (Hwang and Shih 2009)

In El-Khatib and Hamilton (2006), an approach which consists of two layers for real-time navigation of a non-holonomic car-like robot in a dynamic environment is described. More specifically, the first layer is a Sugeno-type fuzzy motion planner of four inputs and one output, which is used to give a clear direction to the robot controller. The second stage is a modified proportional navigation-based fuzzy controller, which is based on the proportional navigation guidance law and it is able to optimize the robot's behavior in real time. This means that it is able to avoid stationary and moving obstacles (such as pedestrians and vehicles) in its local environment obeying specific kinematics constraints. The proposed system consists of two subsystems, a fuzzy motion planner (FMP) and a modified proportional navigation (PN) based fuzzy controller, inspired by human routing. The fuzzy motion planner uses Takagi-Sugeno fuzzy inference for the rule evaluation, resulting in the output of a control function for the system depending on the values of the inputs. The proportional navigation method is a guidance law, which seeks to null the line of sight changing rate (LOS) by making the controlled system (robot) turn rate be directly proportional to the rate of turn of sight (i.e. it seeks to nullify the angular velocity of the line of sight (LOS) angle). For the intelligent behavior of the fuzzy controller, the direction control behavior of the robot in response to an obstacle was also incorporated. As was indicated, the main advantage of this system is the simplicity of its design which makes it suitable for hardware implementation and extensibility as it does not rely on any specific robotic platform, while it is able to use also linguistic representation, which allows the capture of human experiences and intuitive reasoning (Fig. 12).

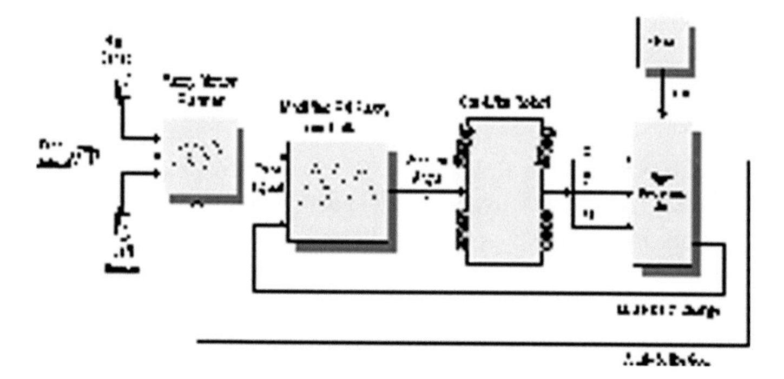

Fig. 12 The complete system of fuzzy motion planner and behavioral fuzzy controller (El-Khatib and Hamilton 2006)

3.2 *Internet-Based Smart Space Navigation Using Fuzzy-Neural Adaptive Control*

In Hwang and Chang (2008) a different approach is presented for a navigation system, which includes a path tracking and an obstacle avoidance apparatus for a car-like wheeled robot (CLWR) within an Internet-based smart-space (IBSS) using fuzzy-neural adaptive control (FNAC). This method relies on two distributed charge-coupled device (CCD) cameras, which capture both the dynamic pose of the CLWR and the obstacle. Based on the control authority of these two CCD cameras, a suitable reference command has been planed, which contains the desired steering angle and angular velocity for the FNAC built into the client computer. The FNAC method that the authors presented in Hwang and Chang (2008) contains also a neural network consisting of a radial basis function (RBFNN) to learn the time-related uncertainties due to the fuzzy-model error, which stem from wireless network delays and CLW slippage (Fig. 13).

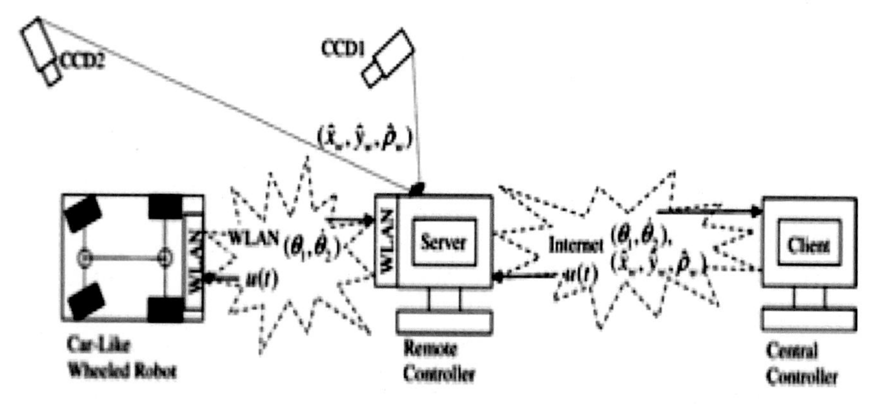

Fig. 13 Block diagram of the overall system (Hwang and Chang 2008)

4 Sensor-Based Methods

4.1 Dynamic Window Approach (DWA)

Another approach presented in Rebai et al. (2007) investigates the use of Dynamic Window Approach (DWA) to solve the high speed autonomous navigation problem for mobile robots in unknown and unstructured environments. Since the DWA algorithm considers periodically a short time interval when computing the next motion command and based on the fact that the obstacles in the closer environment of the robot impose restriction on the translational and rotational velocities, authors define a Dynamic Window (DW) in order to limit the accelerations executable by the motors. In addition, in order to reduce the time of the motion command selection, they use the sensory data from the environment directly in the obstacle avoidance process without the grid cells building in the velocity space, while in order to determine the Distance To Collision (DTC), they have adopted an analytic solution for polygonal robot. Regarding the experimental results of the algorithm, the obstacle avoidance tests using the extended DWA for different environments (simple and cluttered) at high speeds indicated a good performance and efficiency.

4.2 Generalized Voronoi Graph (GVG) Theory

Another approach which relies on a sensor based algorithm for car-like robot based on GVG theory is presented in Quan et al. (2011). For generating the completed GVG, the car-like robot goes through each edge and vertex of GVG in two tangent directions. In addition, the authors proposed backward motion for direction changes at boundary points, also with favorable results (no collision) in unknown environments.

In Quan et al. (2011) a new algorithm is presented that enables a car-like robot to explore an unknown planar workspace, based on Generalized Voronoi Graph (GVG) theory. More specifically, since GVG is a set of points in the plane equidistant to two obstacles, the robot of the proposed system has three degrees of freedom and hence the authors defined a rod-GVG edge as the set of the points equidistant to three obstacles.

In Gall et al. (2010), an intelligent scaled car-like mobile robot that possesses the capability of autonomous driving in an extra-road environment and fully autonomous parking on standard parking lots is presented. In particular, authors describe a low weight and low cost complex mobile robot that is able to navigate across a previously unknown terrain combining some mechanical, sensorial, computing and communication modules (rather than implementing a new sophisticated algorithm). An algorithm associated with the autopilot of the system was also implemented, in order to make the mobile robot completely autonomous; many of these functions

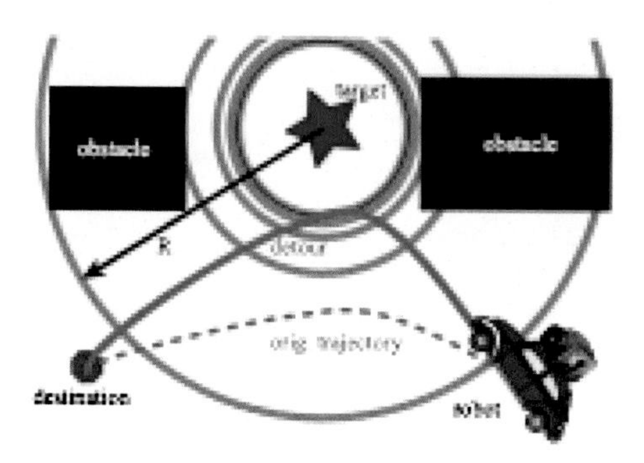

Fig. 14 A schematic representation of a car-like robot making a detour from a path towards its primary destination to opportunistically gather additional information about a secondary target (indicated by a *blue star*) once the presence of the latter has been detected at distance R (Grady et al. 2012)

are written in MATLAB, and therefore available for analysis and modification using open-source modules (xPC toolbox).

The robot described in Grady et al. (2012) is equipped with a sensor that can alert it if an anomaly appears within some range while the robot is moving. In that case, the robot tries to deviate from its computed path and gather more information about the target without incurring considerable delays in fulfilling its primary mission, which is to move to its final destination. The originality of this approach is to take a "semi-corrective" action, i.e. deviating while attempting to further define the problem, akin to a car stepping out of its lane when flashing lights appear ahead—not changing lanes yet, just gaining a view of the obstacle. This model relies on a sampling-based planner called SYCLOP, which works by automatically defining a decomposition of the workspace, creating an adjacency and abstraction graph, and searching that graph for a high-level guide. Then, a low-level planning layer computes the actual dynamically feasible paths and informs the upper layer for how to assign informative weights to the edges of the abstraction graph (Fig. 14).

4.3 Navigation in Dynamic Environments Using Trajectory Deformation

A different approach is presented in Delsart and Fraichard (2008), where authors present a new trajectory deformation scheme in order to improve path deformation. During the course of execution, the still-to- be-executed part of the motion is continuously deformed in response to sensor information (internal and external) acquired on-line, thus accounting for the incompleteness and inaccuracies of the a priori world model (Fig. 15).

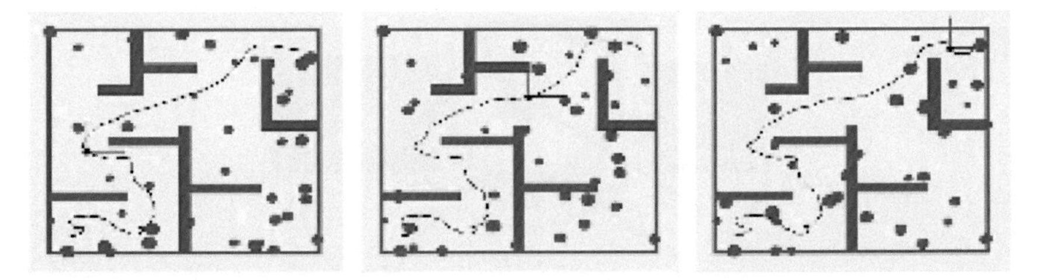

Fig. 15 Double integrator system: the snapshots depict the path at different time instant (Delsart and Fraichard 2008)

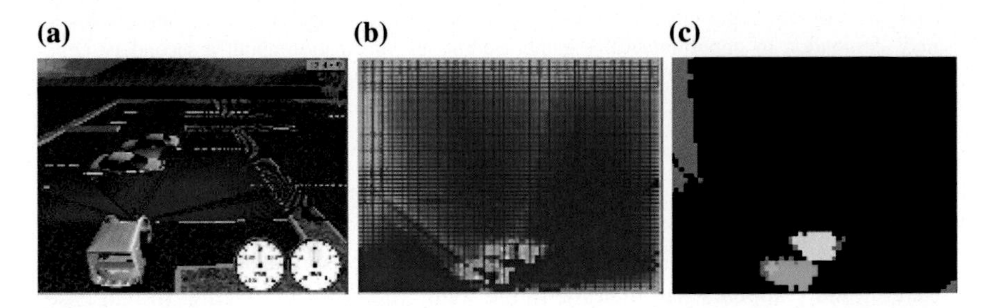

Fig. 16 Simulated detection of two cars crossing each others. **a** Simulated environment: the robot equipped with a laser range finder detects a car moving from left to right and a second car moving from right to left. **b** Dynamic occupancy grid: *red* is high, *blue* is low probability of occupation. The space behind the cars has low probability of occupation. **c** Clustering: different colours characterise objects and occluded or free space (Fulgernzi et al. 2007)

4.4 Probabilistic Velocity Obstacle (PVO)

In Fulgernzi et al. (2007), the Probabilistic Velocity Obstacle (PVO) provides a probabilistic estimation of the occupied free space around the robot and of the velocity with which the objects are moving. The observations of the mobile robot update a 4D probabilistic occupancy grid (incl. space and velocity), and the probability of collision in time is estimated for each reachable velocity of the robot. The proposed system shows that is able to take directly into account limited range and occlusions, uncertain estimations of velocity and position of the obstacles, allowing the robot to navigate safely toward the goal (Fig. 16).

5 SLAM-Based Methods

5.1 On-line Path Following

In Rezaei et al. (2004) the authors address the problem of on-line path following for a car working in unstructured outdoor environments. More specifically, the partially

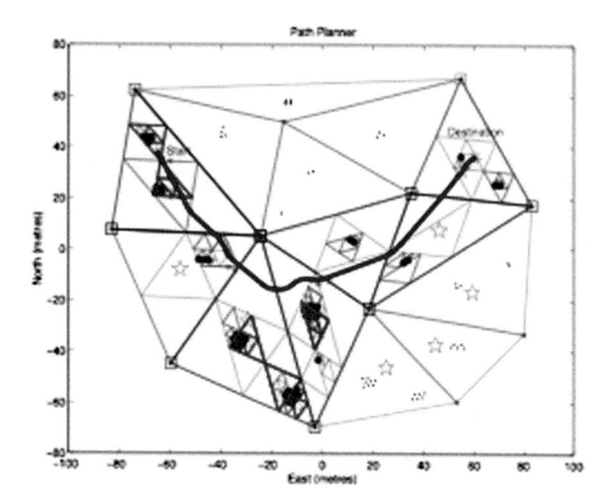

Fig. 17 An example of a generated path (Pepy and Lambert 2006)

known map of the environment is updated and expanded in real time by a Simultaneous Localization and Mapping (SLAM) algorithm. This information is used to implement global path planning based on a new method which constructs a cost graph using the D* search algorithm. In this stage, uncertainty is incorporated in the cost function, and since the continuity of the path is crucial for car type robots, the algorithm chooses only the continuous-curvature local paths. Finally, an improved feedback linearization control algorithm is used to guide the car along this computed reference path (Fig. 17).

5.2 *The CyCab: A Car-Like Robot Navigating Autonomously and Safely Among Pedestrians*

In Pradalier et al. (2005) authors present a bi-steerable car, which allow steering by turning either the back or the front wheels. Using this car, they address the integration of the four essential autonomy abilities (i.e. simultaneous localisation and environment modelling, motion planning and motion execution) into a single application. Then they build a kind of simplified occupancy grid on the environment and they apply the motion planner adopted for the CyCab, expressed as a Bayesian inference problem. Bayesian methods are also used for trajectory tracking (Fig. 18).

In Pradalier et al. (2004) a new kind of public transportation system is presented, which relies on a particular double-steering kinematic structure (as described above). The authors in this work address the integration of these four essential autonomy abilities into one application, applying a reactive execution of planned motion. In addition, they address the fusion of controls, issued from the control law and the obstacle avoidance module, using probabilistic techniques. The planner first

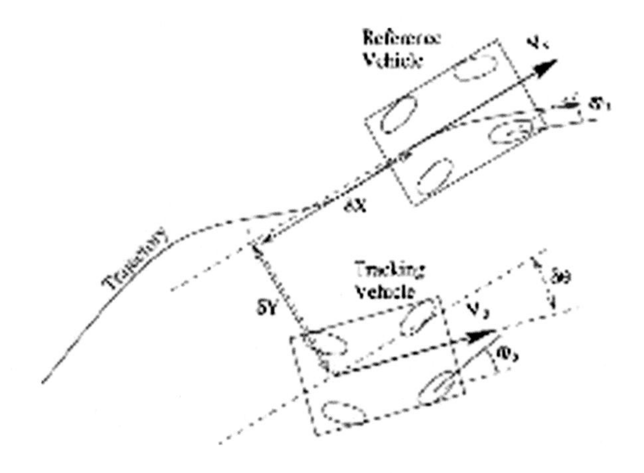

Fig. 18 Variables involved in trajectory tracking behaviour, using Bayesian inference (Pradalier et al. 2005)

builds a collision-free path without taking into account the non-holonomic constraints of the system. Then, this path is approximated by a sequence of collision-free feasible sub-paths computed by a suitable steering method and then is smoothed properly.

5.3 V-Slam

In Lategahn et al. (2011) authors propose a dense stereo V-SLAM algorithm that estimates a dense 3D map representation which is more accurate than raw stereo measurements. The proposed system is composed of two main parts. First a sparse V- SLAM system based on an EKF is calculated, which takes the resulting pose estimates in order to compute a locally dense representation from dense stereo correspondences. The state vector of the EKF contains all landmark positions, the current camera pose and a subset of past camera poses. To tackle the computational complexity problem inherent to EKF SLAM, authors utilize a sub mapping method called conditionally independent sub maps. After incorporating new observations and updating the EKF state vector a new camera pose is obtained. This allows the dense part to be continuously updated.

5.4 SLAM-Based Turning Strategy in Restricted Environments

In Cheein et al. (2010) a strategy to turn a car-like mobile robot in a restricted environment using a Simultaneous Localization and Map Building (SLAM) algorithm is presented. More specifically, in the first step of the proposed method, the environment's information and the vehicle's pose (position and orientation) estimation is provided to the vehicle by a SLAM algorithm, which is implemented on an Extended Kalman Filter (EKF), extracting the lines and corners (convex and concave) from the environment. In the next phase, a turning algorithm, which is based on a semi-circle trajectory, following with direction switching, plans from the vehicle's initial pose the first semi-circle trajectory with respect to the environment until it reaches a neighborhood of the closest geometric map feature provided by the SLAM system state. Then, a next semi-circle trajectory is planned in the opposite direction to the previous trajectory. The proposed algorithm continues until the vehicles reaches the desired orientation, while a kinematic trajectory controller drives the vehicle through the generated paths.

5.5 L-Slam

Authors in Petridis and Zikos (2010) present a new SLAM method, called L-SLAM. It is a low dimension version of the FastSLAM family algorithms, which reduces the dimensionality of the particle filter that FastSLAM algorithms use, while achieving better accuracy with less or the same number of particles. The key idea they used is to sample only the robot's orientation on each particle, in contrast to the FastSLAM algorithms that sample the orientation along with the position of the robot.

6 Conclusions and Future Work

As we have seen, for any mobile device, the ability to navigate in its environment is important. Avoiding dangerous situations such as collisions and unsafe conditions (temperature, radiation, exposure to weather, etc.) comes first, but if the robot has a purpose that relates to specific places in the robot environment, it must find those places. As a result, mobile robots capable of moving in a dynamical and uncertain environment is an important issue in real-world applications. The problem that how to find an optimal real-time collision-free path with a limited sensing range in the presence of dynamically moving objects is arising naturally. The optimal solution should take motion constraints into consideration (including boundary conditions and kinematic constraint), explicitly handle dynamically moving objects, and be analytical.

Based on the previous mentioned methods, it is evident that this technology is well promising for the future. While the human-machine interface is not yet at a transparent level, the degree of autonomy available after a machine has been program is now approaching that once considered purely science fiction. Things that could be done in the future, related to the previous algorithms, are for example to optimize the current techniques using more state-of-the-art methods, testing the navigation algorithms to have a measure of its performance in more complex and realistic scenarios, or even considering for instance that the knowledge about the future behaviour of a robot is less reliable in the distant future, so it could be interesting to monotonically decrease the influence of the obstacles with respect to time.

6.1 Future Directions in Autonomous Robot Navigation and Obstacle Perception

One of the ways in which autonomous robot perception could be improved, especially for safe navigation of urban environments, is by object perception. For the majority of this review, all obstacles have been treated as essentially equal (i.e. rough patches to be avoided). However, as the reader may have suspected at some point, not all obstacles are equal. In fact, some obstacles present quite opposite problems to the optimization routine. A patch of ground for example is a "rough patch" that, although preferably avoided, could in theory be traversed if the cost-to-go function (i.e. "roughness-to-go") were to deem a trajectory through that path to be necessary. However, in other situations, especially in urban environments, "traversing" an obstacles is absolutely not an option. One obvious case of an obstacle that may not under any circumstances be traversed is a pedestrian. Pedestrians must be avoided at all costs, including the cost of potentially never reaching the end destination or (from a programmatic point of view) never being able to calculate possible trajectories leading to the destination. This would be the case in a hypothetical situation where a never-ending stream of pedestrians is crossing a street.

One of the recent entries into the US Department of Defense—sponsored annual competitions for autonomous robot navigation was the Stanford car dubbed "Junior" (2013). As reported (Levinson et al. 2011), "Junior" was able to very accurately tell the difference between people, cars, animals, signs, and roads. This was largely thanks to a novel laser and sensor calibration scheme that involved a great deal of machine learning in the original situation in the navigation environment. This was an example of a case for which existing information about the visual environment was used to estimate or interpolate parameters for that environment at later times points, by looking primarily at the aspects of the environment that change. This calibration scheme is illustrated in Fig. 19.

Another means by which machine vision is becoming more sophisticated in the perception of objects, is in a sense by moving in the opposite direction to how Stanford's "Junior" progressed from earlier autonomous vehicles. Whereas "Junior" was able to use more specific, fine-grained features of the environment, (Maddern and Vidas 2012) is able to detect whether it is nighttime or daytime outside, and base interpretation of features, obstacles density paths, and corresponding trajectories on this information. For example, an accurate assessment of the position of the sun (as well as other light sources, during the night) allows for the detection of

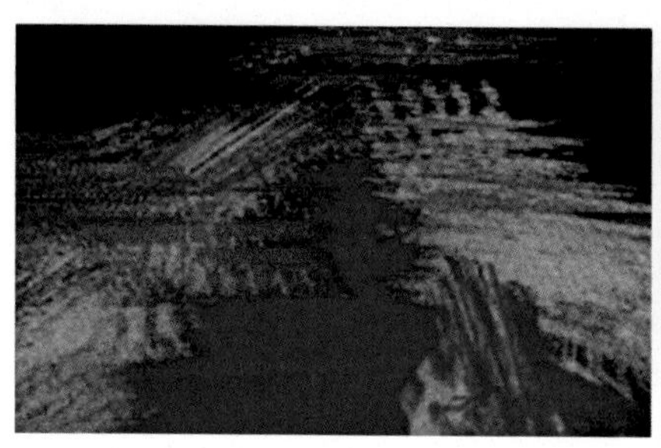

Fig. 19 The evolution of obstacle perception—obstacle discrimination. The Stanford "Junior" autonomous vehicle (Levinson et al. 2011) is able to turn a density "cloud" of obstacle perception (*left*) into a refined, crisp image (*right*) with sufficient details to make out the identity of different obstacles, provided that an initial sensor calibration is performed. Once this calibration has been performed at the beginning of the vehicle's trip or trajectory, it does not need to be repeated until the vehicle is transported by carrier and placed in a new environment. The refined image allows for certain obstacles (such as pedestrians in a crowded urban environment) to be identified and avoided at all costs, in favor of traversing less important obstacles if need be (such as curbs, stairs, etc.)

shadows with greater accuracy. Evidently, shadows can be traversed provided there are not hidden obstacles. To this author's knowledge, no methods that delve into predicting obstacles hidden in shadows have been developed to date.

As one can see, the richness of the visual information is increasingly being taken advantage of by autonomous vehicles, thanks in large part to these vehicles' increasingly powerful artificial intelligences. As a result, the rate of acquisition of this data is becoming increasingly "thirsty", and autonomous vehicles are traveling faster and faster. "Junior" Niclass et al. (2013), for example, can travel up to 35 miles per hours in a crowded urban environment (slowing or stopping where necessary, of course, to avoid pedestrians and other key obstacles). However, the faster that autonomous vehicles go, the more error is introduced to their sensors thanks simply to some basic principles of optics. For example, bending of light occurs even at moderate (highway) speed. As a results, in Niclass et al. (2013) the authors present a sensor capable of recording single-photon time of flight information based on correlation with other photons.

6.2 Future Directions in Applications of Autonomously-Navigating Robots

With all the above literature review and discussed devoted to sensors, techniques, problem types, and optimization algorithms for autonomous vehicle perception of obstacles and navigation by optimization of trajectories around obstacles, little has been said thus far about the actual applications of autonomously-navigating robots. What is the interest in, and what are therefore some possible applications of, these increasingly intelligent and self-aware road and off-road travelers? Below is a discussion describing a variety of different existing and emerging applications of such robots.

An obvious, although far from universally accepted or even much considered, application for autonomous vehicles is for the transportation of people. Some autonomous vehicles already transport people. However, there is potential for autonomous personal and public vehicles to largely replace the manually-operated equivalents of today and yesterday. Because there is not yet an "internet of things" (IoT), a term used below that refers to the potential future in which all objects are connected to the internet via tiny wireless sensors, directions and features of roads and in particular traffic conditions (for example detours due to construction) are often not updated into mobile road navigation apps on many drivers' cell phones or GPS units. Therefore, at least for the foreseeable future (until there is a veritable IoT or at least higher-integrity, more reliable set of traffic/road conditions comprehensively and instantaneously updated in real time—no 15-min delay allowable) autonomous vehicles would have to be able to read road signs just as any human driver would. A recent study Mathias et al. (2013) provides a method for achieving rapid, in-transit machine-vision sign reading. However, the authors performed their training as well as validation under conditions of fair lighting, unlike what may often be the case even with headlights illuminated.

One possible applications autonomously-navigating robots is as traffic-monitoring "drone" vehicles. These vehicles would patrol highways, and collect information about traffic density, and other environmental factors such as temperature, humidity, and surface conditions (e.g. precipitation accumulation). These drone vehicles would be networked to an information hub, either a higher-level computer or a human operator and traffic surveyor. Relaying information to a central information hub would enable high-resolution, real-time information about traffic to be distributed to passengers. This would be accomplished in multiple possible ways, for example by allowing the information to be accessible to mobile phone apps. One recently-proposed means of distributing the information gathered by robotic drone vehicles is by using cloud computing Whaiduzzaman et al. (2014). In this model, cloud computing would also be used to allow communication (and thereby formation of consensus data interpretation and analysis) between different robots. Cloud computing is fast, easily accessible, and cheap.

In order to address issues of security related to autonomous vehicles, especially those with a multitude of sensors containing possible sensitive information (but generally without the size, complexity, or infrastructure to effectively protect against virus or rogue cyber intrusions), some sophisticated theoretical as well as practical design steps have already been taken. As explained in Ho et al. (2012), while denial of service (DoS) style attacks (or sophisticated cyber-attacks originating from multiple points simultaneously) on static networks remains a problem, the technological development of wireless sensors has in many cases led to the adoption of a mobile, robotic platform. In parallel, the possibility of DoS by a mobile, malignant node arises. This article is the first to describe this problem explicitly, describing the unique advantages to DoS agents that mobility brings, and to propose a solution for overcoming these new advantages.

The article Ho et al. (2012) mentions several means by which malicious, mobile nodes could disrupt wireless sensor networks (WSNs) that would be impossible without mobility. A mobile node, if equipped with robotic arms, could move up to a node in the WSN, pick it up, and move it. This disruption in position would throw flags in the WSN security routine, and may even automatically cut out the node from the WSN, leaving it easy pretty for the malicious node. Malignant, mobile nodes could also move to as many different positions as possible, searching for weaknesses (i.e. spots where their positions would be more likely to be accepted as characteristic of a "safe" node), jamming communications, and moving nodes. In addition, the diversification of attack paths would make traceback impossible, without a prior assumption that the nodes were both mobile and hostile.

Although purely theoretical, the article does propose several strategies for combatting the threat of a mobile, malignant node or swarm of nodes. The article focuses on the case wherein the WSN is static, and only the malignant nodes are mobile. A straightforward means of detecting a mobile malignant node would be to keep a list of neighbors, leveraging the fact that the WSN is static. However, this would place severe constraints on the topology of the WSN, as a pre-defined set of neighbors would have to be supplied to the base node as a unique key, for each node. The authors suggest using an adjustable threshold maximum time limit between signals from a neighboring node, with the assumption that, beyond this threshold, the neighbor would be considered as potentially mobile (and therefore malignant). The more nodes flag the same outside node as malignant according to this criterion, the more likely is the base node to pass a judgment of "malignant".

References

Balakirsky, S., Dimitrov, D.: Single-query, bi-directional, lazy roadmap planner applied to car-like robots. In: IEEE International Conference on Robotics and Automation (ICRA), pp. 5015–5020 (2010)

Baturone, I., Gersnoviez, A.: A simple neuro-fuzzy controller for car-like robot navigation avoiding obstacles. In: IEEE FUZZ, pp. 1–6 (2007)

Burns, B., Brock, O.: Single-query motion planning with utility-guided random trees. In: IEEE International Conference on Robotics and Automation (ICRA), pp. 3307–3312 (2007)

Cheein, F.A.A., Carelli, R., De la Cruz, C., Bastos-Filho, T.F.: SLAM-based turning strategy in restricted environments for car-like mobile robots. In: International Conference on Industrial Technology (ICIT), pp. 602–607 (2010)

Delsart, V., Fraichard, T.: Navigating dynamic environments using trajectory deformation. In: IEEE International Conference on Intelligent Robots and Systems (IROS), pp. 226–233 (2008)

Devaurs, D., Simeon, T., Cortes, J.: Parallelizing RRT on large-scale distributed-memory architectures. IEEE Trans. Robot. Autom. **29**, 571–579 (2013)

El-Khatib, M.M., Hamilton, D.J.: A layered fuzzy controller for nonholonomic car-like robot motion planning. In: IEEE International Conference on Mechatronics, pp. 194–198 (2006)

Espinoza, J.L., Sánchez, A., Osorio, M.A.: Exploring unknown environments with RRT-based strategies. In: International Conference on Artificial Intelligence, pp. 1150–1159 (2006)

Fulgernzi, C., Spalanzani, A., Laugier, C.: Dynamic obstacle avoidance in uncertain environment combining PVOs and occupancy grid. In: IEEE International Conference on Robotics and Automation (ICRA), pp. 1610–1616 (2007)
Fulgenzi, C., Tay, C., Spalanzani, A., Laugier, C.: Probabilistic navigation in dynamic environment using rapidly-exploring random trees and Gaussian processes. In: Intelligent Robots and Systems (IROS), pp. 1056–1062 (2008)
Gall, R., Troster, F., Mogan, G.: On the development of an experimental car-like mobile robot. In: International Conference on Optimization of Electrical and Electronic Equipment (OPTIM), pp. 734–739 (2010)
Garrido, S., Blanco, D., Moreno, L., Martın, F.: Improving RRT motion trajectories using VFM. In: IEEE International Conference on Mechatronics (ICM), pp. 1–6 (2009)
Garrido, S., Moreno, L., Blanco, D., Martın, F.: Smooth path planning for non-holonomic robots using fast marching. In: Int. J. Robot. Autom. 154–176 (2011)
Grady, D.K., Moll, M., Hegde, C., Sankaranarayanan, A.C., Baraniuk, R.G., Kavraki, L.E.: Multi-objective sensor-based replanning for a car-like robot. In: International Symposium on Safety, Security, and Rescue Robotics (SSRR), pp. 1–6 (2012)
Ho, J.W., Wright, M., Das, S.K.: Distributed detection of mobile malicious node attacks in wireless sensor networks. Ad Hoc Netw. **10**(3), 512–523 (2012)
Hwang, C.-L., Chang, L.-J.: Internet-based smart-space navigation of a car-like wheeled robot using fuzzy-neural adaptive control. IEEE Trans. Fuzzy Syst. **16**(5), 1271–1284 (2008)
Hwang, C.-L., Shih, C.Y.: A distributed active-vision network-space approach for the navigation of a car-like wheeled robot. IEEE Trans. Industr. Electron. **56**(3), 846–855 (2009)
Jaillet, L., Yershova, A., LaValle, S.M., Simeon, T.: Adaptive tuning of the sampling domain for dynamic-domain RRTs. In: IEEE International Conference on Intelligent Robots and Systems, pp. 2851–2856 (2005)
Jaillet, L., Cortés, J., Siméon, T.: Transition-based RRT for path planning in continuous cost spaces. In: IEEE International Conference on Robots and Systems, pp. 2646–2652 (2012)
Karaman, S., Frazzoli, E.: Sampling-based optimal motion planning for non-holonomic dynamical systems. In: IEEE International Conference on Robotics and Automation (ICRA), pp. 5041–5047 (2013)
Kim, J., Esposito, J.M., Kumar, V.: An RRT-based algorithm for testing and validating multi-robot controllers. In: Conference on Robotics Science and Systems, pp. 249–256 (2005)
Kuwata, Y., Fiore, G.A., Teo, J., Frazzoli, E., How, J.P.: Motion planning for urban driving using RRT. In: International Conference on Intelligent Robots and Systems (IROS), pp. 1681–1686 (2008)
Lategahn, H., Geiger, A., Kitt, B.: Visual SLAM for autonomous ground vehicles. In: IEEE International Conference on Robotics and Automation (ICRA), pp. 1732–1737 (2011)
Levinson, J., Askeland, J., Becker, J., Dolson, J., Held, D., Kammel, S., Thrun, S., et al.: Towards fully autonomous driving: systems and algorithms. In: IEEE Intelligent Vehicles Symposium (IV), pp. 163–168 (2011)
Maddern, W., Vidas, S.: Towards robust night and day place recognition using visible and thermal imaging. RSS 2012: beyond laser and vision: alternative sensing techniques for robotic perception (2012)
Mathias, M., Timofte, R., Benenson, R., Van Gool, L.: Traffic sign recognition—how far are we from the solution? In: IEEE International Joint Conference on Neural Networks (IJCNN), August 2013, pp. 1–8
Niclass, C., Soga, M., Matsubara, H., Kato, S., Kagami, M.: A 100-m range 10-frame/s 340 96-pixel time-of-flight depth sensor in 0.18-CMOS. IEEE J. Solid-State Circ. **48**(2), 559–572 (2013)
Pepy, R., Lambert, A.: Safe path planning in an uncertain-configuration space using RRT. In: Intelligent Robots and Systems (IROS), pp. 5376–5381 (2006)
Pepy, R., Lambert, A., Mounier, H.: Path planning using a dynamic vehicle model. In: ICTTA, pp. 781-786 (2006)

Petridis, V., Zikos, N.: L-SLAM: reduced dimensionality FastSLAM algorithms. In: IEEE International Joint Conference on Neural Networks (IJCNN), pp. 1–7 (2010)
Petti, S., Fraichard, T.: Safe navigation of a car-like robot in a dynamic environment. In: European Conference on Mobile Robots (2005)
Phillips, J.M., C.S. Draper Laboratories: Guided expansive spaces trees: a search strategy for motion- and cost-constrained state spaces. In: International Conference on Robotics and Automation (ICRA), vol. 4, pp. 3968–3973 (2004)
Pradalier, C., Hermosillo, J., Koike, C., Braillon, C., Bessihre, P., Laugier, C.: An autonomous car-like robot navigating safely among pedestrians. In: IEEE International Conference on Robotics and Automation (ICRA), vol. 2, pp. 1945–1950 (2004)
Pradalier, C., Hermosillo, J., Koike, C., Braillon, C., Bessière, P., Laugier, C.: The CyCab: a car-like robot navigating autonomously and safely among pedestrians. In: Robotics and Autonomous Systems, pp. 51–67 (2005)
Quan, Y., Lee, J.Y., Changsoo, H.: Sensor-based navigation algorithm for car-like robot to generate completed GVG. In: International Conference on Control, Automation and Systems (ICCAS), pp. 1442–1447 (2011)
Rebai, K., Azouaoui, O., Benmami, M., Larabi, A.: Car-like robot navigation at high speed. In: Robotics and Biomimetics, pp. 2053–2057 (2007)
Rezaei, S., Guivant, J.E., Nebot, E.M.: Car-like robot path following in large unstructured environments. In: IEEE IROS, pp. 2468–2473 (2004)
Rodriguez, S., Tang, X., Lien, J.-M., Amato, N.M.: An obstacle- based rapidly-exploring random tree. In: International Conference on Robotics and Automation, pp. 895–900 (2006)
Tahirovic, A., Magnani, G.: A roughness-based RRT for mobile robot navigation planning. In: IFAC World Congress, vol. 18, pp. 5944–5949 (2001)
Whaiduzzaman, M., Sookhak, M., Gani, A., Buyya, R.: A survey on vehicular cloud computing. J. Netw. Comput. Appl. **40**, 325–344 (2014)
Yershova, A., Jaillet, L., Simeon, T., LaValle, S.M.: Dynamic-domain RRTs: efficient exploration by controlling the sampling domain. In: IEEE International Conference on Robotics and Automation (ICRA), pp. 3856–3861 (2005)
Ziegler, J., Werling, M.: Navigating car-like robots in unstructured environments using an obstacle sensitive cost function. In: IEEE Intelligent Vehicles Symposium, pp. 787–791 (2008)

Computing a Similarity Coefficient for Mining Massive Data Sets

M. Coşulschi, M. Gabroveanu and A. Sbîrcea

Abstract Large amounts of data can be found today in all areas as a result of various processes like e-commerce transactions, banking or credit card transactions, or web navigation user sessions (recorded into web server logs). The development and implementation of algorithms able to process huge amounts of data have become more affordable due to cloud computing and the *MapReduce programming model*, which, in turn, enabled the development of some open-source frameworks, such as *Apache Hadoop*. Based on the values obtained by computing the Jaccard similarity coefficients for two very large graphs, we have analysed in this paper the connections and influences that certain nodes have over other nodes. Also, we have illustrated how the Apache Hadoop framework and the MapReduce programming model can be used for a large amount of computations.

Keywords Big data · Virtualization · Hadoop · Mapreduce · Jaccard similarity

1 Introduction

Nowadays, the majority of all domain companies have to handle a significant amount of data, which they must process during their every-day activities. The first businesses, which have been impacted by the exponential growth of data were search engines like Yahoo! and Google, as well as social networks like Facebook or Twitter. For example, lots of data is regularly generated on Facebook—there are

M. Coşulschi (✉) · M. Gabroveanu · A. Sbîrcea
Department of Computer Science, University of Craiova, 13 A. I. Cuza Street, 200585 Craiova, Romania
e-mail: mirelc@central.ucv.ro

M. Gabroveanu
e-mail: mihaiug@central.ucv.ro

A. Sbîrcea
e-mail: sbirceaadriana@yahoo.com

G.A. Tsihrintzis et al. (eds.), *Intelligent Computing Systems*,
Studies in Computational Intelligence 627, DOI 10.1007/978-3-662-49179-9_15

over 300 million users of this social network, 30 million users are updating their Facebook statuses every day, while many of these active users are uploading lots of pictures and videos or sharing content like web links, blog posts, photos, notes etc. In 2009, the amount of videos and photos uploaded on Facebook was over 1 billion each month.

Big data did not appear out of nowhere lately, since most companies already had started to collect their strategic data one way or another: e.g. pharmaceutical companies have huge amounts of information collected during their research, stored in local data warehouses and made available through their private intranets. Surveillance cameras, RFIDs, all kind of sensors used in physical security systems, all have produced data that could and are indeed stored.

Big data processing cannot be performed by people or by the existing tools designed to process small sets of data, like database management systems or content management systems, because they are no longer efficient in case the amount of data exceeds a certain limit. Classical databases are useful especially for structured-type data, while big data does not mean only structured data, but also semi-structured or unstructured data. The amount of unstructured data is growing faster than the amount of structured data, about 80 % of the public data turning out to be unstructured data.

As previously mentioned, big data is produced by various sources, such as:

- EBay was using in April 2009 two data warehouses, Terradata and Greenplum, the former having 6.5 PB, 17 trillion already stored records, and collecting more than 150 billion new records/day, resulting into an ingest rate exceeding 50 TB/day.[1]
- Facebook estimated in May 2009 an amount of 2.5 PB of user data, with more than 15 TB of new data per day.[2]
- Google was processing about 20 PB per day in 2008.
- Ancestry.com stores over 2.5 PB.[3]

The necessity of developing algorithms, which can process a significant amount of data, becomes more pressing given the data age we live in (White 2012). The implementation of such algorithms has become easier due to the cloud-computing concept. The ability to store, aggregate, combine data and then use the results to perform a deep data analysis has now become more accessible, as trends such as *Moore's Law*[4] in computing, its equivalent in digital storage and cloud computing, continue to lower costs and other technology barriers.[5]

[1]Source http://www.dbms2.com/2009/04/30/ebays-two-enormous-data-warehouses/.

[2]Source http://www.dbms2.com/2009/05/11/facebook-hadoop-and-hive/.

[3]Source http://blog.familytreemagazine.com/insider/Inside+Ancestrycoms+TopSecret+Data+Center.aspx.

[4]Moore's law is the observation that, throughout the history of computing hardware, the number of transistors on integrated circuits has doubled approximately every two years.

[5]McKinsey Global Institute: Big data: The next frontier for innovation, competition, and productivity (2011).

Those tools have to leverage people's ability for answering big questions by inference through large amounts of data. Data centers have emerged as distributed information systems for massive data storage and processing, providing many online applications and infrastructure services through its large number of servers (Ding et al. 2012).

In this chapter, we have analyzed the connections and influences that some nodes have over others using the Jaccard similarity coefficients for two very large graphs, computed with the help of a cluster of machines and a MapReduce application managed by a Hadoop distribution.

2 Related Work

Mining of massive datasets or *big data analytics* are two frequently encountered hot topics today. For example, there are many conferences, presentations, demos etc., all having in the center of their universe mathematical methods and algorithms for solving new problems arisen from analyzing big data. These topics cover the process of analyzing very large amounts of data with emphasis on data mining and machine learning methods.

The development and implementation of algorithms able to process huge amounts of data has become more affordable due to cloud computing (Armbrust et al. 2010) and the *MapReduce programming model*,[6] which enabled the development of some open-source frameworks, such as *Apache Hadoop*.

A very important role in cloud computing evolution is played by Amazon, which offers the most accessed online library worldwide. This company introduced two cloud computing services, initially intended only for internal use, and which, in time, have become very popular within the IT community, thanks to their clever design, which makes web-scale computing easier for developers: `EC2`[7]—for computations and `S3`[8]—for storage, to manage their resources.

MapReduce has gained widespread popularity, mainly due to the development of Apache Hadoop. MapReduce applications are extremely broad, being tested on general distributed computing frameworks with security applications, such as botnet detection, spam classification (Caruana et al. 2011) and spam detection (Indyk et al. 2013). The authors of the last paper experimented their algorithms on the WEBSPAM-UK2007 dataset, which we have also considered in this work.

In Kunegis et al. (2009) the authors analyze the corpus of user relationships of the Slashdot technology news site. The paper explores, among other characteristics, link-level characteristics, such as distances and similarity measures.

---

[6]MapReduce—The Programming Model and Practice, Sigmod 2009 Tutorial. http://research.google.com/archive/papers/mapreduce-sigmetrics09-tutorial.pdf.

[7]Amazon Elastic Compute Cloud—http://aws.amazon.com/ec2/.

[8]Amazon Simple Storage Service—http://aws.amazon.com/s3/.

Many similarity measures having been defined and studied, researchers are now aware of their advantages and disadvantages due to the various comparative studies performed (Rajaraman and Ullman 2012). One of these measures, the Jaccard similarity index, has various applications (Bank and Cole 2008; Blundo et al. 2012; Engen et al. 2011; Mulqueen et al. 2001). In Macha et al. (2011), the Jaccard coefficient was chosen among different similarity measures for evaluating their proposed Rank Based Fingerprinting (RBF) localization algorithm. In another work (Leydesdorff 2008), Leydesdorff reports the results of an empirical comparison of a number of direct and indirect similarity measures: the Spearman rank correlations between the association strength (referred to as the probabilistic affinity or the activity index), the cosine, and the Jaccard index. Measures are applied to a smaller data set than the one considered in this paper, consisting of the co-citation frequencies of 24 authors.

Our implementation of the Jaccard similarity index computing is inspired from the work of Bank and Cole (2008).

3 Cloud Computing

Cloud computing is, according to a special document published under the auspices of NIST,[9] "a model for enabling ubiquitous, convenient, on-demand network access to a shared pool of configurable computing resources (e.g., networks, servers, storage, applications, and services) that can be rapidly provisioned and released with minimal management effort or service provider interaction" (Mell and Grance 2011).

Another definition can be encountered in an official document emerged from *European Commission*[10]:

> a 'cloud' is an elastic execution environment of resources involving multiple stakeholders and providing a metered service at multiple granularity for a specified level of quality (of service).

As a conclusion, cloud computing can be seen as a new approach for using IT services, with the advantage of accessing them with less costs and usage complexity, maintaining at the same time a high level of scalability and reliability. In Fig. 1, one can see the cloud computing architecture from a logical view.[11]

The best service behavior from a client point of view is a cloud computing service offering the access to resources (hardware and software) with the

[9]National Institute of Science and Technology—http://www.nist.gov/index.html.

[10]European Commission Expert Group Report: The Future of Cloud Computing (2010) http://cordis.europa.eu/fp7/ict/ssai/docs/cloud-report-final.pdf.

[11]http://en.wikipedia.org/wiki/Cloud_computing.

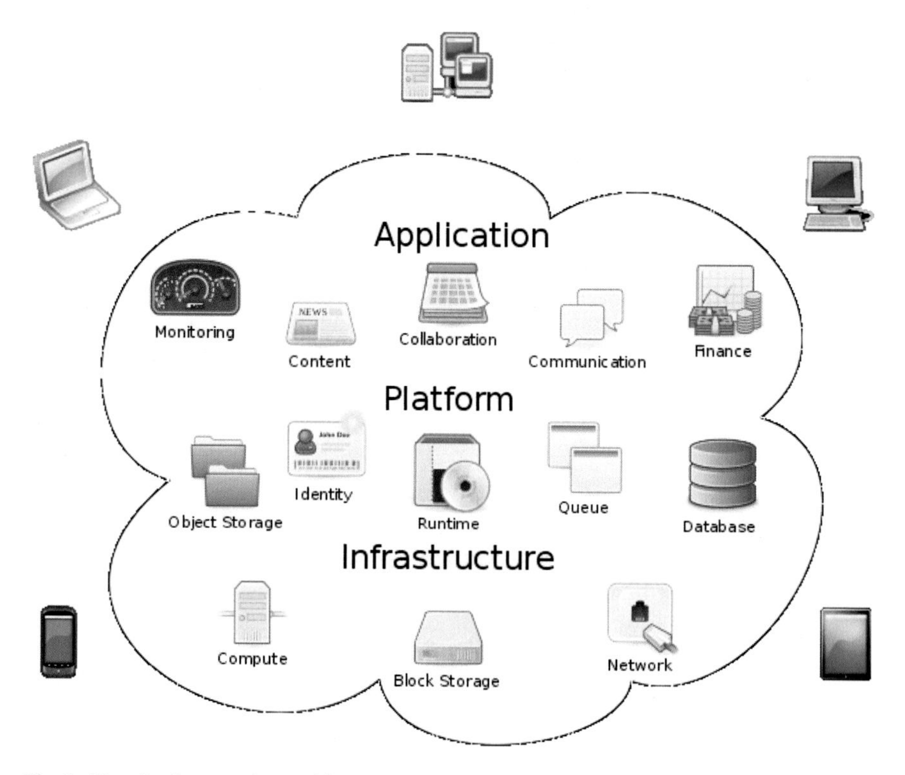

Fig. 1 The cloud computing architecture

appearance to him as being infinite. This requires an automated allocation and management of resources in order to handle the requests. Virtualization is a solution to this problem becoming common that scales well. Also, for security reason virtualization is a strong mechanism.

When running multiple virtual machines in cloud computing, the CPUs, main memory, network devices and hard-disks have to be shared by the VMs launched on the same physical server. Although the CPUs and main memory can be shared well, the bottlenecks occur at the network and disk I/O level. Here, the engineers of the hardware architecture and operating system developers still have to work in order to improve the virtualizations of interrupts and I/O channels.

The main *deployment models* for the clouds are:

1. *Public cloud*—in this deployment model, cloud's resources are made available for the general public;
2. *Community cloud*—community cloud is operated by a community, allowing the member organisations to share the cloud's infrastructure. The member organisations have in common the same interests, objectives, goals.

3. *Hybrid cloud*—is an in-between model from *public cloud* to *private cloud*;
4. *Private cloud*—the service is available only for a single entity, being developed and maintained by the same actor.

The cloud infrastructure lies on three *service models*:

1. *Infrastructure as a Service* (IaaS)—cloud providers offer the infrastructure for cloud users, usually as virtual machine (see Footnote 10) (see *Amazon S3,*[12] *SQL Azure,*[13] *Amazon EC2*[14]);
2. *Platform as a Service* (PaaS)—cloud providers offer to cloud users an entire development platform composed of the operating system, the programming language environment, the database and the web server (see *Google App Engine,*[15] *Windows Azure*[16]);
3. *Software as a Service* (SaaS)—cloud providers offer implementations of various software applications to cloud users (see *Google Docs,*[17] *Microsoft Office 365*[18]).

Some researchers argue (Armbrust et al. 2010) that even the cost of the pay-as-you-go service is more expensive than the amortization cost of a computer server over a certain period, the benefits being overwhelmed by the overprovisioning or underprovisioning risk transfer and the property of service elasticity.

Overprovisioning appears when the service operators estimate a certain workload (resource usage) for a period of time and, for some reasons, like seasonal demand, the resources remain underutilized with certain amount of money loss—the amount of money spent for renting resources which were partially used.

Underprovisioning is characterized by a saturation of the resources as a result of the demand burst due to some external events. In such situations, resources are not able to fulfill the users' requests, concretized in user rejection who are highly likely not to come back again. It is difficult to assess losses produced by the excess users turned down, because we cannot know with certain precision the number of users that will abandon the service.

The elasticity is the property of adding or removing a large quantity of hardware or software resources, even at a fine grain level.

In conclusion, the characteristics that make cloud computing so attractive are short-term usage, no upfront cost and infinite capacity on demand (Armbrust et al. 2010).

[12]http://aws.amazon.com/s3/.

[13]http://www.windowsazure.com/en-us/home/features/data-management/.

[14]http://aws.amazon.com/ec2/.

[15]https://developers.google.com/appengine/.

[16]http://www.windowsazure.com.

[17]http://docs.google.com.

[18]http://www.microsoft.com/office365/.

3.1 Virtualization

Among various kind of virtualization, a hosted *Virtual Machine* is installed as an application program, the virtualizing software being executed on top of the existing operating system. This operating system is responsible for providing virtualizing software with device drivers and low-level services. *VMware GSX server* is an implementation of a hosted VM (Smith and Nair 2005; Sugerman et al. 2002).

The architecture of the *VMware ESX server* is different from the architecture of the *VMware GSX server*. The former provides a virtual intermediate layer between the hardware and the virtual machines, while the latter provides a virtual layer between the host operating system and the virtual machines. This can be achieved by having its own kernel, called *vmkernel*, with microkernel design, and, as a consequence, the ESX server runs directly on the host server without any third-party operating system. The virtual layer introduced by ESX server abstracts the system hardware into a pool of logical resources that can be allocated to the VMs. The computational performances of VMs running under ESX server are superior to the VMs running under the GSX server.

4 Hadoop

Hadoop is an open-source project from *Apache Software Foundation*, its core function consisting of *MapReduce* programming model implementation. This platform was created for solving the problem of very large amounts of data processing, a mixture of complex and structured data. It is used especially when data analytics applications requiring a large number of computations have to be executed.

The key Hadoop property consists of enabling the execution of applications on thousands of nodes and petabytes of data. A typical enterprise configuration is composed of tens or thousands of physical or virtual machines interconnected through a fast network. The machines have to run on POSIX compliant operating systems, while a Java virtual machine with a version newer than 6 has to be installed and be functional.

Doug Cutting was the creator of the Hadoop framework, two papers published by Google researchers having a decisive influence on his work: *The Google File System* (Ghemawat et al. 2003) and *MapReduce: simplified data processing on large clusters* (Dean et al. 2004). In 2005 he started to use an implementation of MapReduce in *Nutch,*[19] an application software for web searching, and very soon the project became independent from *Nutch* under the codename *Hadoop*, aiming to offer a framework for distributed running of many jobs within a cluster.[20]

[19]http://nutch.apache.org/.

[20]Nutch presentations. http://wiki.apache.org/nutch/Presentations.

The language used for the Hadoop framework development is Java. The framework is providing the users a Java API for developing MapReduce applications. In this way, Hadoop becomes very popular in the community of scientists whose name is linked to the big data processing. Yahoo! is an important player in this domain, the company having for a longer period a major role in the design and implementation of the framework. Another big player is Amazon,[21] a company currently offering a web service[22] based on Hadoop that is using the infrastructure provided by the *Elastic Compute Cloud—EC2.*[23] Meanwhile, Hadoop becomes very popular, and for sustaining this allegation it is sufficient just to mention the presence of other important players such as IBM, Facebook, The New York Times, Netflix,[24] Hulu,[25] in the list of companies deploying Hadoop based applications (Irving 2010).

The Hadoop framework hides the details of processing the jobs, leaving to the developers the liberty to concentrate on application logic. The framework has the following specific features:

1. *accessibility*—it can be executed on large clusters or on cloud computing services;
2. *robustness*—it was designed to run on commodity hardware; a major aspect considered during its development was its tolerance to frequent hardware failures and crash recovery;
3. *scalability*—it scales easily for processing increased amount of data by transparently adding more nodes to the cluster;
4. *simplicity*—allows developers to write specific code in a short time.

There is a quite large application domain for Hadoop. Some examples include online e-commerce applications for providing better recommendations for clients looking for certain products or finance applications for risk analysis and evaluation with sophisticated computational models, whose results cannot be stored in a database (Ding et al. 2012).

Basically, Hadoop can be executed on each major OS: Linux, Unix, Windows, Mac OS. The only platform officially supported for production is Linux.

The Hadoop architecture is depicted in Fig. 2. The main project has three parts:

1. *Hadoop Common*—contains utilities necessarily for other Hadoop partial projects;
2. *HDFS* (*Hadoop Distributed File System*)—represents the distributed file system offering the possibility to store data and to retrieve it when needed;
3. *Hadoop MapReduce*—a framework used for distributed processing of large datasets in clusters.

[21] http://www.amazon.com.

[22] http://aws.amazon.com/elasticmapreduce/.

[23] http://wiki.apache.org/hadoop/AmazonEC2.

[24] https://signup.netflix.com/global.

[25] http://www.hulu.com/.

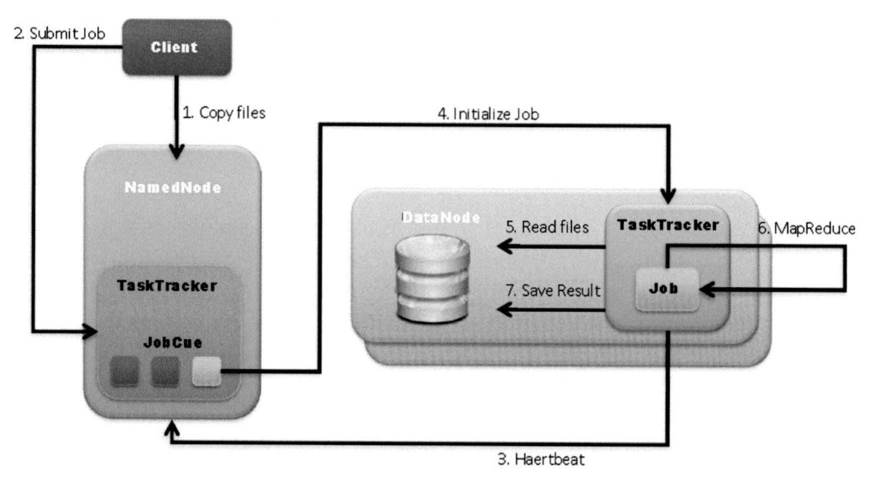

Fig. 2 The Hadoop architecture (Hildebrandt 2010)

The Hadoop framework can be used in one of the following ways:

- *Local* (*Standalone*)—there are no daemons launched, all processes being executed on a local machine. It is used for development, testing and debugging.
- *Pseudo-distributed*—Hadoop daemons are running on the local machine, but on separate virtual machines, simulating a real cluster. The inter-machine communications represented by network traffic are absent;
- *Fully-Distributed*—Hadoop daemons are executed on a real cluster.

The framework is implementing a master-slave architecture used both for distributed data storage and for distributed computations.

Hadoop framework running on a completely configured cluster is composed of a set of daemons or internal modules executed on various machines from the cluster. Those daemons have specific tasks, some of them running only on a single machine, others having instances running on several machines. The Hadoop daemons are (Lam 2010; White 2012):

- *NameNode*—the most important daemon. *NameNode* daemon is the HDFS main controller, coordinating the slave *DataNode* daemons for performing I/O operations. This is the only critical failure point from the cluster;
- *DataNode*—each slave machine from the cluster will host a *DataNode* daemon reading and writing HDFS blocks from the files stored in the local file system;
- *Secondary NameNode*—*SNN*—monitors the health status of HDFS cluster. Each cluster has one *NameNode* and a *Secondary NameNode* running on separate machines, the main difference between the two daemons is that the last one does not receive realtime changes inside the HDFS;

- *JobTracker*—ensures the synchronization between the application and the Hadoop framework. *JobTracker* establishes the execution plan in concordance with the files which will be processed, assigns nodes to each task and monitors all running tasks. Each Hadoop cluster has a single *JobTracker* daemon, running on a server as a master node;
- *TaskTracker*—is responsible for performing the tasks assigned by *JobTracker*. Each slave node has a single *TaskTracker* that can start several JVMs for running more tasks in parallel.

HDFS is the storage system used by the Hadoop applications. HDFS decides how and where to replicate the data blocks and distributes them on the cluster nodes where the computations took place; the goal is to have faster, reliable computations.

Hadoop splits the files stored in the distributed file system into several blocks, in order to facilitate the transport through the network. Those blocks are stored in HDFS, blocks from the same file being stored on 2–3 different machines (see Fig. 3). Each block is stored as a file on the local file system of the OS where *DataNode* is hosted. Usually the block's size is 64 or 128 Mb.

A sample Hadoop cluster architecture is depicted in Fig. 4. For a small cluster, a *Secondary NameNode* daemon can be hosted on a slave node, whereas, on a large cluster, it is customary to separate the *NameNode* and *JobTracker* daemons on different machines. In Fig. 4, we have a master node running *NameNode* and

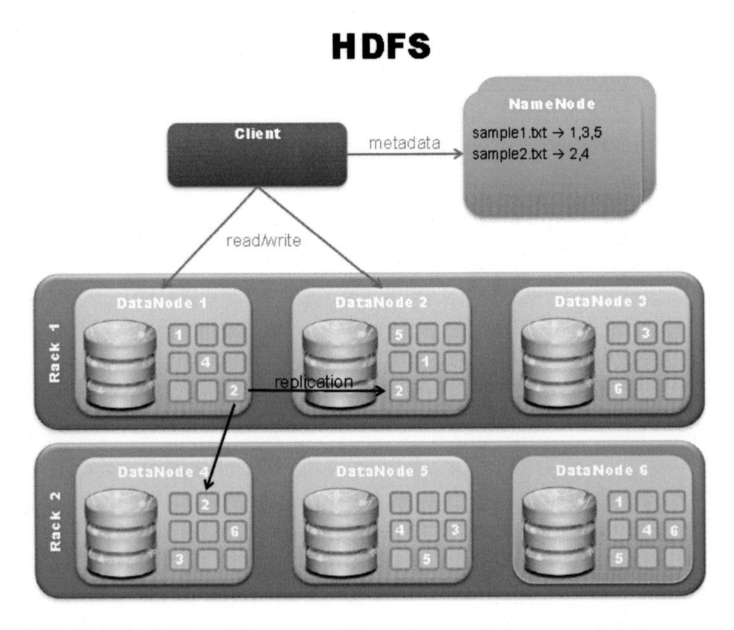

Fig. 3 The Hadoop HDFS (Hildebrandt 2010)

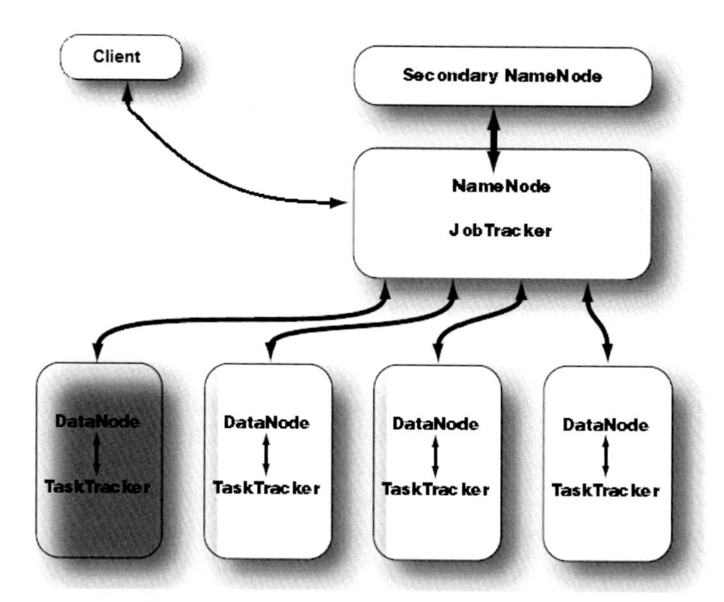

Fig. 4 Hadoop cluster architecture

JobTracker daemons and an independent node hosting the *Secondary NameNode* daemon, in case of failure of the master node.

Each slave machine hosts two daemons, a *DataNode* and a *TaskTracker*, in order to run computational tasks on the same node where their input data is stored.

5 Hadoop Cluster

We have decided to use a preconfigured *VMware* image[26] from a Yahoo tutorial.[27,28]

As stated before, there are three supported modes for running a Hadoop cluster[29]:

[26]http://ydn.zenfs.com/site/hadoop/hadoop-vm-appliance-0-18-0_v1.zip.

[27]Yahoo! Hadoop Tutorial. http://developer.yahoo.com/hadoop/tutorial/.

[28]Hadoop Virtual Image Documentation. http://code.google.com/intl/ro-RO/edu/parallel/tools/hadoopvm/index.html.

[29]http://hadoop.apache.org/common/docs/r0.20.2/quickstart.html.

- local (standalone) mode,
- pseudo-distributed mode,
- fully-distributed mode.

We have tested our application in all three modes in order to dose the grade of the configuration difficulty from easy to complex. For the first mode, the local (standalone) one, Hadoop is configured by default to run in a non-distributed mode. We have used this mode for debugging the application.

The second mode, the pseudo-distributed mode, is closer to the *real*-like behavior, Hadoop running also as a single node. The major difference between those two initial modes relates to the observation that, in the former mode, there is only a Java process running in the Java Virtual Machine, while in the latter, all daemons run in separate processes, each one for a distinct Java Virtual Machine. The pseudo-distributed name comes from the fact that, although all processes have separate space for memory, they are running on the same machine.

The fully-distributed mode is the one recommended for processing large amounts of data in a properly tuned cluster. In order to perform computation on various quantities of big data analytics in reasonable time, we have to properly adjust the size and resources allocation. This can be done by choosing a proper number of machines with specified RAM memory and space on the hard-disk.

5.1 Design

All machines in the cluster have been prepared in the same way in order to reduce the maintenance time. They have identical hardware and software configurations, with the exception of a few characteristics related to the role and identity details for each node, which must be different (Table 1).

We have allocated 6 *virtual machines* as follows (see Fig. 1): the name of each machine is `hadoopn`, where `n` is a digit from the set $\{1, 2, \ldots, 6\}$, while the associated IP address is 10.100.20.*n*.

Table 1 Allocation of daemons inside the specified cluster

Component	Machine host names					
	Hadoop1	Hadoop2	Hadoop3	Hadoop4	Hadoop5	Hadoop6
MapReduce monitor	X					
MapReduce worker		X	X	X	X	X
HDFS monitor	X					
HDFS worker		X	X	X	X	X

The decisive reason for where each type of daemon (*worker* or *monitor*) will be hosted relates to the fact that, while a *monitor* is using more RAM for monitoring and coordination activities of the *workers*, a *worker* is more computationally oriented and is therefore using more CPU and I/O bandwidth, requiring less RAM implied by the distributed architecture of the cluster.

6 MapReduce

The MapReduce programming model was described for the first time by Jeffrey Dean and Sanjay Ghemawat in a paper published in 2004, entitled *MapReduce: Simplified Data Processing on Large Clusters* (Dean and Ghemawat 2004). This programming model was designed and introduced by Google researchers earlier, but its implementation was not published, remaining private. A job in MapReduce splits the input data into many independent blocks of data, which can be processed independently.

MapReduce was designed around two important concepts, the *mapper* and the *reducer*, while the lists of key-value pairs are the basic data structure for inter-process communications. Mapper and reducer are designed as two functions implemented by the user, having the following signatures:

- Map: $(k1, v1) \to [(k2, v2)]$
- Reduce: $(k2, [v2]) \to [(k3, v3)]$.

A mapper is created for every map task and a reducer is created for every reduce task. The mapper is applied to every input pair and generates output pairs, which represent the input data for the reducer. A grouping operation is performed between the two tasks, therefore the output from the first one reaches the second one sorted and grouped by key. Usually, a reducer receives data from more than one mapper, the number of reducers being thus smaller than the number of mappers. The map and reduce instances are distributed across multiple machines from the cluster, allowing parallel processing. The types of keys and values can be of any type, user-defined or not user-defined. There is one rule that applies for the mapper's output and reducer's input: the types of keys and values from the mapper's output should match with the types of keys and values of the reducer's input.

A few examples of algorithms developed using MapReduce consist of *pi*, *distributed grep*, *inverted index*, *count of URL frequency* computation etc. (Borthakur 2009; Lin 2010; Rajaraman and Ullman 2012; White 2012).

One of the key features of the MapReduce programming model is that it allows everyone to develop fast algorithms for big data processing (Zikopoulos et al. 2011).

7 Jaccard Similarity

From a simple point of view, measures of similarity denote the closeness among two or more objects. An important class of measures of similarity is represented by distance measures. Let A be a set of elements, called *space*. A distance measure defined on A is a function $d(x,y)$, $d : A \times A \rightarrow \mathbb{R}$, having as arguments two elements from this space and returning as output a real number. A function is called a *distance measure* if it satisfies the following four axioms:

(a) $d(x,y) \geq 0$ (the distance between two elements is always a positive number).
(b) $d(x,y) = 0 \Leftrightarrow x = y$.
(c) $d(x,y) = d(y,x)$ (symmetry).
(d) $d(x,y) \leq d(x,z) + d(z,y)$ (known as the triangle inequality).

The similarity notion is very flexible, so there are many distance measure formulas, such as the *Euclidean distance*, the *Jaccard distance*, the *Cosine distance*, the *Edit distance*, the *Tanimoto distance*, the *Hamming distance*, the *Soergel distance* etc. (Rajaraman and Ullman 2012). Some of them are described below:

1. the *Euclidean distance*—$d_E(A,B) = \sqrt{\sum_{i=1}^{n}(a_i - b_i)^2}$;
2. the *Cosine distance*—$C(A,B) = \frac{|A \cap B|}{\sqrt{|A| \cdot |B|}}$;
3. the *Manhattan distance* (Minkowski's L_1 distance)—$d_M(A,B) = \sum_{i=1}^{n} |a_i - b_i|$;
4. the *Soergel distance*—$d_S(A,B) = \frac{\sum_{i=1}^{n} |a_i - b_i|}{\sum_{i=1}^{n} \max\{a_i, b_i\}}$;
5. the *Tanimoto distance*—$d_T(A,B) = \frac{\sum_{i=1}^{n} a_i + \sum_{i=1}^{n} b_i - 2\sum_{i=1}^{n} \min\{a_i, b_i\}}{\sum_{i=1}^{n} a_i + \sum_{i=1}^{n} b_i - \sum_{i=1}^{n} \min\{a_i, b_i\}}$ or $d_T(A,B) = \frac{\sum_{i=1}^{n} (\max\{a_i, b_i\} - \min\{a_i, b_i\})}{\sum_{i=1}^{n} \max\{a_i, b_i\}}$.

A very useful measure, the *Jaccard index* (*Jaccard similarity coefficient*) evaluates the similarity between two data sets as the ratio between the cardinal of the intersection of two data sets and the cardinal of the reunion of the same two data sets:

$$J(A,B) = \frac{|A \cap B|}{|A \cup B|}. \tag{1}$$

Jaccard index is not a distance measure and it can be applied to objects having binary attributes.

The function defined as the difference between 1 and the Jaccard index is a real measure distance, known as the *Jaccard distance*. It measures the *degree of dissimilarity* between two data sets, defined as the complementary of the *Jaccard coefficient*:

$$J_{\delta}(A,B) = 1 - J(A,B) = \frac{|A \cup B| - |A \cap B|}{|A \cup B|}. \quad (2)$$

Let's consider two objects, A and B, described by n binary attributes. The Jaccard index measures, in this case, the similarity degree between A and B from the attributes point of view:

- w_{11}—the total number of attributes from A and B having the same values 1;
- w_{01}—the total number of attributes having values 0 for object A and having values 1 for object B;
- w_{10}—the total number of attributes having values 1 for object A and having values 0 for object B;
- w_{00}—the total number of attributes from A and B having the same values 0.

Thus, the formula for the Jaccard index (or the Jaccard similarity coefficient) becomes:

$$J(A,B) = \frac{w_{11}}{w_{01} + w_{10} + w_{11}}. \quad (3)$$

A few examples of how the Jaccard similarity coefficient can be used are the documents' similarity, the online shopping, the movie ratings, the similarity of the users from social networks etc.

8 Experimental Results

The first experiments were performed on a large set of data, named WEBSPAM-UK2007.[30] From this set we have used the *hostgraph*, which represents a very large collection of annotated spam/non-spam hosts, containing 114,529 hosts, numbered from 0 to 114,528. The input file for the MapReduce algorithm implementation contains the hosts graph, which summarizes the URL to URL links by converting from multiple links on different web pages into a single link among two hosts. A line from the input file has the following generic structure:

$$\texttt{dest1} : \texttt{nlinks1}\ \texttt{dest2} : \texttt{nlinks2}, \ldots, \texttt{destk} : \texttt{nlinksk}$$

where `dest_` represents the destination host and `nlinks_` represents the number of page-to-page links between the host having the *id* equal with the line number and the `dest_` host. When a host does not have any link with another host, its corresponding line is empty.

[30] "Web Spam Collections". http://chato.cl/webspam/datasets/ Crawled by the Laboratory of Web Algorithmics, University of Milan, http://law.dsi.unimi.it/. URLs retrieved March 2014.

Table 2 Values of the Jaccard index for sample pairs of nodes from the WEBSPAM-UK2007 graph

$Node_1$	$Node_2$	Jaccard index
100012	100002	0.016528925
100021	100002	0.002851711
100024	100012	0.023121387
100024	100021	0.018365473
100059	100007	0.1
100059	100012	0.006264462
100061	100024	0.012987013
100062	100010	0.03125
100062	100018	0.045454547
100064	100021	0.001897533
100083	100077	0.33333334
100088	100010	0.03448276
100107	100104	0.25
100110	100007	0.1
100110	100012	0.008264462
100110	100092	0.030303031

We have processed it in order to obtain a new file containing the adjacency list of the graph of hosts. Every line from the original file is parsed having as a result a set of lines, each one corresponding to the pair [current host—every host on the original line].

The application was run on a virtual cluster consisting of six virtual machines. The master machine hosted the *NameNode*, the *Secondary NameNode* and the *JobTracker* daemons, while the *DataNode* and the *TaskTracker* daemons ran on the other five machines of the cluster, the slaves. Every virtual machine had 1 GB of RAM, a 1.99 GHz dual core processor and a hard drive with a capacity of 25 GB. The virtual machines were launched and managed using *VMWare ESX* 5.0.[31] The Hadoop version installed on these machines was *Hadoop 0.18.0* (Coşulschi et al. 2012; Coşulschi et al. 2014).

The size of the input file for the application was 16 MB, while the size of the result file was over 2 GB. A few lines from the output file are listed in Table 2.

The Jaccard index histogram is represented in Fig. 5, where one can see that most nodes from the host graph have the Jaccard similarity coefficient included in the [0.0, 0.1) interval. This interval contains over 70,000,000 indexes, thus it can be concluded that most of our graph nodes are completely different. As regards to the identical nodes from the similarity point of view, there are more than 150,000 indexes having values 1, while over 50,000 are almost identical, with the Jaccard coefficient included in [0.9, 1.0) interval. In order to further explore the [0.0, 0.1) interval, it was split into smaller intervals, the results being displayed in Fig. 6.

[31]http://www.vmware.com/products/esxi-and-esx/.

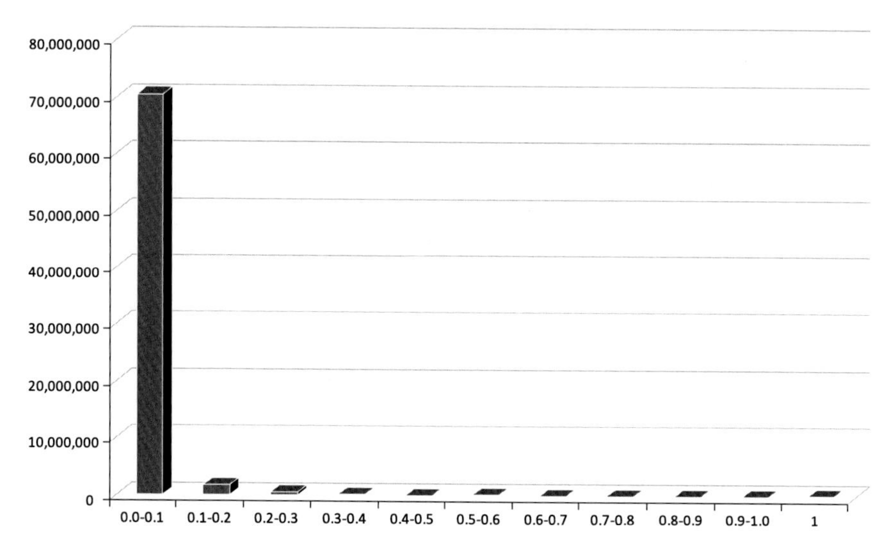

Fig. 5 The Jaccard index histogram for the WEBSPAM-UK2007 data collection

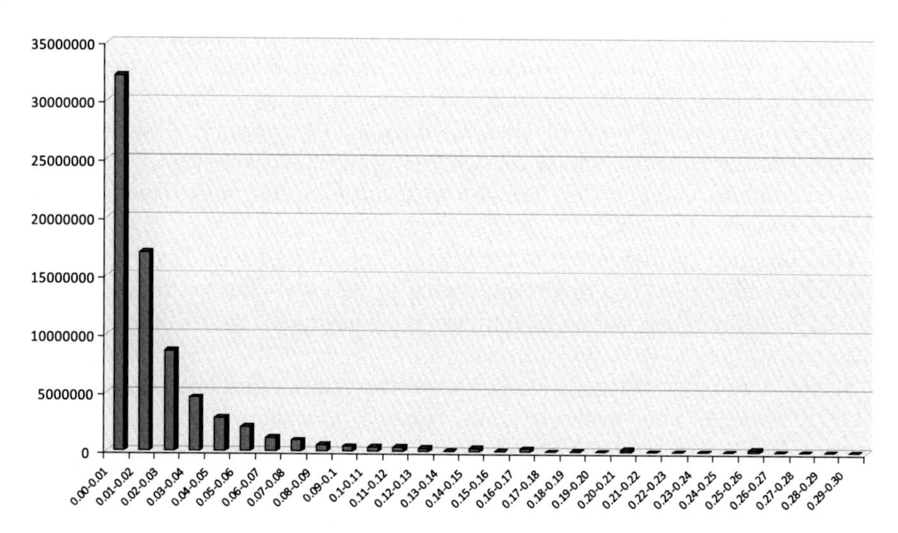

Fig. 6 The Jaccard index values distribution over [0.0, 0.3) interval for the WEBSPAM-UK2007 dataset

In this figure, one can notice that most of the sites from the chosen dataset have one or two common links for every 100 adjacent sites. The distribution of the Jaccard index is very heterogeneous due to the chosen dataset. Lower values are obtained depending on the number of sites that have the Jaccard similarity coefficient values from 0.13 to 0.14 (for sites having 13–14 links in common on every 100 neighbors).

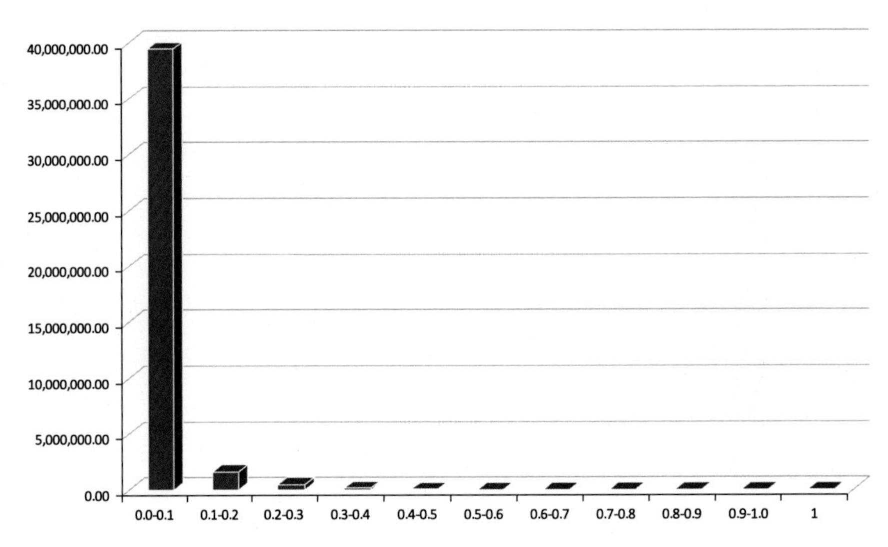

Fig. 7 The Jaccard index values distribution for the Slashdot graph

The second test was performed on a different dataset, i.e. Slashdot, which was collected in February 2009.[32] Slashdot is a technology-related news website,[33] offering a Slashdot Zoo feature that enables users to tag each other as friends or foes. The corresponding graph contains friend/foe links between the Slashdot users. The file containing the dataset has on each line a pair of nodes, separated by space, with the structure $node_1$ $node_2$. The dataset contains 82,168 nodes and 948,464 edges (Leskovec et al. 2009).

The histogram corresponding to the sequence of values for all Jaccard indexes can be seen in Fig. 7. The results are similar to the ones obtained from the first dataset, the majority pairs of nodes from the graph having the values of the Jaccard index included in the [0.0, 0.1) interval.

Similarly, in order to have a closer look at the values distribution, the interval [0.0, 0.3) was split into smaller intervals. Figure 8 presents the histogram for this interval. As a conclusion, the Slashdot graph has many pairs of nodes with the corresponding values of the Jaccard index belonging to the [0.0, 0.1) interval, which means that most of its nodes are very different from the point of view of the similarity functions used: e.g. they have little nodes in common. The lower frequency values of histogram are located in the [0.24, 0.25) and [0.29, 0.3) intervals (for friends having 24–25 and 29–30 links in common on every 100 neighbors). A few values of the dataset are listed in Table 3.

[32]http://snap.stanford.edu/data/soc-Slashdot0902.html.

[33]http://slashdot.org/.

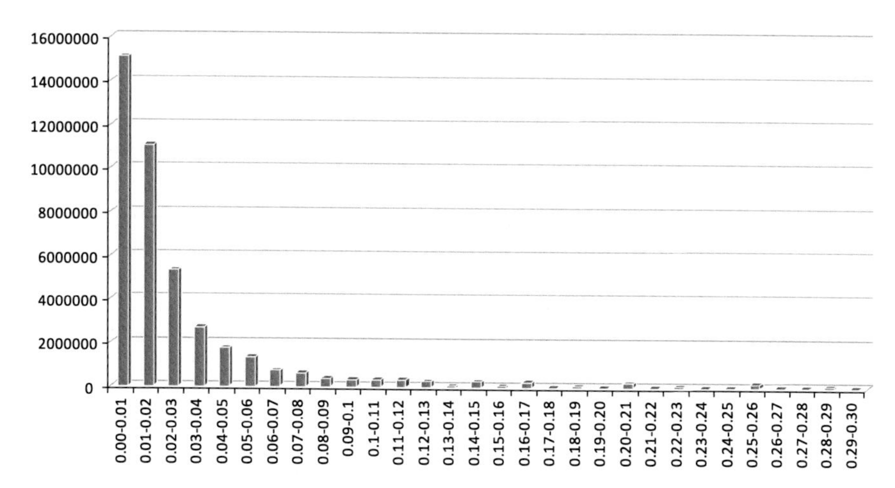

Fig. 8 The Jaccard index values distribution over [0.0, 0.3) interval for the Slashdot data set

Table 3 Values of the Jaccard index for sample pairs of nodes from the Slashdot graph

$Node_1$	$Node_2$	Jaccard index
1	0	0.04359673
10	1	0.005291005
100	1	0.020338982
10000	1	0.005181347
10002	10001	0.04347826
10003	100	0.018072288
10003	10001	0.004651163
10004	0	0.011111111
10004	100	0.015957447
10004	10001	0.014705882
10005	1	0.005291005
10005	10001	0.060606062
10005	10004	0.012658228
10006	100	0.009615385
10006	10002	0.011494253
10006	10005	0.042105265
10007	0	0.004385965
10002	10002	0.041666668
10007	10005	0.05882353
10008	100	0.011278195
10008	10002	0.006849315
10008	10005	0.012820513
10009	0	0.0041841003
1001	10000	0.022727273
1001	10003	0.01369863
1001	10009	0.024390243

9 Conclusion

As regards to the WEBSPAM-UK2007 data collection, let's consider two hosts u_1 and u_2, with A the set of neighboring hosts with u_1 (hosts that have at least a web page pointed by a link from a web page from u_1) and B the set of neighboring hosts with u_2. Jaccard index for the hosts u_1 and u_2 represents the measure of overlap that sets A and B shares. In other words, this value represents the number of common neighboring hosts reported to the total number of neighboring hosts.

If two hosts u_1 and u_2 have the Jaccard index ≈ 1, then:

1. we can conclude that their degree as *hub* is similar, considering only the number of hosts they refer to [we consider the *hub* notion as introduced for the HITS algorithm (Kleinberg 1999)];
2. it could be a good measure for classifying hosts as spam/non-spam (if a host was already classified as *spam*, then all the hosts having the Jaccard index $\approx$1, are good candidates for the same labeling);
3. the two hosts are connected from the semantic point of view (if they have a lot of neighboring hosts in common, then the information that they are hosting is similar).

On the other side, a Jaccard index $\approx$0 means that the two hosts have less in common from the semantic point of view.

Regarding the Slashdot dataset, the set of neighbors of a node u can be interpreted as a group of people with an already familiar u. Thus, the Jaccard index is an indicator of the level of commonly known people. A pair of people having the value of the Jaccard index greater than 0.9, means that there are strong connections between them at the professional level or that the respective people have common interests. Also, a level lower than 0.1 can be seen as an indicator that the respective two people have little in common.

We can create a new graph on the basis of the Jaccard index with the same set of vertices, while the set of edges consist of edges between two elements u and v if and only if the Jaccard index computed for the two nodes has a value greater than a threshold value ($J(u, v) > \alpha$, where e.g. $\alpha = 0.9$).

Based on the values obtained by computing the Jaccard similarity coefficients for two very large graphs, we have analyzed the connections and influences that certain nodes have over other nodes. Also, we have illustrated how the Apache Hadoop framework and the MapReduce programming model can be used for high volume of computations. All the tests were performed on a distributed cluster, in order to obtain the previously described results.

References

Armbrust, M., Fox, A., Griffith, R., Joseph, A., Katz, R., Konwinski, A., Lee, G., Patterson, D., Rabkin, A., Stoica, I., Zaharia, D.: A view of cloud computing. Commun. ACM **53**(4), 50–58 (2010)

Bank, J., Cole, B.: Calculating the Jaccard similarity coefficient with map reduce for entity pairs in Wikipedia. http://www.weblab.infosci.cornell.edu/papers/Bank2008.pdf (2008)

Blundo, C., De Cristofaro, E., Gasti, P.: EsPRESSo: efficient privacy-preserving evaluation of sample set similarity, In: 7th ESORICS Workshop on Data Privacy Management (DPM 2012) (2012)

Borthakur, D.: Hadoop architecture and its usage at facebook. http://borthakur.com/ftp/hadoopmicrosoft.pdf (2009)

Caruana, G., Li, M., Qi, M.: A MapReduce based parallel SVM for large scale spam filtering. In: 8th International Conference on Fuzzy Systems and Knowledge Discovery (FSKD), IEEE, vol. 4, pp. 2659–2662 (2011)

Coşulschi, M., Gabroveanu, M., Slabu, F., Sbîrcea, A.: Experiments with computing similarity coefficient over big data. In: 5th International Conference on Information, Intelligence, Systems and Applications (IISA 2014), pp. 112–117. IEEE (2014)

Coşulschi, M., Gabroveanu, M., Sbîrcea, A.: Running Hadoop applications in virtualization environment. Ann. Univ. Craiova Math. Comput. Sci. Ser. **39**(2), 322–333 (2012)

Dean, J., Ghemawat, S.: MapReduce: simplified data processing on large clusters. In: Proceedings of the 6th Conference on Symposium on Operating Systems Design & Implementation (OSDI04), vol. 6, pp. 137–150 (2004)

Ding, Z., Guo, D., Chen, X., Luo, X.: Performing MapReduce on data centers with hierarchical structures. Int. J. Comput. Commun. **7**(3), 432–449 (2012)

Engen, S., Grøtan, V., Sæther, B.-E.: Estimating similarity of communities: a parametric approach to spatio-temporal analysis of species diversity. Ecography **34**, 220–231 (2011)

Ghemawat, S., Gobioff, H., Leung, S.-T.: The Google file system. In: Proceedings of the 19th ACM Symposium on Operating Systems Principles (SOSP03), pp. 29–43. ACM (2003)

Hildebrandt, E.: Distributed computing the Google way, Java Forum Stuttgart and Herbstcampus (2010). http://www.soa-at-work.com/2010/09/distributed-computing-google-way.html

Indyk, W., Kajdanowicz, T., Kazienko, P., Plamowski, S.: Web spam detection using MapReduce approach to collective classification. In: International Joint Conference CISIS/ICEUTE/SOCO Special Sessions, vol. 189. Springer (2013)

Irving, B.: Big data and the power of Hadoop, Yahoo! Hadoop Summit (2010)

Kleinberg, J.: Authoritative sources in a hyperlinked environment. J. ACM **46**(5), 604–632 (1999)

Kunegis, J., Lommatzsch, A., Bauckhag, C.: The slashdot zoo: mining a social network with negative edges. In: Proceedings of World Wide Web Conference, pp. 741–750 (2009)

Lam, C.: Hadoop in Action. Manning Publications (2010)

Leskovec, J., Lang, K., Dasgupta, A., Mahoney, M.: Community structure in large networks: natural cluster sizes and the absence of large well-defined clusters. Internet Math. **6**(1), 29–123 (2009)

Leydesdorff, L.: On the normalization and visualization of author co-citation data: Salton's Cosine versus the Jaccard index. J. Am. Soc. Inform. Sci. Technol. **59**(1), 77–85 (2008)

Lin, J., Dyer, C.: Data-intensive text processing with MapReduce. Morgan & Claypool Publishers (2010)

Machaj, J., Piché, R., Brida, P.: Rank based fingerprinting algorithm for indoor positioning. In: International Conference on Indoor Positioning and Indoor Navigation (IPIN), pp. 1–6 (2011)

Mell, P., Grance, T.: The NIST Definition of Cloud Computing. National Institute of Science and Technology (2011)

Mulqueen, C.M., Stetz, T.A., Beaubien, J.M., O'Connell, B.J.: Developing dynamic work roles using Jaccard similarity indices of employee competency data. Ergometrika **2**, 26–37 (2001)

Rajaraman, A., Ullman, J.D.: Mining of Massive Datasets. http://www.mmds.org/. Cambridge University Press (2012)

Smith, J.E., Nair, R.: The architecture of virtual machines. Computer **38**(5), 32–38 (2005)

Sugerman, J., Venkitachalam, G., Lim, B.H.: Virtualizing I/O devices on VMware workstation's hosted virtual machine monitor. In: Proceedings of the General Track: 2002 USENIX Annual Technical Conference, pp. 1–14 (2001)

White, T.: Hadoop: The Definitive Guide. Storage and Analysis at Internet Scale, 3rd edn. O'Reilly Media/Yahoo Press (2012)

Zikopoulos, P., Eaton, C., DeRoos, D., Deutsch, T., Lapis, G.: Understanding Big Data: Analytics for Enterprise Class Hadoop and Streaming Data. McGraw-Hill (2011)

A Probe Guided Crossover Operator for More Efficient Exploration of the Search Space

K. Liagkouras and K. Metaxiotis

Abstract Crossover operators play very important role for the entire performance of evolutionary algorithms, as through the recombination of the fittest solutions, guide the population towards higher fitness regions of the search space. This study proposes a new Probe Guided Crossover (PGC) operator for the more efficient exploration of the search space. The proposed recombination operator is applied to three well-known Multiobjective Evolutionary Algorithms (MOEAs) namely the Non-dominated Sorting Genetic Algorithm II (NSGAII), the Strength Pareto Evolutionary Algorithm 2 (SPEA2) and the Indicator Based Evolutionary Algorithm (IBEA) and its performance is evaluated in comparison with the Simulated Binary Crossover (SBX). The proposed methodology is tested with the assistance of five test instances from the Li-Zhang (LZ) set of test functions. The experiments show that the PGC operator generates better results with confidence than the SBX operator.

Keywords Multiobjective optimization · Evolutionary algorithms · Crossover

1 Introduction

Evolutionary algorithms can efficiently solve multiobjective optimization problems (MOPs) by obtaining near optimal solution sets (Liagkouras and Metaxiotis 2014, 2015; Metaxiotis and Liagkouras 2012). A number of operators are being used by the multiobjective evolutionary algorithms (MOEAs) in order to evolve a set of candidate solutions (Koza 1992; Liagkouras and Metaxiotis 2013; Metaxiotis and Liagkouras 2013). Distinguished position among the various types of operators utilized by the MOEAs possess the recombination operator, as one of main

K. Liagkouras (✉) · K. Metaxiotis (✉)
Decision Support Systems Laboratory, Department of Informatics, University of Piraeus, 80, Karaoli & Dimitriou Str., 18534 Piraeus, Greece
e-mail: kliagk@unipi.gr; kmetax@unipi.gr

G.A. Tsihrintzis et al. (eds.), *Intelligent Computing Systems*,
Studies in Computational Intelligence 627, DOI 10.1007/978-3-662-49179-9_16

mechanisms responsible for the exploration of the search space (Zitzler et al. 2000). In MOEAs recombination operators are being used to generate offspring solutions, by exchanging parts from a selected pool of parent solutions (Zeng et al. 2010). One of the most popular real coded recombination operators that has been applied to a considerable number of MOEAs, is the Simulated Binary Crossover (SBX) (Deb and Agrawal 1995). In this paper we revisit the classical SBX and propose a probe guided version of the SBX operator that allows the more efficient exploration of the search space.

The remainder of the paper is structure as follows. A description of the Simulated Binary Crossover (SBX) is given in Sect. 2. Then, in Sect. 3, a detailed description of the proposed Probe Guided Crossover (PGC) operator is presented. The experimental environment is presented in Sect. 4. Section 5 presents the performance metrics. In Sect. 6 a series of experiments involving multiobjective optimization problems with complicate Pareto sets are conducted and in Sect. 7, the relevant results are discussed. Finally, Sect. 8 concludes the paper.

2 Simulated Binary Crossover (SBX)

The need for solving continuous search space optimization problems led the researchers to develop real-coded genetic operators. One of the most well-known and extensively applied, real-coded crossover operator is the simulated binary crossover (SBX) that was introduced by Deb and Agrawal in (1995). The SBX uses a probability distribution around two parents to create two children solutions. The probability distribution used by the SBX is similar in principle to the probability of creating children solutions in crossover operators used in binary-coded algorithms. In SBX as introduced by Deb and Agrawal (1995) each decision variable x_i, can take values in the interval: $x_i^{(L)} \leq x_i \leq x_i^{(U)}$, $i = 1, 2, \ldots, n$. Where $x_i^{(L)}$ and $x_i^{(U)}$ stand respectively for the lower and upper bounds for the decision variable i. In SBX, two parent solutions $p^{(1)}$ and $p^{(2)}$ generate two children solutions $c^{(1)}$ and $c^{(2)}$ as follows:

1. Calculate the spread factor β:

$$\beta = 1 + \frac{2}{p^{(2)} - p^{(1)}} \min\left[(p^{(1)} - p^{(l)}), (p^{(u)} - p^{(2)})\right]$$

2. Calculate parameter a:

$$\alpha = 2 - \beta^{-(\eta_c + 1)}$$

3. Create a random number u between 0 and 1.

$$u \rightarrow [0,\ 1];$$

4. Find a parameter β_q with the assistance of the following polynomial probability distribution:

$$\beta_q = \begin{cases} (au)^{1/(\eta_c+1)} & \textit{if } u \leq \frac{1}{a}, \\ \left(\frac{1}{2-au}\right)^{1/(\eta_c+1)} & \textit{otherwise} \end{cases}$$

The aforementioned procedure allows a zero probability of creating any children solutions outside the prescribed range $[x^{(L)}, x^{(U)}]$. Where η_c is the distribution index for SBX and can take any nonnegative value. In particular, small values of η_c allow children solutions to be created far away from parents and large values of η_c allow children solutions to be created near the parent solutions.
5. The children solutions are then calculated as follows:

$$c^{(1)} = 0.5\left[\left(p^{(1)} + p^{(2)}\right) - \beta_q\left|p^{(2)} - p^{(1)}\right|\right]$$
$$c^{(2)} = 0.5\left[\left(p^{(1)} + p^{(2)}\right) + \beta_q\left|p^{(2)} - p^{(1)}\right|\right]$$

The probability distributions as shown in *step 4*, do not create any solution outside the given bounds $[x^{(L)}, x^{(U)}]$ instead they scale up the probability for solutions inside the bounds.

3 Probe Guided Crossover (PGC) Operator

Crossover consists of swapping chromosome parts between individuals but is not performed on every pair of individuals. The crossover frequency is controlled by a crossover probability (P_c) (Arumugan et al. 2005). The efficiency of any recombination operator is judged by its ability to move progressively towards the higher fitness regions of the search space.

In this study we propose a new crossover operator named Probe Guided Crossover (PGC) operator due to its ability to probe efficiently the search space. As we mentioned earlier the PGC is a variation of the classical SBX operator. Thus, the two operators share some common elements. Indeed, the first two steps are common for both methods. In particular, as shown below, first we calculate the spread factor β and then the parameter α in the same manner as the SBX.

However, in *step 3* we follow a different strategy. As shown in Sect. 2 that illustrates the SBX operator, a random number $u \in [0, 1]$ is generated. If $u \leq 1/a$, it samples to the left hand side (region between $p^{(L)}$ and $p^{(i)}$), otherwise if $u > 1/a$ it samples to the right hand side (region between $p^{(i)}$ and $p^{(U)}$), where $p^{(i)}$ is the ith parent solution.

In PGC at this particular point as shown below we follow a different methodology. Specifically, instead of generating a random number $u \in [0, 1]$, we generate two random numbers, $u_L \in [0, 1/a]$ to sample the left hand side and a random number $u_R \in (1/a, 1]$ to sample the right hand side of the probability distribution. From the aforementioned process emerge two values of β_q, the β_q^L that samples the left hand side of the polynomial probability distribution and the β_q^R that samples the right hand side of the polynomial probability distribution. Next, as shown below in *step 5* with the assistance of β_q^L and β_q^R are formulated two variants for each child solution. Specifically, $c_L^{(1)}$ and $c_R^{(1)}$ are the two variants that emerge by substituting the β_q^L and β_q^R to $c^{(1)}$. Respectively $c_L^{(2)}$ and $c_R^{(2)}$ are the two variants that emerge by substituting the β_q^L and β_q^R to $c^{(2)}$.

Then, by substituting to the parent solution vector at the position of the selected variable to be crossovered, respectively the $c_L^{(1)}$ and $c_R^{(1)}$ we create two different child solution vectors (*csv*), the $csv_L^{(1)}$ and $csv_R^{(1)}$. Thanks to the generated $csv_L^{(1)}$ and $csv_R^{(1)}$ we are able to perform fitness evaluation for each one of the corresponding cases. As soon as we complete the fitness evaluation process, we select the best child solution between the two variants $c_L^{(1)}$ and $c_R^{(1)}$ with the assistance of the Pareto optimality framework. The same procedure is followed for $c_L^{(2)}$ and $c_R^{(2)}$. The proposed methodology allows us to probe more efficiently the search space and move progressively towards higher fitness solutions. Whenever, there is not a clear winner i.e. strong or weak dominance, between the $c_L^{(1)}$ and $c_R^{(1)}$, or respectively between the $c_L^{(2)}$ and $c_R^{(2)}$ the generation of a random number allows the random choice of one of the two alternative child solutions.

The procedure of computing children solutions $c^{(1)}$ and $c^{(2)}$ from two parent solutions $p^{(1)}$ and $p^{(2)}$ under the Probe Guided Crossover (PGC) operator is as follows:

1. Calculate the spread factor β:

$$\beta = 1 + \frac{2}{p^{(2)} - p^{(1)}} \min\left[(p^{(1)} - p^{(l)}), (p^{(u)} - p^{(2)})\right]$$

2. Calculate parameter *a:*

$$\alpha = 2 - \beta^{-(\eta_c + 1)}$$

3. Create 2 random numbers $u_L \in [0, 1/a]$ and $u_R \in (1/a, 1]$.

$$u_L \rightarrow [0, 1/\alpha];$$
$$u_R \rightarrow (1/\alpha, 1];$$

4. Find 2 parameters β_q^L and β_q^U with the assistance of the following polynomial probability distribution:

$$\beta_q^L = (au_L)^{1/(\eta_c+1)}, \quad u_L \in [0, 1/a],$$
$$\beta_q^R = \left(\frac{1}{2-au_R}\right)^{1/(\eta_c+1)}, \quad u_R \in [1/a, 1]$$

5. Thus, instead of a unique value for $c^{(1)}$ and $c^{(2)}$, we obtain two evaluations for each child solution that correspond to β_q^L and β_q^R respectively:

$$c_L^{(1)} = 0.5\left[\left(p^{(1)} + p^{(2)}\right) - \beta_q^L\left|p^{(2)} - p^{(1)}\right|\right]$$
$$c_R^{(1)} = 0.5\left[\left(p^{(1)} + p^{(2)}\right) - \beta_q^R\left|p^{(2)} - p^{(1)}\right|\right]$$
$$c_L^{(2)} = 0.5\left[\left(p^{(1)} + p^{(2)}\right) + \beta_q^L\left|p^{(2)} - p^{(1)}\right|\right]$$
$$c_R^{(2)} = 0.5\left[\left(p^{(1)} + p^{(2)}\right) + \beta_q^R\left|p^{(2)} - p^{(1)}\right|\right]$$

6. We perform fitness evaluation for each variant child solution, by substituting the candidate solutions into the parent solution vector.
7. We select the best variant between the $c_L^{(1)}$ and $c_R^{(1)}$, based on the Pareto optimality framework. The same procedure is followed for $c_L^{(2)}$ and $c_R^{(2)}$.
8. Whenever, there is not a clear winner i.e. strong or weak dominance, between the $c_L^{(1)}$ and $c_R^{(1)}$, or respectively between the $c_L^{(2)}$ and $c_R^{(2)}$ the generation of a random number allows the random choice of one of the two alternative child solutions.

4 Experimental Environment

All algorithms have been implemented in Java and run on a 2.1 GHz Windows Server 2012 machine with 6 GB RAM. The jMetal (Durillo and Nebro 2011) framework has been used to compare the performance of the proposed, Probe Guided Crossover (PGC) operator against the Simulated Binary Crossover (SBX) operator with the assistance of three state-of-the-art MOEAs, namely the NSGAII, the SPEA2 and the IBEA. In all tests we use, binary tournament and polynomial mutation (PLM) (Deb and Agrawal 1995) as, selection and mutation operator, respectively. The crossover probability is $P_c = 0.9$ and mutation probability is $P_m = 1/n$, where n is the number of decision variables. The distribution indices for the crossover and mutation operators are $\eta_c = 20$ and $\eta_m = 20$, respectively. Population size is set to *100*, using *25,000* function evaluations with *100* independent runs.

5 Performance Metrics

5.1 *Hypervolume*

The Hypervolume (Zitzler et al. 2007) of a set of solutions measures the size of the portion of objective space that is dominated by those solutions collectively. The hypervolume is an indicator of both the convergence and diversity of an approximation set (Zitzler et al. 2003). Thus, given a set S containing m points in n objectives, the hypervolume of S is the size of the portion of objective space that is dominated by at least one point in S. The hypervolume of S is calculated relative to a reference point which is worse than (or equal to) every point in S in every objective. The greater the hypervolume of a solution the better considered the solution.

5.2 *Spread*

Deb et al. (2002) introduced the spread of solutions (Δ) as another indicator of the quality of the derived set of solutions. Spread indicator examines whether or not the solutions span the entire Pareto optimal region. First, it calculates the Euclidean distance between the consecutive solutions in the obtained non-dominated set of solutions. Then it calculates the average of these distances. After that, from the obtained set of non-dominated solutions the extreme solutions are calculated. Finally, using the following metric it calculates the nonuniformity in the distribution.

$$\Delta = \frac{d_f + d_l + \sum_{i=1}^{N-1} \left| d_i - \bar{d} \right|}{d_f + d_l + (N-1)\bar{d}}$$

where d_f and d_l are the Euclidean distances between the extreme solutions and the boundary solutions of the obtained nondominated set. The parameter $\bar{d}$ is the average of all distances d_i, $i = 1, 2, \ldots, (N-1)$, where N is the number of solutions on the best nondominated front. The smaller the value of Spread indicator, the better the distribution of the solutions. Spread indicator takes a zero value for an ideal distribution of the solutions in the Pareto front.

5.3 *Epsilon Indicator I_ε*

Zitzler et al. (2003) introduced the epsilon indicator (I_ε). There are two versions of epsilon indicator the multiplicative and the additive. In this study we use the unary additive epsilon indicator. The basic usefulness of epsilon indicator of an approximation set A ($I_{\varepsilon+}$) is that it provides the minimum term ε by which each point of the real front R in the objective space can be shifted by component-wide addition, such that the resulting transformed approximation set is dominated by A. The additive epsilon indicator is a good measure of diversity, since it focuses on the worst case distance and reveals whether or not the approximation set has gaps in its trade-off solution set.

6 Experimental Results

A number of computational experiments were performed to test the performance of the proposed Probe Guided Crossover (PGC) operator for the solution of five test instances of the Li and Zhang (LZ) test suite (Li and Zhang 2009). The performance of the proposed PGC operator is assessed in comparison with the Simulated Binary Crossover (SBX) operator with the assistance of three well-known MOEAs, namely the Non-dominated Sorting Genetic Algorithm II (NSGAII), the Strength Pareto Evolutionary Algorithm 2 (SPEA2) and the Indicator Based Evolutionary Algorithm (IBEA). The evaluation of the performance is based on a variety of metrics that assess both the proximity of the solutions to the Pareto front and their dispersion on it.

6.1 The Li and Zhang (LZ) Test Suite

The Li and Zhang (LZ) (Li and Zhang 2009) test suite introduces a class of continuous multiobjective optimization test instances with complicated Pareto sets. In this study we use five test instances of the Li and Zhang (LZ) test suite (Li and Zhang 2009) in order to evaluate the performance of the proposed crossover operator. In particular we use the LZ 1-2, 5-6 and 9 test instances. The LZ 1-2, 5 and 9 are bi-objective instances. The LZ6 has three objectives and its Pareto Set (PS) is nonlinear. The LZ 1-2, 5 have convex Pareto Front (PF) shapes, but their Pareto Set (PS) shapes are non-linear in the decision space. On the contrary, the LZ9 examines a test instance with concave Pareto Front.

Li-Zhang's function N.1 (LZ1) problem:
Li-Zhang's function N.2(LZ2) problem:

$$Min = \begin{cases} f_1(x) = x_1 + \frac{2}{|J_1|}\sum_{j\epsilon J_1}\left(x_j - x_1^{0.5\left(1.0+\frac{3(j-2)}{n-2}\right)}\right)^2, \\ f_2(x) = 1 - \sqrt{x_1} + \frac{2}{|J_2|}\sum_{j\epsilon J_2}\left(x_j - x_1^{0.5\left(1.0+\frac{3(j-2)}{n-2}\right)}\right)^2, \\ where\ J_1 = \{j|\, j\ is\ odd\ and\ 2 \le j \le n\} \\ \quad and\ J_2 = \{j|\, j\ is\ even\ and\ 2 \le j \le n\} \\ The\ decision\ space\ \Omega = [0,\ 1]^n \end{cases}$$

$$Min = \begin{cases} f_1(x) = x_1 + \frac{2}{|J_1|}\sum_{j\epsilon J_1}\left(x_j - sin\left(6\pi x_1 + \frac{j\pi}{n}\right)\right)^2, \\ f_2(x) = 1 - \sqrt{x_1} + \frac{2}{|J_2|}\sum_{j\epsilon J_2}\left(x_j - sin(6\pi x_1 + \frac{j\pi}{n})\right)^2, \\ where\ J_1 = \{j|\, j\ is\ odd\ and\ 2 \le j \le n\} \\ \quad and\ J_2 = \{j|\, j\ is\ even\ and\ 2 \le j \le n\} \\ The\ decision\ space\ \Omega = [0,\ 1] \times [-1, 1]^{n-1} \end{cases}$$

Li-Zhang's function N.5 (LZ5) problem:

$$Min = \begin{cases} f_1(x) = x_1 + \frac{2}{|J_1|}\sum_{j\epsilon J_1}\left\{x_j - \left[0.3x_1^2 cos\left(24\pi x_1 + \frac{4j\pi}{n}\right) + 0.6x_1\right]cos\left(6\pi x_1 + \frac{j\pi}{n}\right)\right\}^2 \\ f_2(x) = 1 - \sqrt{x_1} + \frac{2}{|J_2|}\sum_{j\epsilon J_2}\left\{x_j - \left[0.3x_1^2 cos\left(24\pi x_1 + \frac{4j\pi}{n}\right) + 0.6x_1\right]sin\left(6\pi x_1 + \frac{j\pi}{n}\right)\right\}^2 \\ where\ J_1 = \{j|\, j\ is\ odd\ and\ 2 \le j \le n\} and\ J_2 = \{j|\, j\ is\ even\ and\ 2 \le j \le n\} \\ The\ decision\ space\ \Omega = [0, 1] \times [-1, 1]^{n-1} \end{cases}$$

Li-Zhang's function N.6 (LZ6) problem :

$$Min = \begin{cases} f_1(x) = \cos(0.5x_1\pi)\cos(0.5x_2\pi) + \frac{2}{|J_1|}\sum_{j\epsilon J_1}(x_j - 2x_2\, sin(2\pi x_1 + \frac{j\pi}{n}))^2, \\ f_2(x) = \cos(0.5x_1\pi)\sin(0.5x_2\pi) + \frac{2}{|J_2|}\sum_{j\epsilon J_2}\left(x_j - 2x_2 \sin\left(2\pi x_1 + \frac{j\pi}{n}\right)\right)^2, \\ f_3(x) = \sin(0.5x_1\pi) + \frac{2}{|J_3|}\sum_{j\epsilon J_3}\left(x_j - 2x_2 \sin\left(2\pi x_1 + \frac{j\pi}{n}\right)\right)^2, \\ where \quad J_1 = \{j|3 \le j \le n,\ and\ j-1\ is\ a\ multiplication\ of\ 3\} \\ \qquad\qquad J_2 = \{j|3 \le j \le n,\ and\ j-2\ is\ a\ multiplication\ of\ 3\} \\ \qquad\qquad J_3 = \{j|3 \le j \le n,\ and\ j\ is\ a\ multiplication\ of\ 3\} \\ The\ decision\ space\ \ \Omega = [0,\ 1]^2 \times [-2,\ 2]^{n-2} \end{cases}$$

Li-Zhang's function N.9 (LZ9) problem:

$$Min = \begin{cases} f_1(x) = x_1 + \frac{2}{|J_1|}\sum_{j\epsilon J_1}\left(x_j - sin\left(6\pi x_1 + \frac{j\pi}{n}\right)\right)^2, \\ f_2(x) = 1 - x_1^2 + \frac{2}{|J_2|}\sum_{j\epsilon J_2}\left(x_j - sin\left(6\pi x_1 + \frac{j\pi}{n}\right)\right)^2, \\ where\ J_1 = \{j|\, j\ is\ odd\ and\ 2 \le j \le n\} and\ J_2 = \{j|\, j\ is\ even\ and\ 2 \le j \le n\} \\ The\ decision\ space\ \ \Omega = [0,\ 1] \times [-1,\ 1]^{n-1} \end{cases}$$

The results in the Tables 1, 2 and 3 have been produced by using jMetal (Durillo and Nebro 2011) framework. Table 1 presents the results of LZ 1-2, 5-6 and 9 test

Table 1 Mean, std, median and IQR for HV, spread and epsilon

Problem: LZ1	NSGAII		SPEA2		IBEA	
	PGC	SBX	PGC	SBX	PGC	SBX
HV. Mean and Std	$6.52e-01_{6.8e-04}$	$6.52e-01_{6.7e-04}$	$6.52e-01_{3.1e-03}$	$6.52e-01_{2.8e-03}$	$6.54e-01_{3.2e-03}$	$6.54e-01_{1.7e-03}$
HV. Median and IQR	$6.52e-01_{1.0e-03}$	$6.52e-01_{1.0e-03}$	$6.52e-01_{4.0e-03}$	$6.53e-01_{2.8e-03}$	$6.55e-01_{2.3e-03}$	$6.55e-01_{1.6e-03}$
SPREAD. Mean and Std	$4.96e-01_{1.1e-01}$	$5.29e-01_{1.3e-01}$	$4.26e-01_{1.5e-01}$	$5.45e-01_{2.0e-01}$	$7.47e-01_{5.0e-02}$	$7.75e-01_{5.0e-02}$
SPREAD. Median and IQR	$4.60e-01_{1.4e-01}$	$4.77e-01_{1.8e-01}$	$3.62e-01_{2.2e-01}$	$5.05e-01_{3.0e-01}$	$7.43e-01_{7.2e-02}$	$7.75e-01_{7.4e-02}$
EPSILON. Mean and Std	$1.89e-02_{4.3e-03}$	$1.89e-02_{1.9e-03}$	$4.97e-02_{1.3e-02}$	$4.98e-02_{1.2e-02}$	$4.84e-02_{1.3e-02}$	$4.79e-02_{1.3e-02}$
EPSILON. Median and IQR	$1.77e-02_{2.6e-03}$	$1.86e-02_{2.7e-03}$	$4.69e-02_{1.7e-02}$	$4.71e-02_{1.7e-02}$	$4.68e-02_{1.7e-02}$	$4.64e-02_{2.0e-02}$
Problem: LZ2	NSGAII		SPEA2		IBEA	
	PGC	SBX	PGC	SBX	PGC	SBX
HV. Mean and Std	$5.08e-01_{4.3e-02}$	$5.02e-01_{4.4e-02}$	$5.03e-01_{4.1e-02}$	$5.06e-01_{3.8e-02}$	$4.92e-01_{4.2e-02}$	$4.85e-01_{5.8e-02}$
HV. Median and IQR	$5.26e-01_{4.9e-02}$	$5.21e-01_{7.4e-02}$	$5.19e-01_{4.8e-02}$	$5.21e-01_{4.0e-02}$	$5.10e-01_{6.4e-02}$	$4.98e-01_{7.3e-02}$
SPREAD. Mean and Std	$1.52e+00_{1.1e-01}$	$1.48e+00_{1.2e-01}$	$1.52e+00_{1.1e-01}$	$1.50e+00_{1.2e-01}$	$1.48e+00_{1.3e-01}$	$1.48e+00_{1.4e-01}$
SPREAD. Median and IQR	$1.50e+00_{1.7e-01}$	$1.48e+00_{1.7e-01}$	$1.50e+00_{1.2e-01}$	$1.48e+00_{1.0e-01}$	$1.50e+00_{4.5e-02}$	$1.50e+00_{7.1e-02}$
EPSILON. Mean and Std	$2.17e-01_{5.5e-02}$	$2.28e-01_{6.0e-02}$	$2.26e-01_{5.3e-02}$	$2.22e-01_{5.4e-02}$	$2.42e-01_{5.9e-02}$	$2.57e-01_{8.6e-02}$
EPSILON. Median and IQR	$1.93e-01_{8.2e-02}$	$2.08e-01_{9.4e-02}$	$2.07e-01_{9.5e-02}$	$2.03e-01_{7.6e-02}$	$2.30e-01_{9.3e-02}$	$2.41e-01_{1.0e-01}$
Problem: LZ5	NSGAII		SPEA2		IBEA	
	PGC	SBX	PGC	SBX	PGC	SBX
HV. Mean and Std	$6.12e-01_{6.2e-03}$	$6.09e-01_{7.8e-03}$	$6.04e-01_{7.7e-03}$	$6.02e-01_{5.9e-03}$	$6.06e-01_{1.2e-02}$	$6.04e-01_{1.4e-02}$
HV. Median and IQR	$6.11e-01_{7.6e-03}$	$6.09e-01_{7.2e-03}$	$6.06e-01_{8.5e-03}$	$6.03e-01_{7.6e-03}$	$6.08e-01_{1.1e-02}$	$6.08e-01_{1.3e-02}$
SPREAD. Mean and Std	$6.36e-01_{6.0e-02}$	$6.36e-01_{7.1e-02}$	$6.17e-01_{9.8e-02}$	$6.23e-01_{8.3e-02}$	$1.08e+00_{8.1e-02}$	$1.07e+00_{6.8e-02}$
SPREAD. Median and IQR	$6.27e-01_{8.0e-02}$	$6.23e-01_{9.1e-02}$	$6.04e-01_{1.3e-01}$	$6.09e-01_{1.1e-01}$	$1.07e+00_{1.3e-01}$	$1.07e+00_{9.9e-02}$
EPSILON. Mean and Std	$1.16e-01_{2.2e-02}$	$1.21e-01_{2.5e-02}$	$1.39e-01_{3.3e-02}$	$1.35e-01_{2.2e-02}$	$1.51e-01_{5.0e-02}$	$1.62e-01_{5.8e-02}$
EPSILON. Median and IQR	$1.20e-01_{2.7e-02}$	$1.19e-01_{2.0e-02}$	$1.32e-01_{2.9e-02}$	$1.33e-01_{2.2e-02}$	$1.29e-01_{6.8e-02}$	$1.42e-01_{6.5e-02}$

(continued)

Table 1 (continued)

Problem: LZ6	NSGAII		SPEA2		IBEA	
	PGC	SBX	PGC	SBX	PGC	SBX
Problem: LZ6	NSGAII		SPEA2		IBEA	
	PGC	SBX	PGC	SBX	PGC	SBX
HV. Mean and Std	$2.25e-01_{3.2e-02}$	$1.54e-01_{4.1e-02}$	$2.57e-01_{2.0e-02}$	$2.46e-01_{2.6e-02}$	$8.51e-02_{9.3e-02}$	$7.39e-02_{8.2e-02}$
HV. Median and IQR	$2.34e-01_{4.3e-02}$	$1.54e-01_{6.6e-02}$	$2.62e-01_{1.9e-02}$	$2.54e-01_{2.2e-02}$	$2.82e-02_{2.0e-01}$	$3.77e-02_{1.4e-01}$
SPREAD. Mean and Std	$9.08e-01_{7.1e-02}$	$9.32e-01_{7.5e-02}$	$6.94e-01_{6.8e-02}$	$7.03e-01_{6.7e-02}$	$1.48e+00_{4.6e-01}$	$1.61e+00_{4.4e-01}$
SPREAD. Median and IQR	$9.07e-01_{8.5e-02}$	$9.38e-01_{9.7e-02}$	$6.98e-01_{8.7e-02}$	$6.99e-01_{8.4e-02}$	$1.77e+00_{9.2e-01}$	$1.84e+00_{3.5e-01}$
EPSILON. Mean and Std	$2.93e-01_{8.0e-02}$	$3.31e-01_{7.2e-02}$	$3.07e-01_{1.1e-01}$	$3.07e-01_{1.0e-01}$	$5.21e-01_{1.6e-01}$	$4.69e-01_{1.4e-01}$
EPSILON. Median and IQR	$2.73e-01_{2.0e-02}$	$3.08e-01_{5.2e-02}$	$2.70e-01_{1.9e-02}$	$2.70e-01_{4.2e-02}$	$3.97e-01_{3.1e-01}$	$3.97e-01_{1.0e-01}$
Problem: LZ9	NSGAII		SPEA2		IBEA	
	PGC	SBX	PGC	SBX	PGC	SBX
HV. Mean and Std	$1.70e-01_{5.1e-02}$	$1.53e-01_{5.5e-02}$	$1.67e-01_{4.9e-02}$	$1.64e-01_{4.6e-02}$	$1.52e-01_{5.7e-02}$	$1.49e-01_{5.6e-02}$
HV. Median and IQR	$1.89e-01_{6.9e-02}$	$1.57e-01_{7.5e-02}$	$1.70e-01_{6.0e-02}$	$1.62e-01_{6.0e-02}$	$1.52e-01_{7.8e-02}$	$1.50e-01_{7.8e-02}$
SPREAD. Mean and Std	$1.66e+00_{1.6e-01}$	$1.64e+00_{1.8e-01}$	$1.69e+00_{1.4e-01}$	$1.66e+00_{1.4e-01}$	$1.68e+00_{1.3e-01}$	$1.68e+00_{1.2e-01}$
SPREAD. Median and IQR	$1.62e+00_{2.8e-01}$	$1.65e+00_{2.9e-01}$	$1.69e+00_{2.4e-01}$	$1.65e+00_{2.3e-01}$	$1.64e+00_{2.3e-01}$	$1.64e+00_{2.0e-01}$
EPSILON. Mean and Std	$2.51e-01_{6.3e-02}$	$2.65e-01_{6.4e-02}$	$2.57e-01_{6.6e-02}$	$2.65e-01_{7.3e-02}$	$2.96e-01_{9.3e-02}$	$2.93e-01_{7.9e-02}$
EPSILON. Median and IQR	$2.32e-01_{9.5e-02}$	$2.54e-01_{1.1e-01}$	$2.38e-01_{8.7e-02}$	$2.51e-01_{1.1e-01}$	$2.91e-01_{1.0e-01}$	$2.92e-01_{1.1e-01}$

functions. Specifically, it presents the mean, standard deviation (STD), median and interquartile range (IQR) of all the independent runs carried out for Hypervolume (HV), Spread (Δ) and Epsilon indicator respectively.

Regarding the HV (Fonseca and Flemming 1998; Zitzler et al. 2007) indicator the higher the value (i.e. the greater the hypervolume) the better the computed front.

Table 2 Boxplots for HV, spread and epsilon

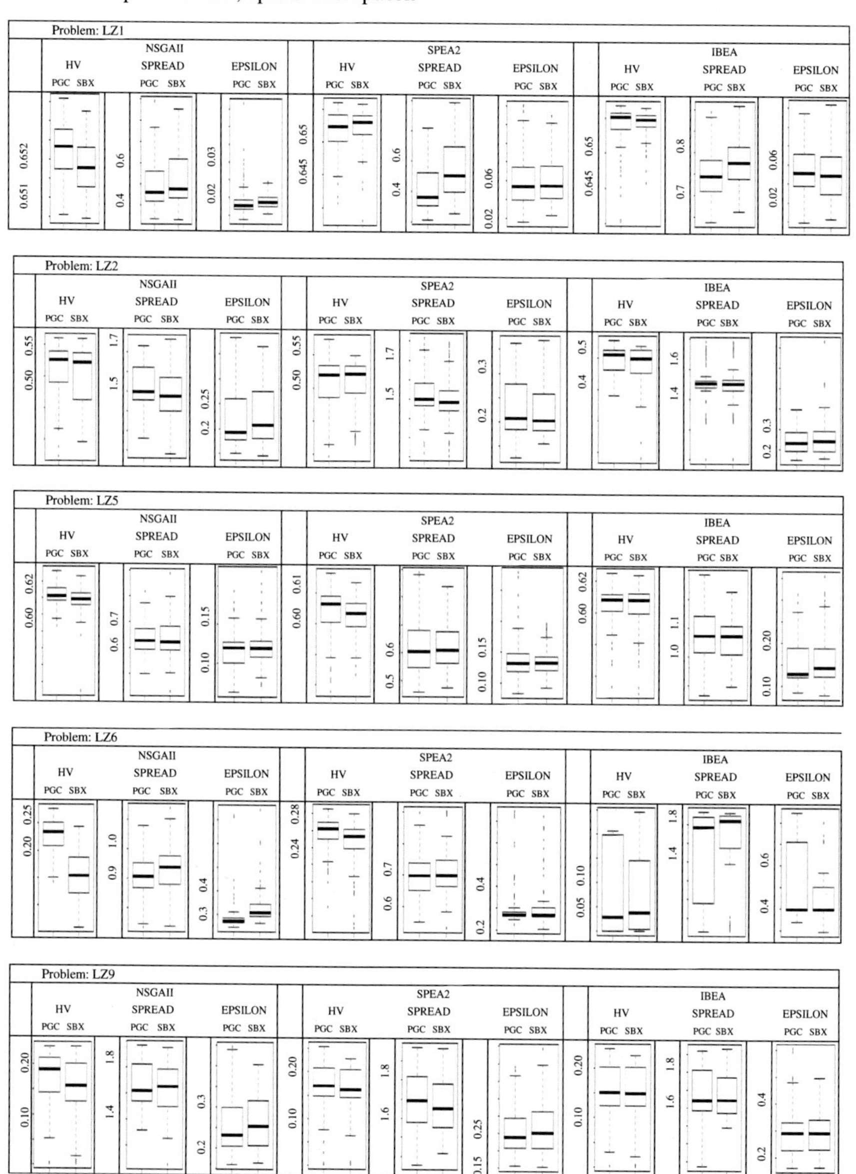

The second indicator the Spread (Δ) (Deb et al. 2002) examines the spread of solutions across the Pareto front. The smaller the value of this indicator, the better the distribution of the solutions. This indicator takes a zero value for an ideal distribution of the solutions in the Pareto front. The third indicator, the Epsilon (Zitzler et al. 2003) is a measure of the smaller distance that a solution set A, needs to be changed in such a way that it dominates the optimal Pareto front of this problem. Obviously the smaller the value of this indicator, the better the derived solution set.

Table 2 use boxplots to present graphically the performance of NSGAII, SPEA2 and IBEA under two different configurations, the PGC and the SBX respectively, for HV, Spread (Δ) and Epsilon performance indicators. Boxplot is a convenient way of graphically depicting groups of numerical data through their quartiles.

Table 3, presents if the results of NSGAII, SPEA2 and IBEA derived under the two different configurations (PGC and SBX) are statistically significant or not. For that reason, we use the Wilcoxon rank-sum test as it is implemented by the jMetal framework (Durillo and Nebro 2011). In Table 3, three different symbols are used. In particular "–" indicates that there is not statistical significance between the algorithms. "↑" means that the algorithm in the row has yielded better results than the algorithm in the column with confidence and "↓" is used when the algorithm in the column is statistically better than the algorithm in the row.

Finally, Table 4 provides the total CPU time, in seconds, needed for each one of the presented tests problems (LZ 1-2, 5-6 and 9) under the typical configuration with the SBX operator and under the proposed PGC. As expected, there is a computational cost associated with the fitness evaluation process of PGC.

We carried out our experiments on a 2.1 GHz Windows Server 2012 machine with 6 GB RAM. On average the proposed PGC has an overhead of 11.14 % in CPU time when applied to the NSGAII, an overhead of 10.56 % when applied to SPEA2 and an overhead of 4.17 % when applied to IBEA.

7 Analysis of the Results

In this section, we analyze the results obtained by applying the Probe Guided Crossover (PGC) operator and the Simulated Binary Crossover (SBX) operator respectively to three well-known MOEAs, namely the NSGAII, the SPEA2 and the IBEA for solving five test instances from the Li-Zhang (LZ) set of test functions (Li and Zhang 2009). The assessment of the performance of the proposed crossover operator is done with the assistance of three well-known performance indicators, namely *Hypervolume*, *Spread* and *Epsilon* indicator.

The results from the experiments regarding the HV indicator show that the PGC operator generates better results with confidence than the SBX operator. In particular, in 46.67 % of the cases the PGC yields better results with confidence than the conventional configuration of the NSGAII, SPEA2 and IBEA with the SBX operator. Also, in 53.33 % of the cases there was not statistical significance between

Table 3 Wilcoxon test for HV, spread and epsilon

Problem			NSGAII with SBX		SPEA2 with SBX		IBEA with SBX
LZ1	HV. Mean and Std	NSGAII with PGC	↑	SPEA2 with PGC	–	IBEA with PGC	↑
	HV. Median and IQR		↑		–		↑
	SPREAD. Mean and Std		↑		↑		↑
	SPREAD. Median and IQR		↑		↑		↑
	EPSILON. Mean and Std		↑		–		–
	EPSILON. Median and IQR		↑		–		–
Problem			NSGAII with SBX		SPEA2 with SBX		IBEA with SBX
LZ2	HV. Mean and Std	NSGAII with PGC	–	SPEA2 with PGC	–	IBEA with PGC	↑
	HV. Median and IQR		–		–		↑
	SPREAD. Mean and Std		–		–		–
	SPREAD. Median and IQR		–		–		–
	EPSILON. Mean and Std		↑		–		–
	EPSILON. Median and IQR		↑		–		–
Problem			NSGAII with SBX		SPEA2 with SBX		IBEA with SBX
LZ5	HV. Mean and Std	NSGAII with PGC	–	SPEA2 with PGC	↑	IBEA with PGC	–
	HV. Median and IQR		–		↑		–
	SPREAD. Mean and Std		–		–		–

(continued)

Table 3 (continued)

Problem			NSGAII with SBX		SPEA2 with SBX		IBEA with SBX
	SPREAD. Median and IQR		–		–		–
	EPSILON. Mean and Std		–		–		↑
	EPSILON. Median and IQR		–		–		↑
Problem			NSGAII with SBX		SPEA2 with SBX		IBEA with SBX
LZ6	HV. Mean and Std	NSGAII with PGC	↑	SPEA2 with PGC	↑	IBEA with PGC	–
	HV. Median and IQR		↑		↑		–
	SPREAD. Mean and Std		↑		–		–
	SPREAD. Median and IQR		↑		–		–
	EPSILON. Mean and Std		↑		–		–
	EPSILON. Median and IQR		↑		–		–
Problem			NSGAII with SBX		SPEA2 with SBX		IBEA with SBX
LZ9	HV. Mean and Std	NSGAII with PGC	↑	SPEA2 with PGC	–	IBEA with PGC	–
	HV. Median and IQR		↑		–		–
	SPREAD. Mean and Std		–		–		–
	SPREAD. Median and IQR		–		–		–
	EPSILON. Mean and Std		↑		–		–
	EPSILON. Median and IQR		↑		–		–

Table 4 Mean and Std total CPU times (in seconds) for the test problems for 100 independent runs

Problem	NSGAII		
	PGC	SBX	Overhead (%)
LZ1	$2.3751e{+}00_{3.1350e-02}$	$2.2412e{+}00_{2.715e-02}$	5.97
LZ2	$3.2013e{+}00_{2.1313e-01}$	$2.9502e{+}00_{2.0659e-01}$	8.51
LZ5	$3.3301e{+}00_{1.4402e-01}$	$3.1505e{+}00_{1.3132e-01}$	5.70
LZ6	$2.1042e{+}00_{2.2230e-02}$	$1.9502e{+}00_{2.3940e-02}$	7.89
LZ9	$3.7106e{+}00_{1.3859e-01}$	$2.9075e{+}00_{1.1950e-01}$	27.62
		Average	11.14
Problem	SPEA2		
	PGC	SBX	Overhead (%)
LZ1	$6.4045e{+}00_{8.6970e-01}$	$6.2865e{+}00_{8.4750e-01}$	1.88
LZ2	$7.6013e{+}00_{8.6021e-01}$	$6.7405e{+}00_{8.4354e-01}$	12.77
LZ5	$8.4127e{+}00_{7.6112e-01}$	$8.1310e{+}00_{7.9750e-01}$	3.46
LZ6	$8.3250e{+}00_{5.1588e-01}$	$7.2545e{+}00_{6.1183e-01}$	14.76
LZ9	$7.3208e{+}00_{8.2992e-01}$	$6.1025e{+}00_{7.5627e-01}$	19.96
		Average	10.56
Problem	IBEA		
	PGC	SBX	Overhead (%)
LZ1	$11.8502e{+}00_{8.4850e-01}$	$11.5765e{+}00_{8.115e-01}$	2.36
LZ2	$11.9053e{+}00_{8.9313e-01}$	$11.3890e{+}00_{8.6659e-01}$	4.53
LZ5	$12.5181e{+}00_{9.1402e-01}$	$12.2505e{+}00_{8.3132e-01}$	2.18
LZ6	$12.0832e{+}00_{8.7230e-01}$	$11.9504e{+}00_{8.5940e-01}$	1.11
LZ9	$12.0196e{+}00_{8.3859e-01}$	$10.8575e{+}00_{8.1950e-01}$	10.70
		Average	4.17

the PGC and SBX. Finally, in none of the examined test functions the SBX outperformed the PGC operator regarding the HV indicator (Fig. 1).

Regarding the *Spread indicator* the PGC operator generates better results with confidence than the SBX operator. Specifically, in 27 % of the cases the PGC yields better results with confidence than the conventional configuration of the examined MOEAs with the SBX operator. Also, in 73 % of the cases there was not statistical

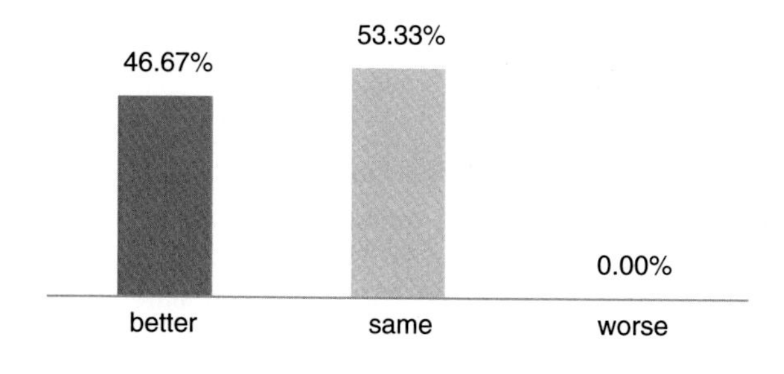

Fig. 1 HV PGC performance compared with SBX

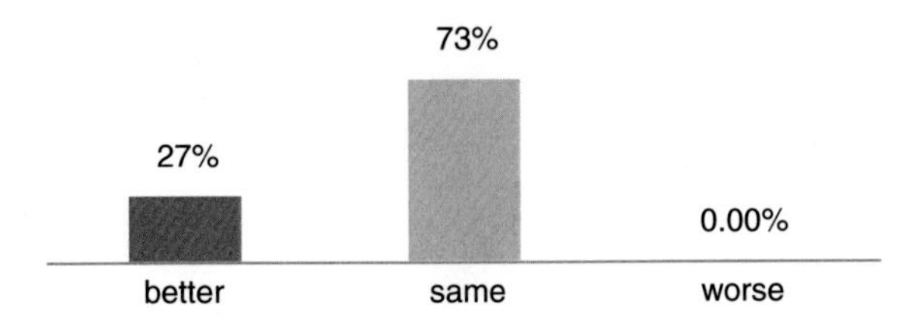

Fig. 2 Spread PGC performance compared with SBX

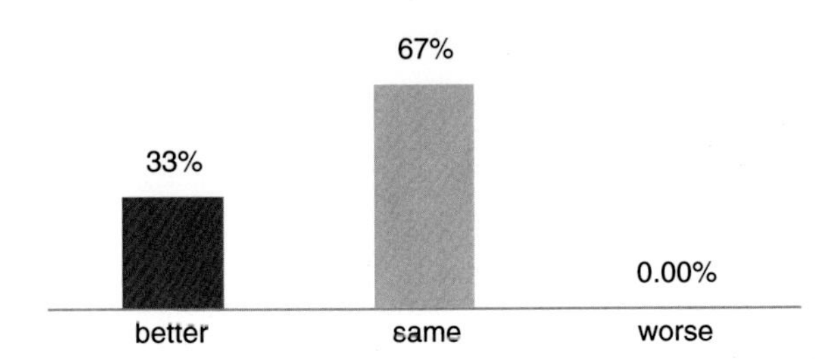

Fig. 3 Epsilon PGC performance compared with SBX

significance between the PGC and SBX. Lastly, in none of the examined test functions the SBX outperformed the PGC operator regarding the Spread indicator (Fig. 2).

Regarding the *Epsilon indicator* the PGC operator again generates better results with confidence than the SBX operator. In particular, in 33 % of the cases the PGC yields better results with confidence than the conventional configuration of the NSGAII, SPEA2 and IBEA with the SBX operator. Also, in 67 % of the cases there was not statistical significance between the PGC and SBX. Finally, in none of the examined test functions the SBX outperformed the PGC operator regarding the Epsilon indicator (Fig. 3).

8 Conclusions

In this paper, we have developed a new Probe Guided Crossover (PGC) operator for real coded evolutionary algorithms. The proposed crossover operator incorporates a probe guided mechanism that allows the efficient exploration of the most promising regions of the search space in terms of fitness value. This mechanism aids the PGC operator to find better solutions and move progressively towards higher fitness regions of the search space.

The PGC operator is applied to three well-known Multiobjective Evolutionary Algorithms (MOEAs) namely the Non-dominated Sorting Genetic Algorithm II (NSGAII), the Strength Pareto Evolutionary Algorithm 2 (SPEA2) and the

Indicator Based Evolutionary Algorithm (IBEA) and its performance is evaluated in comparison with the Simulated Binary Crossover (SBX). The efficiency of the proposed PGC operator is established with the assistance of five test instances from the Li-Zhang (LZ) set of test functions.

We utilize three different performance metrics in order to assess the performance of the proposed recombination operator. According to the relevant results the PGC operator outperforms the SBX operator, on a number of test problems with complicate Pareto sets. The examined MOEAs with the proposed recombination method have repeatedly found solutions closer to the true optimal solutions than when the SBX applied to the same MOEAs.

References

Arumugan, M.S., Rao, M.V.C., Palaniappan, R.: New hybrid genetic operators for real coded genetic algorithm to compute optimal control of a class of hybrid systems. Appl. Soft Comput. **6**(1), 38–52 (2005)

Deb, K., Agrawal, R.B.: Simulated binary crossover for continuous search space. Complex Syst. **9** (2), 115–148 (1995)

Deb, K., Pratap, A., Agarwal, S., Meyarivan, T.: A fast and elitist multiobjective genetic algorithm: NSGA, II. IEEE Trans. Evol. Comput. **6**(2), 182–197 (2002)

Durillo, J.J., Nebro, A.J.: jMetal: a java framework for multi-objective optimization. Adv. Eng. Softw. **42**, 760–771 (2011)

Fonseca, C., Flemming, P.: Multiobjective optimization and multiple constraint handling with evolutionary algorithms—part II: application example. IEEE Trans. Syst. Man Cybern. **28**, 38–47 (1998)

Koza, J.R.: Genetic Programming. MIT Press, Cambridge (1992)

Li, H., Zhang, Q.: Multiobjective optimization problems with complicated Pareto sets, MOEA/D and NSGA-II. IEEE Trans. Evol. Comput. **13**(2), 284–302 (2009)

Liagkouras, K., Metaxiotis, K.: An elitist polynomial mutation operator for improved performance of MOEAs in computer networks. In: 22nd International Conference on Computer Communications and Networks (ICCCN), pp. 1–5 (2013). doi:10.1109/ICCCN.2013.6614105

Liagkouras, K., Metaxiotis, K.: A new probe guided mutation operator and its application for solving the cardinality constrained portfolio optimization problem. Expert Syst. Appl. **41**(14), 6274–6290 (2014)

Liagkouras, K., Metaxiotis, K.: Efficient portfolio construction with the use of multiobjective evolutionary algorithms: best practices and performance metrics. Int. J. Inf. Technol. Decis. Making **14**(3), 535–564 (2015). doi:10.1142/S0219622015300013

Metaxiotis, K., Liagkouras, K.: A fitness guided mutation operator for improved performance of MOEAs. In: IEEE 20th International Conference on Electronics, Circuits, and Systems (ICECS), pp. 751–754 (2013). doi:10.1109/ICECS.2013.6815523

Metaxiotis, K., Liagkouras, K.: Multiobjective evolutionary algorithms for portfolio management: a comprehensive literature review. Expert Syst. Appl. **39**(14), 11685–11698 (2012)

Zeng, F., Decraene, J., Hean Low, M.Y., Hingston, P., Wentong, C., Suiping, Z., Chandramohan, M.: Autonomous bee colony optimization for multi-objective function. In: Proceedings of the IEEE Congress on Evolutionary Computation (2010)

Zitzler, E., Brockhoff, D., Thiele. L.: The hypervolume indicator revisited: on the design of Pareto-compliant indicators via weighted integration. In: Conference on Evolutionary Multi-Criterion Optimization (EMO 2007), pp. 862–876. Springer, Berlin (2007)

Zitzler, E., Deb, K., Thiele, L.: Comparison of multiobjective evolutionary algorithms: empirical results. Evol. Comput. **8**(2), 173–195 (2000)

Zitzler, E., Thiele, L., Laumanns, M., Fonseca, C.M., Da Fonseca, V.G.: Performance assessment of multiobjective optimizers: an analysis and review. IEEE Trans. Evol. Comput. **7**(2), 117–132 (2003)

Printed in the United States
By Bookmasters